Advances in Polar Ecology

Volume 5

Series editor
Dieter Piepenburg
Alfred Wegener Institute, Helmholtz Centre for Polar and Marine Research, Bremerhaven, Germany

Advances in Polar Ecology

In recent years, the polar regions have received increased scientific and public interest. Both the Arctic and Antarctic have been recognized as key regions in the regulation of the global climate and polar ecosystems have been identified to be particularly susceptible to the ongoing environmental changes. Consequently, the international efforts in polar research have been enhanced considerably, and a wealth of new findings is being produced at a growing rate by the international community of polar researchers. The aim of the book series Advances in Polar Ecology is to foster the progress in the scientific knowledge about the polar and sub-polar regions of both hemispheres by contributing to the fast and wide-ranging dissemination of novel scientific information gained from recent studies of sea, freshwater and land biota among polar researchers, environmental managers and policy makers. Advances in Polar Ecology's broad ecology-oriented scope encompasses environmental and biological research of both recent and past ecosystems. The Series offers an outlet to publish contributions (monographs, edited works, conference proceedings, etc.) addressing scientific topics that need more comprehensive and in-depth analysis than the length of typical journal articles allow for. These include, but are not limited to, thorough accounts of the current status of active research areas of wide importance that outline promising perspectives for future research. An international editorial board oversees the rigorous peer review that all contributions will be subjected to.

More information about this series at http://www.springer.com/series/10290

Yuri Mikhalev

Whales of the Southern Ocean

Biology, Whaling and Perspectives of Population Recovery

Springer

Yuri Mikhalev
Institute of Marine Biology of National
Academy of Ukraine
Odessa, Ukraine

ISSN 2468-5712 ISSN 2468-5720 (electronic)
Advances in Polar Ecology
ISBN 978-3-030-29251-5 ISBN 978-3-030-29252-2 (eBook)
https://doi.org/10.1007/978-3-030-29252-2

This Springer imprint is published by the registered company Springer Nature Switzerland AG.
The registered company address is: Gewerbestrasse 11, 6330 Cham, Switzerland

Annotation

This monograph examines the distribution and migration patterns of whales of the Southern Ocean. It defines distinct populations on the basis of phenes, as well as whale breeding zones, which are located in the adjacent to the Southern Ocean waters at lower latitudes. The book records the presence of a new species of killer whale in the Southern Ocean – Orcinus nana. Prenatal growth patterns, pregnancy and lactation duration, and mean sizes of newborn whales are determined. Methods for the graphic recording of registering structures are described, and an original method for their decoding is proposed to determine animal age. The age of sexual and physical maturity and life expectancy are determined. Earlier unknown "pair formations" on the lower jaw of baleen whales and sperm whales are described, together with their macro, histological, and electronic microscopic structure.

The impact of the extermination of whales on the Southern Ocean ecosystem is examined, recommendations for control of the current state of whale populations are given, and perspectives of whale population recovery are estimated. Regions that could be used as testing areas for whale registration method are defined.

The book is intended for biologist-cytologists, ecologists, and other specialists interested in cetaceans and for biology students.

Marine Mammal Council

MARINE MAMMAL COUNCIL

36 Nachimovsky av.

117851, Moscow, Russia

Tel (fax): 7-495-1247579

E-mail: mmc@ocean.ru

СОВЕТ ПО МОРСКИМ МЛЕКОПИТАЮЩИМ

Россия, 117851, Москва,

Нахимовский пр-т. 36

Tel (fax): 7-495-1247579

http://2mn.org

Having examined *Whales of the Southern Hemisphere: Biology, Whaling and Perspectives of Population Recovery*, written by doctor of biology and professor of the Institute of Marine Biology of Ukrainian National Academy of Science (Odessa, Ukraine), Yuri Mikhalev, the Marine Mammal Council (Russia) concluded that the book is about a very important issue, contains high-level scientific research, and should be published.

Yuri Mikhalev is a member of the Council and has worked in the area of whale biology since 1964. He participated in six cruises of Soviet whaling fleets *Slava*, *Sovetskaya Ukraina*, *Yuri Dolgoruky*, and scientific-research boat "Bodry-25" as a member of a scientific team. Before starting his teaching career, he was a chief at the marine mammal laboratory at the Odessa branch of AzCherNIRO. Based on a huge amount of scientific data, he published more than 100 papers. His first thesis was about the biology of the Southern Ocean fin whale reproduction (Scientific Council of All-Union Institute of Sea Fishery and Oceanography, Moscow, 1972). He completed two doctorate theses concerning the biology of Southern Hemisphere whales, focusing on "environmental protection and rational use of natural resources" (Scientific Council of the Institute of Ecology Problems and Evolution, Russian Academy of Science, Moscow, 1997) and "zoology" (Scientific Council of the Institute of Zoology, National Academy of Science of Ukraine, Kiev, 2005). He was awarded the "150 Years of Antarctica Discovery" Medal for his research on the Southern Ocean cetaceans.

Yuri Mikhalev actively opposed illegal whaling and the falsification of scientific data by whaling fleet authorities. He did much to save the true whaling data that the government tried to destroy and hide from the public.

Since 1994, he has been an independent expert of the Scientific Committee of International Whaling Commission (IWC).

This book is not a compilation of other scientists' research. Rather, it includes only issues that the author himself studied. Nevertheless, partly as a result of discussions with other researchers, many aspects of whaling in the Southern Hemisphere, as well as whale distribution, growth, reproduction biology, and recovery capability, are carefully examined in this book. The book also analyzes the influence of whaling on the Southern Ocean ecology, and the author forecasts perspectives of whale population recovery. The book suggests recommendations for management and suggests regions of the world's oceans that could be used as research areas for improving scientific methods of estimating abundance. The solutions proffered by Yuri Mikhalev on many issues are original, and they, as well as his hypotheses, will provoke discussion such that this book will be useful for both specialists and beginners. I believe it could be read with interest even by anyone who is interested in cetacean biology.

Chairman of the Council, Member-Correspondent of *Russian Academy of Science.*

A. V. Yablokov

Acknowledgments

As a biologist, I am obliged to the chair of zoology of the Kishinev State University under the direction of highly skilled zoologists – the rector of the university and managing chair of zoology, Victor Sergeevich Chepurnov; the dean of biological faculty, Magda Sadykovich Burnashev; and the professor in charge of our course, Lyudmila Viktorovna Chepurnova. I am grateful to destiny that during my externship at AzCherNiRo (the Azovo-Black Sea Scientific Research Institute of Fishery Economics), she introduced me to Yury Petrovich Altukhov (subsequently the director of the USSR AS Institute of Genetics), in coauthorship with whom my first paper on jack mackerels of Black Sea was published.

I have warm memories of the director of L.S. Berg Museum of the Kishinev University, Alexandra Matveevna Didusenko. Like all other graduates of our chair, who went to the different corners of the Soviet Union, I considered it my duty to supply the museum with new exhibits. The cooperation with Aleksandra Matveevna lasted for many long years. I was extremely touched upon receiving the letter from her daughter with the sad news of Alexandra Matveevna's death, in which she informed me that some hours prior to her death she wrote to me that she saw me as the next director of the museum. It was Alexandra Matveevna who recommended me for work at the Odessa AzCherNiRo Laboratory (later transformed into the Odessa VNIRO branch on whaling and after that into the AzCherNiRo Branch). I am also sincerely grateful to the director of this laboratory, Arcady Vasilevich Krotov, who sent me to the whaling flotilla scientific group and after defending my master's thesis appointed me the head of the laboratory of marine mammals. Arcady Vasilevich was a quick-witted person with delicate humor and great life experience. Conversations with him on the chessboard have taught me a lot.

During the Odessa period of my life, I have had the good luck to form my views under the influence of great people and scientists including the head of zoology and chair of Odessa I.I. Mechnikov National University, Professor Ivan Ivanovich Puzanov ("the last encyclopedist of the twentieth century") and the outstanding geneticist, Alexander Aleksandrovich Malinovsky; they both desperately struggled against "lysenkovschina," the governmental biological ideology of that time.

I have greatly appreciated my conversations/consultations with most outstanding cytologists, Professor Avenir Grigor'evich Tomilin and Professor Vyacheslav Alekseevich Zemsky, and especially with a corresponding member of the Russian Academy of Sciences, Alexey Vladimirovich Yablokov – his criticism helped me not to make too many mistakes.

I would like to express special gratitude to my colleagues in the struggle for factual scientific material, to the former employees of the laboratory of marine mammals, and to the participants of scientific groups in the Southern Ocean voyages of the "Soviet Ukraine" whaling flotilla, Vladimir Pavlovich Savusin, Valery Leonidovich Zinchenko, and Sergey Genrihovich Bushuev, for their courage and scientific honesty.

With gratitude, I remember the kindness extended to me by the captain-director of the *Slava* whaling fleet, Anatoly Stepanovich Labunets; the captain of the scientific research whaling vessel "Bodry-25," Anatoly Vasilevich Grebenshchikov; and sailors of the *Slava* and "the Soviet Ukraine," Alexander Podymov, Yury Savelyev, Nikolay Nerezov, and many others, without whose help, it would have been impossible to work on deck and collect scientific material.

I can say without exaggeration that neither the two theses for my doctor's degree nor this book could have been written without the help of my sons, Vladimir and Igor Mikhalyev, who created a special PC data processing program, translated my articles into English, and helped with international correspondence. *I devote this work to them!*

While working on this book, it was necessary to solve a number of practical problems, which would not have been possible without the help of Anatoly Grigorevich Popovich, Vladimir Nikolaevich Burkanov, Lev Mihajlovich Muhametov, and Sergey Valerevich Keljushok's disinterested aid. Thank you, my friends! Additionally, I am grateful to the US Marine Mammal Commission. I would like to thank the two WWF Russia colleagues, Dr. Konstantin Zgurovsky, the senior advisor of the Marine Program, and Dr. Andrey Vinnikov, the director of the Marine Program, for their tireless efforts and support in publishing this book. I also would like to thank Mr. Stuart Prior, the honorary consul of Belarus in New Zealand, without his help, this book would be never published. Also, I am grateful to Dr.

Eisert (research scientist at the University of Canterbury in Christchurch, New Zealand) and Olga Suvorova-Stuart, who helped to maintain correspondence with foreign colleagues. I would like to thank limited companies "Utrish dolphinarium" and "Odessa dolphinarium" that have borne a part of the financial costs.

Very special thanks to Rosy Whelan. The English edition of the book would not have been possible without her great help.

Introduction

Whales, baleen and toothed, being the top links in trophic chains, play a major role in the ecosystems of the world's oceans. Their huge size has always excited people's imagination: legends and myths were created. Whales became the heroes of literary works and the fine arts. At the same time, whales, with their huge weight, became a desired target for humans. From whales, food; medical, veterinary, and animal products; and sperm and baleen oil have been produced. Particular interest was paid to oil from the bones, from which high-quality lubricants for precision mechanisms and devices were prepared. Fresh and frozen meat from baleen whales has been consumed by humans, and second-grade edible products have been consumed by animals, e.g., on fur farms. Whale liver was used as an edible product and to manufacture medical preparations – primarily, vitamin A concentrate. Bone meal was added to animal fodder (Bodrov and Grigoriev 1963). More complete utilization of whale raw materials took place in Japan (Komatsu and Misaki 2005).

Together with whale catches, the study of whales took place. In the eighteenth to nineteenth centuries, special attention was given to cetacean anatomy, with research continuing into the twentieth century. Among morphological research, cetaceans were considered in V.E. Sokolova's monograph (1973, pp. 242–269) *Integument of Mammals*. Scientists of the department studying marine mammals at the I.I. Schmalhausen Institute of Zoology, under the direction of G.B. Agarkov, published the monographs *Morphology of Dolphins* (Agarkov et al. 1974) and *Functional Morphology of Cetaceans* (Agarkov et al. 1979). The physiology of respiration is considered in A.Z. Kolchinsky, N.N. Mankovsky, and A.G. Misjura's book *Breath and Oxygen Modes of the Dolphin Organism* (1980).

Detailed, general reports on the biology of whales began to appear in the twentieth century. The first monograph description of whales was that of F.E. Beddard, *A Book of Whales* (Beddard 1900). Detailed studies of large baleen whales were described by N.A. Mackintosh and J. F. Wheeler in the book *Sothern Blue and Fin Whales* (Mackintosh and Wheeler 1929). Monographs describing first sperm whales and then humpback whales were also published in *Discovery Reports* by L.H. Matthews (Matthews 1937, 1938).

The postwar period 1950s–1980s proved very fruitful. Among domestic works, the first that must be mentioned is M.M. Sleptsov's monograph *Cetaceans of the Far East Seas* (Sleptsov 1955). As fundamental works, *Cetacea* (1957) and *Mammals of the USSR and Adjacent Countries* by A.G. Tomilin (1962) must be singled out. During the same period, E. Slijper's book *Whales* (Slijper 1962) was published abroad. The collection *Whales, Dolphins, and Porpoises* under K.S. Norris's editorship (Norris, 1966) and *Biology of Sea Mammals* under G.T. Andersen's editorship (Andersen 1969) were published a little later. More attention is paid to cetaceans in handbooks, encyclopedias, and guides including "Cetaceans" in *Guide to the Mammals of the USSR* by N.A. Bobrinsky, B.A. Kuznetsov, and A.P. Kuzyakin (1965, pp. 185–205), *Handbook of Marine Mammals* under P.A. Moiseyev's editorship (Ivashin, Popov, Tsapko, 1972), and the short description of cetaceans in the book *Marine Mammals* by V.A. Arsenyev, V.A. Zemsky, and I.A. Studenetskaya (1973, 15–101). The suborder of toothed whales is described in detail in the second volume *Mammals of the Soviet Union: Pinnipeds and Toothed Whales* by V.G. Heptner, K.K. Chapsky, V.A. Arsenyev, and V.E. Sokolov (1976, pp. 413–660).

With some delay, *The Atlas of Marine Mammals of the USSR* was published in 1980 under V.A. Zemsky's editorship; the work undertaken by prominent VNIRO experts began in the 1950s. The book *Baleen Whales* by V.E. Sokolov and A.A. Arsenyev was published in 1994. Several monograph descriptions of whales species have also been printed: S.E. Kleinenberg, A.V. Yablokov, V.M. Bel'kovich, and M.N. Tarasevich (1964) published *The White Whale*, A.A. Berzin (1971) published *The Sperm Whale*, and, abroad, D.V. Rice and A.A. Wolman (1971) published *The Life History and Ecology of the Gray Whale (Eschrichtius robustus)*. An overview of the results of whale research was provided in the book *Whales and Dolphins* (Yablokov et al. 1972).

Considerable attention was also given to the popularization of information on cetaceans. Among the books of this type, I will mention: *In the Country of Whales and Penguins* by V.A. Arsenyev and V.A. Zemsky (1954); *Whales of the Antarctic* by V.A. Zemsky (1962); *Our Friend the Dolphin* by V.M. Bel'kovich, S.E. Kleinenberg, and A.V. Yablokov (1967); the Russian translation of book *Whale* under L.H. Mathue editorship (1973); the Russian translation in 1962 of the 1851 novel by the American author, Herman Melville, *Moby Dick or The Whale*.

During that period, seminars of scientific groups from the whaling fleets were regularly held. All-union meetings on marine mammals, international conferences, the reports made at those meetings, and collections of articles were issued. After the disintegration of the Soviet Union, Russia restored those traditions: at 2-year intervals, international conferences on marine mammals of the Holarctic have been held, and the proceedings have been published.

More than a quarter of a century has passed since the commercial whaling in the world's oceans was all but stopped; the Russian whaling fleet has been utilized for other purposes or recycled. The future of whales, whose stocks were catastrophically reduced, continues to disturb mankind. It is necessary to comprehend, sum up, and define their prospects. First of all, the question is whether all species and their populations can ever be restored to their original numbers. There are strong grounds for anxiety, as the published facts of large-scale illegal whaling (Yablokov 1994; Zemsky et al. 1994a, 1995, 1996; Mikhalev 1997a, 1997b; Tormosov et al. 1998; Yablokov, etc., 1998; Kasuya 1999, 2002; etc.) became known to a wide range of experts and the public. Unfortunately, other countries, which conducted whaling for a considerably longer time than the Soviet Union and no less intensively, are not in a hurry to open their archives and publish the actual numbers of whales caught! At present, the only attempt I know of is the Japanese researcher, T. Kasuya (Kasuya 1999, 2002), who discovered illegalities in the catching of whales at onshore stations in Japan.

To begin working on the theory and practice of the restoration of whale populations and then to develop rational methods of exploitation (to conduct whale *management*), a full audit of all our information about the original size of whale populations is required. Primarily, it is necessary to ascertain the true numbers of whale catches and what damage was caused to populations by illegal catches.

For various economic and political reasons, solving this problem has not been easy. Successors of the old Soviet system, having this sin on their conscience, have concealed and destroyed archival materials.[1] "Specialists" of these departments are trying to mislead the public by means of speculative polemic.

These factors continue to dictate the necessity of returning to history, to the analysis of whaling methods, and the means of regulating the catch. In this book, the primary focus is on whales that were commercially caught in the Southern Ocean. Whales are feeding in this food-rich zone of the World Ocean during warm time of the year. At cold time, most of the whales with few exceptions leave the Southern Ocean and move to lower latitudes in breeding zones, which are located to the north of the ocean border – line of water convergence. Some species move even to the tropic waters and cross them. For example, pygmy blue whales, humpback whales, Bryde's whales, and some of sperm whales were noticed even in the Arabian Sea. It is not proved yet that the whales migrate over there from the Southern Ocean though. From this perspective, sperm whales are interesting. Their males are twice larger than females, and with few exceptions, they go to the feeding areas in the Southern Ocean, reaching very high latitudes. But small females and average size males stay

[1] It is difficult to believe, but they literally destroyed records. It is authoritatively known that VNIRO's expert, M.V. Ivashin, in the early 1990s spent 2 days burning primary sources on whale catches in the institute courtyard.

for breeding and feeding in the lower latitudes. The females are frequently noticed to the north of convergence zone, but just few were recorded above 40 degrees south.

So whales of the Southern Ocean that are part of its ecosystem are closely linked to the ecosystems of waters in adjacent lower latitudes, and when necessary, the whales from the waters are described in the monograph. Sometimes, the comparison is made between whales of the Southern Ocean and the Northern Hemisphere including its polar zone.

Research Materials Reported

- At conferences of Scientific Committee of the International Whaling Commission: Puerto Vallarta (Mexico), 1994; Galway (Ireland), 1995; Dublin (Ireland), 1995; Aberdeen (Scotland), 1996; Bournemouth (England), 1997; Muscat (Oman), 1998; Grenada (Republic of Grenada), 1999; Adelaide (Australia), 2000; London (England), 2001; Shimonoseki (Japan), 2002; Berlin, 2003; Sorrento (Italy), 2004; and Madera (Portugal), 2009.
- At all-union meetings on the study of marine mammals: Kaliningrad, 1969; Makhachkala, 1972; Kiev, 1975; Simferopol, 1978; Astrakhan, 1982; Arkhangelsk, 1986; and Svetlogorsk (Kaliningrad region), 1990.
- At the international conferences on the study and protection of marine mammals: Istanbul (Turkey), 1994, and Golitsyno (Russia), 1995. At the international scientific conferences on marine mammals of the Holarctic: Arkhangelsk (Russia), 2000; Listvyanka (the Irkutsk region, Russia), 2002; Koktebel (Crimea), 2004; St. Petersburg (Russia), 2006; Odessa (Ukraine), 2008; Istanbul (Turkey), 2009; Kaliningrad (Russia), 2010; Stralsund (Germany), 2010; Oakland (New Zealand), 2010; Suzdal (Russia), 2012; Saint Petersburg (Russia), 2014; Astrakhan (Russia) 2016; Copenhagen (Denmark), 2017; and Réunion Island, Saint Denis (France), 2017.
- At republic, regional, and university conferences: scientific conference of the Kishinev State University, 1965; conferences of young scientists of the Odessa State University, 1969; conferences of young biologists and physicians, Odessa, 1971; conferences of young scientific AzCherNIRO (nowadays YugNIRO), Kerch, 1973; at the conference "Wildlife Management and Rational Use of Natural Resources of the South of Ukraine," Simferopol, 1977; at the inter-university conference on "The Protection of Fish Stocks and Increase in Efficiency of Reservoirs of the Southern Zone of the USSR," Kishinev, 1969; at the accounting scientific conference of the Odessa Pedagogical University, Odessa, 1984; republican scientific-practical conferences, Odessa, 1989 and 1990; at the Anniversary scientific-practical conference devoted to the 130 anniversary of L.S. Berg, Bendery, 2006; etc.

Data and Research Methods

The basic area of research is the waters of the Southern Ocean up to the coast of Antarctica and also adjacent waters of the Atlantic and Indian Oceans to 30°N. The basic data for the research came from the commercial catches of whales in the Southern Ocean.

The data are materials collected by the author as a member of scientific groups on whaling fleet voyages, *Sovetskaya Ukraina* (1964–1965), *Slava* (1965–1966), and *Yuri Dolgoruky* (1966–1967 and 1968–1969), and on the scientific and research whaling vessel "Bodry-25" (1973–1974 and 1974–1975). The data collected for our program by scientific participants on the whaling fleets are also included.

Actual (true) data about the catch of whales are reported, which greatly differ from those reported to the International Whaling Commission (IWC) by the management of the fleets and the inspection service of the Ministry of Fisheries of the USSR. The data on marking of 21,456 whales and 759 return labels taken both from the IWC database and from whale logs collected by biologists on the whaling fleets have been corrected. Personally, I marked 97 whales, mainly in the Pacific sector of the Southern Ocean.

The results of weighing 1250 whales and 888 embryos are included. Weights were taken using spring dynamometers.

Prenatal growth curves of whales were developed on the basis of the sizes (zoological length) of 16,443 embryos from an early stage of development to prenatal fetuses. A series of numerous (standard) measurements of 655 embryos and 550 whales were taken. The characteristics of the variety of body shapes and colors were defined from 404 embryos. The age of 18,770 toothed and 12,800 baleen whales was determined. Using decalcified polished sections and polished sections of teeth from toothed whales, and earplugs of baleen whales, more than 2000 recording structure graphs were drawn. Whale distribution charts were compiled from the catch coordinates of more than 200,000 animals.

Weighting of killer whale (left) and bottlenose whale (right)

Using computing programs, biometric, approximate calculus, and visual analytic methods were employed (Rokitsky 1961, 1964; Plokhinsky 1961, 1978; Bailey 1970; Urbakht 1963; Pogorelov 1968; etc.). An estimation of the reliability of qualitative indicators was made following Wilcoxon (Urbakht 1963, 1964). The degree of distinctions of populations and local groups was estimated by the criteria of similarity and identity (Zhivotovsky 1979, 1982).

Decalcification of teeth by formic acid was made following Bow and Purday (Bow and Purday 1966). The calcium content in tooth sections was determined by an atomic-absorptive method (Dubin et al. 1982). The fat content of whale milk was determined using the *Butera* metric method. For the histological examination of teeth, earplugs, the pituitary gland, and "pair formations," staining with silver nitrate, hematoxylin, and eosin was used. The equipment used for graphical records of the lamination of teeth and earplugs (recording structure graphs) included a domestic M-201 profilograph, an English Talyline-4 profilograph, an IFO-451 microphotometer, and plethysmographs with a special sensor/gauge of our own design.

For the convenience of describing and understanding the material presented, the techniques used are explained in more detail in the chapters in which they are applied. Data from the literature and the data provided by other researchers are stipulated in corresponding places.

The following species are examined in more detail:

Cetacean group – *Cetacea* Brisson, 1762
Suborder baleen whales – *Mysticeti* Flower, 1864
Blue whale – *Balaenoptera musculus* Linnaeus, 1758
Pygmy blue whale – *Balaenoptera brevicaudis* Zemsky, 1972
Fin whale – *Balaenoptera physalus* Linnaeus, 1758
Sei whale – *Balaenoptera borealis* Lesson, 1828
Bryde's whale – *Balaenoptera brydei* Olsen, 1913
Eden's whale – *Balaenoptera edeni* Anderson, 1879
Minke whale – *Balaenoptera acutorostrata* Lacépède, 1804
Humpback whale– *Megaptera novaeangliae* Borowski, 1781
Southern right whale – *Eubalaena glacialis* Muller, 1776
Pygmy right whale – *Caperea marginata* Gray, 1846
Gray whale – *Eschrichtius gibbosus* Erxleben, 1777
Suborder of toothed whales – *Odontoceti* Flower, 1867
Sperm whale – *Physeter macrocephalus* Linnaeus, 1758
Pygmy sperm whale – *Kogia breviceps* de Blainville, 1838
Killer whale – *Orcinus orca* Linnaeus, 1758
Short-bodied killer whale – *Orcinus nana* Mikhalev, 1981
False killer whale – *Pseudorca crassidens* Owen, 1846
Bottlenose whale – *Hyperoodon ampullatus* Forster, 1770
Cuvier's beaked whale– *Ziphius cavirostris* G. Cuvier, 1823
Pilot whale – *Globicephala melaena* Traill, 1809
Beluga – *Delphinapterus leucas* Pallas, 1776
Risso's dolphin – *Grampus griseus* G. Cuvier, 1812
White-beaked dolphin – *Lagenorhynchus albirostris* Gray, 1846
White-sided dolphin – *Lagenorhynchus obliquidens* Gill, 1865
Vaquita – *Phocoena sinus* Norris and McFarland, 1958
Finless porpoise – *Neophocoena phocoenoides* G. Cuvier, 1829
Striped dolphin – *Stenella coeruleoalbus* Meyen, 1833

Contents

About the Author

Yuri Alekseyevich Mikhalev was graduated as biologist-zoologist from the Department of Biology at Kishinev State University in 1963, and was hired as Senior Scientist in the Sea Mammal Laboratory in Odessa branch of AzCherNIRO. He took part as a scientific group member in six cruises of Soviet whaling fleets "Slava" (1965–1966), "Sovetskaya Ukraina" (1964–1965), "Yuri Dolgorukiy" (1966–1967), and scientific-research vessel "Bodriy-25" (1973–1974 and 1974–1975).

In 1970, he became a chief of the Sea Mammal Laboratory in Odessa branch of AzCherNIRO. He is the author of more than 100 papers; most of them were published abroad. His first thesis was about biology of Antarctic fin whale reproduction (Scientific Council of All-Union Institute of Sea Fishery and Oceanography, Moscow, 1972). His two doctorate theses were about biology of whales of Southern Hemisphere (specifically "environment protection and rational use of nature resources," Scientific Council of Institute of Ecology Problems and Evolution, Russian Academy of Science, Moscow, 1997, and "zoology," Scientific Council of Institute of Zoology, National Academy of Science of Ukraine, Kiev, 2005). He was awarded with 150 years of Antarctica discovery medal for his research of Antarctic cetaceans.

When the mammal laboratory was closed in 1982, he started to work as a Professor in the Departments of

Anatomy and Physiology at State Pedagogical University (Odessa, Ukraine). He is currently working at the Institute of Marine Biology, Ukrainian Academy of Science, Odessa.

Since 1994, he has been an independent expert of Scientific Committee of International Whaling Commission (IWC), and since 1995, he has been a member of the Marine Mammal Council (Moscow, Russia).

Chapter 1
Historical Review of Whaling

Y. Mikhalev, *Whales of the Southern Ocean*, Advances in Polar Ecology 5,
https://doi.org/10.1007/978-3-030-29252-2_1

- Whaling in the Southern Ocean and adjacent waters
- Periods of whaling
- Attemps to regulate whaling
- Pecularities of Soviet whaling

From prehistoric times, natives hunted whales for food in the shallow coastal waters of Europe, the Chukchi Peninsula, Kamchatka, the Japanese islands and Alaska. The catch levels were minimal and did not affect the health of whale populations.

Rapid expansion of whaling areas and catch levels began in the Middle Ages. This initiative belonged to the Basques, who, in a relatively short time, expanded their catch areas from the Bay of Biscay up to Spitsbergen and Newfoundland, where right and, probably, gray whales were taken. Subsequently, whalers from different countries developed whaling in the Davis Strait, Baffin Bay and the whole European Russian North. From the ninth century, Northern Russian princes received taxes in the form of whale products from Saami and Finnish people (Zenkovich 1955). On the Arctic coast, hunters from Novgorod also caught whales and walruses. M.V. Lomonosov, noted that the Novgorod chronicles specified that princes and their retinues took part in, and often supervised, whaling expeditions. Pechenga monastery conducted whaling at Spitsbergen, known as Grumant to Russian Pomors. The hunters from the Starostin kin were well-known; these Novgorod natives traded on Grumant for more than 400 years. The last member of the Starostin family died in Saint Petersburg in 1875 (Zenkovich 1955).

Catching a right whale. Engraving in Wellington museum, New Zealand

In the sixteenth century, foreign whalers also started to visit the waters of the Russian North (Sidorov 1867), and until the end of that century, the Novgorod authorities imposed taxes on the whaling trade. The domestic industry stopped with the reign of Peter the Great, who ruled from 1682 to 1725. Funded by the state treasury, the first Russian commercial enterprise, the "Kola Peninsula Whaling Trade" (Veberman 1914) was founded. With the accession of Katherine the Great in 1762, the state monopoly was stopped and private whaling allowed. In 1786 the "Onega River Whaling Company" of Count Vorontsov, a relative of the well-known Odessa General-Governor, was founded. In 1803, the "White Sea Company" was founded with Emperor Alexander I among the shareholders. Later more companies – "Goebel's Whaling Trade", "Whaling Trade Company on the Murman", "Sheremetiev's Whaling Trade" -were created. However, the companies were not profitable, lost everything and ceased to exist.

In the seventeenth century, Spaniards, the English, Dutch, Germans, Norwegians and Americans were actively engaged in whaling in Arctic Waters, the English and Dutch having seized the initiative from the Basques. At that time, more than a thousand sailing ships were engaged in whaling (Zenkovich 1955; Sidorov 1867). Basically, right whales were taken. There are no precise data on whaling of that time, but from 1669 till 1787 whalers of all countries (mainly Dutch) took no less than 200,000 bowhead whales (also known as the Greenland whale) i.e. on average about 1700 animals were taken annually. In due course, stocks of this species were considerably reduced, which led to the search for new whaling areas.

Relatively recently whaler used hand harpoons new Azorean Islands

In the eighteenth and nineteenth centuries, whaling occurred throughout the Bering Strait, the Chukchi Sea, the East Siberian Sea, and the Northern Pacific. As to the Russian Far East, natives of Kamchatka and the Chukchi Peninsula had hunted whales since time immemorial, using them for food, fuel, construction material etc. From 1794, a number of companies was created in Russia: Russian-American, Shelekhov's, Pigota and Partners, Russian-Finnish, etc. (Elfsberg 1863; Lindholm 1888; Tikhomirov 1894). "The Russian-Finnish company" was established in 1850. Merchant O.V. Lindholm from Vladivostok conducted a whale trade from 1864. The first whaling steamships, constructed in Norway, appeared in Russia by the end of that century: the "Gennady Nevelskoy", "Nikolay" and "Georgy". By 1903, "Count G.G. Keyserling's and Co Pacific whaling fishing joint-stock company" even had a rather powerful 5000 t floating whaling base "Michael". At the end of the nineteenth and the beginning of the twentieth century, foreign industrialists appeared in the Russian waters. The governor of the Kamchatka area, Monomakhov, wrote in his report for 1912–1913: "the trade of whales and walruses is visibly decreasing every year due to extermination of these animals by Americans".

In 1923, the Soviet power granted a concession contract to a Norwegian businessman, Christiansen, until January 1938, which allowed the catching of all species of whales within a 12-mile strip from Cape Serdtse-Kamen to Cape Lopatka. His company, "Vega", had four whale catchers and a 9000 t factory, "Comodoren 1". The company was obliged to train citizens of the USSR how to conduct whaling.

In 1930, the government appropriated funds to construct a floating factory and purchase four whaler catchers. The steamship "Glen Ridge", constructed in 1919, was bought in the USA. It was a two-deck dry-cargo ship, 115 m long and 16 m wide, with 10,573 t displacement. The dry-cargo ship was converted into the whaling factory, the "Aleut" in the docks of Kronstadt. Whaler catchers were ordered in Norway for the "Aleut". The "Aleut" fleet arrived in Vladivostok in 1933 and started whaling. The fleet was written off in 1968.

In the twentieth century, whaling started in the Southern Hemisphere, while gray whales continued to be taken in the North Pacific. The initiative in whaling was seized by Americans, with 735 American whalers conducting business in 1876. The English followed them. Sperm whales became the object of the catch, rather than southern right whales. From 1800–1889, American whalers made about 14,000 voyages (Best 1983). As calculated by A. Starbuck (1878), 225,521 sperm whales were taken by all countries from 1804–1876, on average 3000 sperm whales annually. By R. Gambell's calculations (Gambell 1983) the Americans took much more – 4600–5100 sperm whales annually. Taking into account that 10–20% of whales (Brand 1940) were lost each year, about 6000 sperm whales were caught, that is about 430,000 sperm whales removed. Life and romanticism of whaling was realistically described by the American classical writer in his novel "Moby Dick, or, the Whale" (Melville 1851).

Apart from sperm whales, right whales continued to be killed. From 1800 to 1875 about 200,000 southern right whales were taken i.e. 2600 annually, by Americans alone. Excessively intensive, unregulated predatory whaling, first by American, English and Dutch whalers, strongly undermined stocks of sperm whales, gray, bowhead and southern right whales. Of the great whales, only minkes were untouched. Then whaling shifted to the Southern Ocean waters.

1.1 Whaling in the Southern Ocean and Nearby Waters

The period of most intensive whaling started at the beginning of the twentieth century. This time the initiative came from the Norwegians. In 1864, Svend Foyn modernized a whaling ship by setting a harpoon cannon on the bow with a harpoon line attached. When a whale was harpooned, the shock of the associated jerk was absorbed by a special brake system. The captured whale was inflated with the help of a compressor, and as a result, whales with negative buoyancy did not sink. Such whaling ships began to spread through the North Atlantic, where the Norwegians take more than 1000 Minke (and, probably, gray) whales annually, before they turned to the Southern Ocean. From 1892 to 1895 the Norwegians made expeditions to the Southern Ocean in their attempt to find new whaling regions, but the first whale (a humpback) they caught was in 1902 in the Magellan Strait.

In 1904, an Argentinean company built a shore station, headed by the Norwegian K. Larsen, on the island of South Georgia. The following season a Norwegian floating factory, the "Admiral", arrived in these waters. The Norwegians were followed by the British, and intensive whaling began in the world's most whale-rich oceans – the Southern Ocean and nearby waters.

Night. The whaling fleet is trapped in ice

Construction of factory ships with a slipway (a sloping platform to haul whales up to the deck), was introduced by P. Sorlle, and contributed to the intensification of whaling in the Southern Ocean. This innovation was first used in 1923 by a

Norwegian whaling company, "The Globe". Since then whales have been flensed on the decks of the factory ships.

Due to the modernization of whaling and the discovery of new whaling grounds, catches started increasing from season to season at an incomparable scale. By 1923/24, 14 fleets from different countries, with 66 catcher boats, worked in the Southern Ocean. There were also six land stations, but their catches were not as great. The number of fleets grew to 41 in 1930/31. Because the stocks of right whales had been dramatically reduced, the comparatively slow (to the rorquals) humpbacks began to be caught intensively. They were most actively hunted from 1908–1916 and 1934–1938, when 4000, 5000 and even 8000 humpback whales were taken per season. Blue whales and fin whales also became target species. The biggest catches were in the 1930s when the combined catch of all countries was 13,000–30,000 blue whales and 5000–28,000 fin whales per season, and the Southern Ocean became the main whaling region (Table 1.1).

As can be seen, the volumes of catch in the Southern Ocean reached 80% of the total world catch. Such intensive whaling flooded the market, which forced the Norwegians to leave the trade for some time. As a result, whaling in the Southern Ocean had reduced a little before the Second World War, and was insignificant during the war.

However, whaling revived after the war with a total of 14 fleets working in the Southern Ocean in the first post-war season of 1945/1946. The following season there were 17 fleets and in the 1950–60s, their number grew to 21 when the Netherlands, Panama, the USSR, RSA, Japan and other countries went whaling. The number of fleets was half of that before the war but the modernized and new whalers were much more powerful. As a result, whaling became more aggressive and whale populations began to reduce dramatically. The necessity of taking measures to limit catches and to control whaling became evident.

Table 1.1 Whaling catches in the world's oceans and in the Southern Ocean, according to IWC data

Whaling season	World's catch database	The Southern Ocean catch	The Southern Ocean percentage (%)
1909/10	12,301	6099	54
1923/24	16,839	7271	43
1930/31	43,130	40,201	93
1931/32	12,988	9572	74
1932/33	28,907	24,327	84
1933/34	32,586	26,089	80
1934/35	39,311	31,808	81
1935/36	44,868	30,991	69
1936/37	51,586	34,579	67
1937/38	54,902	46,039	84
1938/39	45,783	38,356	84
1939/40	37,705	32,900	87
1940/41	23,638	16,365	69

1.2 Periods of Whaling

This short review allows us to distinguish several whaling periods (epochs), which differ from one another (Mikhalev 2002a). Historically, the longest is the epoch of coastal whale hunting for the subsistence of local people, followed by the start of commercial whaling in the Middle Ages. Then, during the growth of capitalism in the eighteenth and nineteenth centuries, the epoch of large-scale whaling arrives. The period from the beginning of the twentieth century until the Second World War differs from the previous epochs because, with more powerful ships the most intensive whaling moved to the Southern Ocean, where an abundance of baleen whales became the primary target. The end of the post-war period can be considered to be 1972, the year when international inspection was introduced and when the Southern Ocean minke whales were the only species allowed to be caught. Then comes a very short period (1972–1987) – the epoch of the reduction and nearly complete cessation of whaling.

There is no need to discuss the prehistoric and early historical whaling periods. Whaling then was comparatively passive, natural and ecologically harmless. Beginning with the Middle Ages, whaling became more and more intensive. More countries started whaling and the USA began playing the leading part. What saved the populations of whales from complete destruction was the constant search for and discovery of new regions of considerable concentrations of whales. Still, technical limitations of whaling at that time and the restricted need for whale products, made it possible to continue whaling over the course of ten centuries.

The reduction in whaling by the end of the nineteenth century was greatly determined by the use of oil products in industry. The cost of whale-oil went down. The development of other branches of industry and the "gold rush" contributed to the decrease in capital and people from whaling (Brand 1940).

During the pre-war period of the twentieth century, practically all of the world's oceans had been exploited. Whaling took on a threatening scale as the number of whales dramatically reduced. During the war, however, that process slowed down. After the war whaling resumed. In addition to the traditional participants, Argentina, Brazil, Chile, Peru and the USSR joined in whaling, although the USSR only started the Southern Ocean whaling. The country had great political authority at that time and it was not connected with negative tendencies. Had the USSR signed the 1946 Convention on the Regulation of Whaling, they could have actively worked out methods and rules for rational, regulated whaling, and organized efficient control over whaling. Had the USSR taken that stand the whaling process and the destiny of whale populations could have been completely different. But what actually happened? Let's have a good look at the period from 1946 till 1972.

If the first period of active whaling lasted for ten centuries, the following one lasted for about two or three hundred years, the pre-war period for less than 40 years and the post-war period for 25 years. The intensity of whaling was in inverse proportion to the length of the epoch. The shortest post-war period had the most negative impact. The 1950s was a period of whaling restoration and growth. The 1960s was the time of the most destructive whaling with the USSR playing the leading part. We'll come to this matter later. First let's examine the history of whaling regulation.

1.3 Attempts to Regulate Whaling

Spontaneous attempts at whaling regulation were first made long ago, e.g. in 1281, Icelandic whalers agreed on how to decide the matter when one whaler harpooned a whale and another whaler found it. At the end of the nineteenth century, a serious situation took place on the Norwegian coast, when the development of whaling led to a conflict between fishermen and whalers. By the king's edict of 1 February 1904, whaling in Finnmark was banned for 10 years. As a result, four out of ten whaling companies were liquidated, five started whaling near Spitsbergen, two near the Faroe Islands and one near Bear Island.

When whalers came to the Southern Ocean, the regulations meant paying fees to rent the shore stations and use the territorial waters in the area of South Georgia, the South Orkneys and the Falkland Islands. Duty was paid on the catch of various whale species. The number of whales allowed to be caught (the norm) was restricted. Catching suckling calves and nursing female whales with calves was banned, and fines had to be paid for unlicensed whaling. However, inspection did not exist, and the business was developing so intensively that these measures had no positive effects. It was more a method of obtaining money from territorial waters rather than the regulation of whaling.

At the beginning of the twentieth century, scientists and the public started worrying about the dramatic decrease in the number of whales. At a conference on wildlife conservation in 1913, a Swiss naturalist, P. Savatsin, predicted the rapid depletion of whale stocks and urged that whaling be limited and regulated (Brand 1940). The research of British and Norwegian scientists in the 1920s confirmed their fears (Harmer 1920). At the same time, the modernized whaling fleet was increasing its power and the volume of catches grew. A prominent English scientist, S. Harmer (Harmer 1920) stressed the need to take efficient measures to regulate whaling. Businessmen also started to worry. In 1923 the association of Norwegian whaling companies addressed a petition to the government to take necessary measures. However, the legislative measures prepared by Norway were not approved by other countries. Special conferences were required for the problem to be thoroughly discussed.

The first conference took place in Paris in 1927. Two leading whaling countries, Norway and Great Britain, agreed on common rules. In 1929, Norway adopted a law prohibiting the catch of right whales, suckling calves and females with calves. Complete use of whales was ordered by law. The following year the law was completed by setting minimum length limits for blue whales and fin whales.

In 1930 the League of Nations held a conference of experts from different countries. In 1931, in Geneva on the basis of Norwegian law, 26 countries with the aid of the League of Nations signed an international agreement restricting the whaling trade. According to the Agreement it was prohibited to take right whales of all types and nursing females with calves. A licensed system of whaling was introduced. Information on whaling results had to be reported to the Bureau of International Whaling Statistics. The agreement came into force in 1936 after it was ratified by all the participating countries.

The second international conference took place in London in 1937. Some rules of whaling were specified: minimum lengths for blue, fin, humpback and sperm whales were determined. Pelagic whaling in the Southern Ocean was limited to 40 degrees South latitude and within the period from 8 December to 7 March. The countries (with the exception of Japan) agreed to introduce inspection of the whaling trade. It should be noted that this did not mean international inspection, but inspection on behalf of the country interested in whaling.

The following year, a conference in London prohibited the hunting of humpback whales in addition to right whales. At the 1939 conference, the necessity of inspection was confirmed, with an additional request that one of the inspectors should be a biologist. At the 1944 conference, for the first time agreement was reached for a limitation of the number of whales taken – up to 16,000 "Blue Whale Units". The term "Blue Whale Units" (BWU) appeared as a result of agreement between the Norwegian and British whaling companies in 1931. The mass of the BWU was agreed as 83.9 tons, with 110 barrels of oil (i.e. 18.7 t). One BWU meant one blue whale, or two fin whales, or two and a half humpback whales, or six sei whales (and later Bryde's whales). Thus the quota was not set for each species separately but was generalized by "BWU", which led to the taking of the largest whales. After the 1972/73 season, the "BWU" approach ceased and separate quotas were determined for each species.

Unfortunately, all these agreements and legislative proposals proved ineffective. They could not have had a positive effect without international inspection. The volume of catches grew (20,000–40,000 whales were taken per season), reaching a maximum in 1937/38 (46,039 whales), mainly the biggest rorquals – the blue and fin whales.

The victory of the antifascist coalition in the war created a favorable political situation for the leading whaling countries to agree on the rules of the whaling trade. In 1946, the International Convention on the Regulation of Whaling was signed in Washington DC and the International Whaling Committee (IWC) was instigated. All the signatories to the convention agreed with the original regulations, which were written in the rules for the conduct of whaling. Unfortunately, humpback whales were allowed to be taken again, but no more than 1250 per season.

Signing of the International Convention in 1946 in Washington, DC

More progressive was an agreement in 1958 when the common quota on baleen whales in the Southern Ocean was distributed between the countries. In addition, the IWC Scientific Committee founded the so-called "Committee of Three" to evaluate the status of the populations of whales. The reliability of such an evaluation could not have been high and relied on a method for the determination of age and on knowledge of the biology of reproduction.

Beginning in 1974, whale stocks were classified into three categories: Initial Management Stocks; Sustained Management Stocks and Protected Stocks. The principle of whaling regulation was based on that of the "the maximum sustainable yield" (MSY). Catch limits were determined for each whale species separately and in fact by for whaling areas within the Southern Ocean. The Southern Ocean waters were divided into six "Areas" for baleen whales and into nine "Divisions" for sperm whales. As whale stocks continued to decline, more radical measures were needed. After 1975/76, the hunting of fin whales was banned and, after 1978/79, sei whales. In 1979, a new area banning commercial whaling was proclaimed by the IWC within 20°E and 130°E, including the African and Asian coasts in the Northern Hemisphere, i.e. all the Indian Ocean and waters adjacent to the Southern Ocean became a whaling "Sanctuary". In the same year, the decision was made to prohibit using factory ships to hunt of whales (apart from Bryde's whales) on the high seas.

Beginning in 1985/86, the IWC proclaimed a moratorium on commercial whaling for all species in all waters. It was a temporary measure to be reconsidered in 1990, but actually commercial whaling discontinued.

Why were all the measures taken ineffective? Why were nearly all whale species reduced and many populations seriously reduced? The main reason was that independent international inspection to ensure the fulfillment of the IWC decisions was not established. With this lack of inspection, illegal catching and the falsification of data reported to the IWC could and did flourish. It is impossible to use these data without serious correction in any scientific study. However, it is not very easy to "separate the wheat from the chaff" – I will examine this in a separate chapter using Soviet whaling practice as an example.

Peter Best was the first scientist who paid attention to the inaccuracy of the information sent to the IWC (Best 1989). On the basis of his contacts with whalers during the period 1957/1958 to 1978/1979, he indicated that not all taken whales were reported to the IWC. Sometimes the species was incorrectly identified, the lengths of the animals taken were reduced, and the embryos found were not mentioned. On other occasions, the lengths of "undersized" whales were enlarged, and the sizes of the embryos approximated. He mentioned also the inaccuracy of data on stomach contents. Best (1988) also referred to two cases when the USSR representative did not report humpback and right whales taken. The article (Best 1989) was discussed at the Scientific Committee of the IWC (SC/40/O 20).

After the collapse of the USSR, a group of scientists led by A. Yablokov started collecting data on the true extent of Soviet whaling. The data were first divulged in November 1993 at the plenary meeting of the tenth biannual conference of the Society of Marine Mammalogy in Galveston (Texas, USA). D. Tormosov and A. Yablokov presented a report and informed participants that the *Yuri Dolgoruky* fleet illegally took 717 right whales, 7207 humpback and 1433 blue whales in the Southern Ocean (Yablokov 1994). The leaders of the fleet at the time reported the following to the IWC: 0 right whales, 152 humpback and 156 blue whales. Later, four scientists of the former USSR: V. Zemsky, A. Berzin, Y. Mikhalev and D. Tormosov, who used to be in charge of laboratories studying cetaceans and participated in scientific groups on the Soviet fleets, reported the actual catch of whales in the Southern Hemisphere. They reported 3349 right, 48,477 humpback and 8439 pygmy blue whales taken by all four Soviet fleets. The leaders of the fleets had informed the IWC of four right whales, 2710 humpback and ten pygmy blue whales (Zemsky et al. 1994b). When these data were published, the scientific committee recommended making all of the Southern Hemisphere a protected area.

Cambridge, IWC Scientific Committee headquarter

The information on falsification of sperm whale data taken in the North Pacific by Soviet pelagic fleets and Japanese shore stations was presented to the IWC Scientific Committee meeting in 1998 (Brownell et al. 1998). The catch of males and females was underestimated 1.3 and 9.6 times, respectively.

T. Kasuya (1999) gave information about large scale falsification of statistics by Japanese whaling companies in the post-war period, including: the underestimation of the total number of whales taken, overstatement of small-sized whale body length, and the fabrication of correspondence between male and female sperm whales taken. Other scientists had also suspected Japanese shore stations of providing false statistics about sperm whales (Balcomb and Goebel 1977; Allen 1980; Cook and De La Mar 1983; Kasuya et al. 1998). On consideration of these facts at the meeting of Scientific committee of the IWC the scientists had to state that "the data by Japanese shore stations are inauthentic".

Spermwhale albinos

1.4 Characteristics of Whaling During the Soviet Period

Let us begin with the Soviet whaling in the Southern Ocean. After the Second World War, the USSR received a former German whaling factory, the "Vikinger" and several steam catcher boats as part of the war reparations. However, during the season of 1945/46, the British retained them and used them in the Southern Ocean under their flag. It was only in the summer of 1946 in Liverpool that they started to prepare the fleet to operate under the Soviet flag. The factory vessel was renamed the *Slava* and Odessa became its home port. It made 20 successful voyages. The whaling fleet administration was created in Odessa and its first head was the captain-harpooner A.N. Purgin, who made voyages with the fleets "Aleut" and then *Slava* (Purgin 1953).

After the repair work, the *Slava* fleet (the whaling factory and eight steam catchers) left Liverpool for its maiden voyage in December 1946 and started whaling in the Southern Ocean waters in January 1947. Among the crew members were those who had had experience on board the "Aleut" in the North Pacific. The novice crew members were former soldiers in the Second World War. Norwegian specialists were invited to be instructors and consultants on whaling, flensing and equipment usage.

Whaling factory "Slava"

In the first cruise pigs were taken. A storm eventually swiped them over board.

On 28 January, Morozov, the first mate of the whaler "Slava-5", discovered the first whale – a fin -, which was taken at 51°25′ South, 11°10′ West. The epoch of Soviet whaling had begun. The expedition took 154 blue whales, 226 fin whales, five sperm whales and one humpback – 386 whales in total. Having gained some experience, they nearly doubled their catch on the second voyage: 185 blue whales, 592 fin whales and 47 sperm whales – 824 whales total. After the first two voyages, the government dispensed with the service of the Norwegians. Catches rose dramatically, unfortunately mainly due to poaching. Inaccurate data began to be sent to the Bureau of International Whaling Statistics, for instance, the catch of 112 humpback whales was not reported. Instead of the 69 fin whales actually taken, a total of 718 was reported to BIWS and instead of the true catch of 97 sperm whales, 173 were reported because their quota had not been reached, and they wanted to retain the quota for the following voyage. This habit of lying was practiced for nearly a quarter of a century.

Appetites grew with every voyage. Management areas and whaling regulations started to be violated ever more seriously. Catches increased with every voyage – up to 2000, 3000, 4000, 5000 and even 6000 whales, mostly humpbacks, although the quota for them was limited to 1500 at that time. In 1957/1958, the *Slava* fleet alone took 2235 humpback whales, but informed the BIWS of only 60. The following season, the fleet took 4039 humpback whales, but reported only 420 to the Bureau.

Some countries (Norway, The Netherlands, Great Britain, the USA, Canada and Australia) had stopped pelagic whaling by the 1960s due to their comparatively low efficiency because of the reduction in whale stocks and thanks to growing efficiency in cattle breeding. The USSR, on the contrary, increased its whaling capacity. In 1959, a whaling factory the, *Sovetskaya Ukraina* was built, followed in 1960 by the *Yuri Dolgoruky* and in 1961 by the *Sovetskaya Rossia*. In 1963 and 1964, the fleets *Vladivostok* and *Dalny Vostok* were sent to the North Pacific. Slower steam catchers were superceded by more powerful diesel ones that could reach 16 knots. The Soviet whaling fleet became the most powerful in the world. With the rapid decrease in the quotas they relied only on poaching.

Whaling factory “Savetskaya Ukraina” in Odessa port

Whaling factory "Yuri Dolgoruky"

At that time, the IWC allowed the Southern Ocean whaling for baleen whales to start no earlier than 12 December and to finish no later than 7 April; the whaling area was restricted to south of 40° South. However, the Soviet fleets started taking baleen whales in September (!) and sometimes even on their way to the Southern Ocean. Most damagingly, the Soviet poachers went on taking humpback whales. In the 1959/1960 season, the Soviet fleet exceeded the quota (720 humpback whales) by 18 times (12,945 whales), and during the following voyage they took 12,529 humpback whales, exceeding the quota by 40 times (!) (Table 1.2).

Humpback whales became rare, and since the 1963/1964 season, the IWC has prohibited them from being hunted. However, nothing could stop the illegal catching by the Soviet fleet as they took what was left of the humpback stocks. After 1968, although only solitary humpback whales were encountered, they were also taken. Together with the near-extermination of humpback whales, animals from the remaining stocks of Southern right whales were also caught (their whaling was first prohibited in 1937 and that decision was ratified by the 1946 Convention). Despite this, during the post-war whaling period in the Southern Hemisphere, 3364 right whales were taken by the Soviet fleets.

It is important to consider blue whales separately. The largest of the whale species had been almost exterminated before it had been properly studied. At the beginning of the 1960s, the existence of another sub-species (Ichichara 1961, 1963, 1964; Zemsky and Voronin 1964; Gambell 1964) or species (Zemsky 1972) was recognized–the pygmy blue whale–a whale whose size is close to that of the fin whale. Now, backdating, according to regions and periods of whaling, we can approximately estimate how many pygmy blue whales were taken amongst the total blue whale catches.

It is important to note that, beginning with the 1965/1966 season, commercial exploitation of all blue whales was prohibited. After this, only 12 blue whale catches were reported to the BIWS. However, the real situation was different.

Soviet whalers continued taking blue whales until 1972. From 1959, they took 10,944 blue whales, 1622 of them "true" ones (1519 reported to BIWS) and 9322 pygmy blue whales, none of which were reported. After blue whales were protected in 1963, Soviet fleets took 543 of them and 4268 pygmy blue whales.

Recall that the number of reported whales was overstated or understated, depending on the situation needed to retain the quota. For example, in 1959/60, 521 fin whales and 350 sei whales were taken. Greatly overstated figures were reported – 3554 fin whales (by 682%) and 1188 sei whales (by 339%). By contrast, in 1965/66, 1560 fin whales and 4489 sei whales were actually taken, but only 973 fin whales (understated by 62%) and 2611 sei whales (understated by 58%) were reported to the IWC.

The true number of sperm whales taken was usually under-reported by two times or more, because females, which weigh much less than the males, were usually caught. Their lengths were smaller than the allowable minimum length limits, and that is why one large male was reported rather than two or three small females. In the 1960s, the Soviet fleet took 71,979 sperm whales, most of which were females (many pregnant, nursing and undersized). A total of 48,432 sperm whales were reported to the IWC, which was 67.3% below the actual level. In fact, Soviet whalers inexorably led to complete over-exploitation of the sperm whale stocks and the closure of whaling of that species.

The table above (Table 1.2) vividly illustrates the scale of poaching and the falsification of the data presented to the IWC. These true data have been obtained from genuine documents and have been reported at the meetings of the IWC Scientific Committee and published (Zemsky et al. 1994b, 1995, 1996; Mikhalev 1997a, b, 2000a, b, 2002b, c; Tormosov et al. 1998, 2000a, b; Yablokov et al. 1998; Zemsky and Mikhalev 2000a, b).

So much hypocrisy can be found in the words of those who had prepared and presented fabricated whaling data and who were the representatives of the Soviet Union at the IWC: "None of the four whaling fleets practically gave falsified data to the IWC, and all quotas were strictly observed"(Nikonorov "Mystery of whales" Moscow, VNIRO Press, 1995–4 with a Preface written by the Director of VNIRO A. Elizarov). In the second chapter (Nikonorov 1995, p. 22) cautiously confessed that the *Slava* surpassed the quota by 32%, and the *Yuri Dolgoruky* by only 4%.

A similar story was promoted in another work by VNIRO scientists I. Nikonorov and R. Borodin ("Fate of whales", 1997), and in the book by R. Borodin ("Whales: regulations and stock status", 1998). The following table illustrated the real situation (Table 1.3).

Table 1.2 Whaling in the South Hemisphere by the USSR, 1959–1972

Season	Blue whale	Blue whale pygmy	Fin whale	Sei whale	Bryde's whale	Hump-back whale	South right whale	Sperm whale
1959–1960	**129**	**0**	**521**	**350**	**0**	**12,945**	**79**	**59**
	526	0	3554	1188	0	720	0	840
1960–1961	**160**	**0**	**1844**	**933**	**0**	**12,529**	**18**	**497**
	301	0	4373	1070	0	302	0	974
1961–1962	**275**	**0**	**3976**	**2556**	**0**	**5507**	**1443**	**2893**
	407	0	4851	2138	0	270	1	1321
1962–1963	**296**	**677**	**2813**	**2639**	**4**	**2925**	**649**	**3487**
	196	0	4197	2498	0	263	0	2220
1963–1964	**164**	**1267**	**2081**	**4149**	**42**	**664**	**349**	**7105**
	74	0	2724	3390	0	0	0	4034
1964–1965	**104**	**3133**	**1529**	**4585**	**574**	**234**	**80**	**8763**
	20	0	1826	3931	0	0	0	5087
1965–1966	**153**	**865**	**1560**	**4489**	**247**	**2059**	**250**	**5851**
	0	0	973	4489	0	0	0	4954
1966–1967	**150**	**270**	**1991**	**5709**	**47**	**1035**	**160**	**6665**
	0	0	1662	1441	3	0	0	5197
1967–1968	**57**	**342**	**1573**	**8135**	**23**	**929**	**4**	**5820**
	0	0	1182	2566	0	0	0	3668
1968–1969	**94**	**580**	**1671**	**6981**	**4**	**2**	**0**	**5441**
	0	0	1199	2281	0	0	0	3907
1969–1970	**40**	**884**	**1082**	**6368**	**40**	**0**	**78**	**7427**
	0	0	1177	2340	1	0	0	5008
1970–1971	**35**	**798**	**1700**	**4369**	**18**	**0**	**2**	**6746**
	0	5	1282	2016	4	0	2	4907
1971–1972	**14**	**529**	**979**	**2308**	**487**	**3**	**0**	**11,225**
	0	3	1431	1141	1	0	0	6319
Total:	**1670**	**9344**	**23,320**	**53,571**	**1486**	**38,832**	**3212**	**71,971**
	1519	8	30,431	28,611	9	1555	3	48,432

Notes: first bold line – real data, second line – data reported to the IWC

As can clearly be seen, the Soviet fleets did not 'strictly observe' the whaling quota. The *Slava* surpassed the quota not by 32%, but by 37.4%. The difference may seem slight, but biological data, dates of capture, positions of capture, true sex data, etc. were misrepresented. In the 1960s, the quota was surpassed by 100%, i.e. twice as many whales were taken as were reported to the IWC! The *Yuri Dolgoruky* exceeded the quota not by 4%, but by 95.4%. In total, speaking only of numbers of animals, the Soviet fleets combined exceeded the quota by an average of 80%. The percentage of falsification was actually considerably higher, as the data were sometimes overstated, and at other times understated. In addition, the dates and areas of whaling, the species, sizes and biological condition of the whales were also fabricated. The data reported to the IWC were absolutely, 100% false.

Table 1.3 Whaling catches by the Soviet Union's fleets in the Southern Hemisphere and data submitted to IWC

Fleet	Seasons	Whales caught	Reported to the IWC	Difference (percent of quota)
Slava	1947/1966	59,378	43,212	16,166 (37.4%)
Slava	1959/1966	25,285	12,643	12,642 (100.0%)
Sovetskaya Ukraina	1959/1972	73,776	41,723	32,053 (76.8%)
Yuri Dolgoruky	1960/1972	46,006	23,543	22,463 (95.4%)
Sovetskaya Rossia	1961/1972	57,984	33,395	24,589 (73.6%)
Total:	1947/1972	237,140	141,893	95,247 (67,1%)
	1959/1972	203,047	111,324	91,723 (82.4%)

Whaling boat with caught whales

That is why, with all due respect to I.P. Golovlev, the former Chief Inspector of the Soviet whaling fleets, who summoned up the courage to inform the public about the illegal character of Soviet whaling (Golovlev 2000, p. 22), I cannot agree with his statement, "I can assert that on average the Soviet whalers took whales in violation of the rules in 60% of cases". As is shown from the information above, the true number of whales taken was exceeded by no less than 80%, and taking into account all other infringements (size and biological state of the animals, terms and regions of whaling, etc.) the data were completely inadequate. That is why the original statistics sent to the IWC must be replaced by the actual data. I also cannot agree with his statement that during the period when Japanese observers were present onboard the Soviet fleets there were no violations (pp. 17–18). The characteristics of whaling in the presence of the Japanese observers will be examined later.

I knew Igor Golovlev as an honest man and it is not improbable that when he was inspecting the fleet he recorded all the infringements and informed the Ministry about them. However, in my and my colleagues' experience as members of scientific teams on the fleets, the inspectors rarely recorded all the violations. All the inspectors together only informed the IWC about two or three infringements.

Whales near whaling factory

In the end, the large-scale poaching of the Soviet whalers led to dramatic decreases in the populations of fin whales, sei whales and sperm whales whose stocks had already been reduced, and to the commercial extermination of right whales and blue whales, which forced the USSR to reduce its whaling capacity and to bear losses. In 1966, the *Slava* fleet was removed from whaling and other fleets followed.

The year 1972 marked a turning point when the world 'Green' movement became involved and the Marine Mammal Protection Act was adopted in the USA. Under pressure from the public (as well as some prominent Soviet scientists – Sokolov, Tomilin, Yablokov), the IWC decided to introduce international inspection of whaling. Even after the introduction of international inspectors, the poachers found a way to continue their illegal catching but at a much lower rate. It was not by chance that Japanese observers were appointed to be on Soviet fleets and Soviet observers on Japanese fleets! Nevertheless, it must be stated that with the introduction of international observers, the illegal catching of blue, right and humpback whales stopped. Fin whale quotas were reduced and then hunting stopped three seasons later, while sei whaling stopped five seasons later. The killing of female sperm whales practically stopped, and then of males as well. Only the Southern Ocean minke whales continued to be taken and the USSR had to make two fleets redundant. In 1975, the fleet *Yuri Dolgoruky* went out of business, the *Sovetskaya Rossia* followed 5 years after.

A new epoch then started during which only the Southern Ocean minke whales were taken; it was also not without its problems. The Japanese industry was interested in Soviet whaling; fresh whale meat was sold to Japan and was reloaded to Japanese vessels at sea. That is why the international observers colluded with 300–350 minke whales over the quota being taken each season. Consequently, the scientific data and statistics reported to the IWC were false. We are preparing the associated material for publication.

The authors of the publications mentioned earlier were trying to promote the idea that the role of Soviet whaling in the near-extermination of whale stocks was not very important. For example, Nikonorov (1995, p. 3) stated, "Soviet whalers started intensive whaling at the end of the 1940s when it was not possible to influence substantially the situation". In the preface, mentioned above, the VNIRO deputy director Elizarov said, "The part played by the USSR in the whaling of the most valuable whales was insignificant. … During the whole whaling period in the Southern Ocean, beginning in 1919/1920 till its closure, 1.4 million whales were taken, and only 210,000 of these by Soviet fleets, i.e. less than 15%" (Nikonorov 1995, p. 22). Underlying their hypothesis were the statements that "the proportion of the whales taken by the former USSR in global whaling is only 12–16%" (Borodin 1996, p. 189; Nikonorov and Borodin 1997, p. 12) and that by the beginning of Soviet whaling "the stocks of humpback, blue and fin whales had already been undermined" (Nikonorov 1995, p. 5).

There is no doubt that Soviet whaling in the Southern Ocean started after the Second World War, and reached its peak in the 1960s. It is true that before the war, the main whale stocks in the Southern Hemisphere had already been overexploited. The question is, why should the USSR have spent money on building new whaling ships, increasing their capacity from one to four fleets, when other countries had already been reducing their whaling fleet due to the low profitability of the whaling business? Had the stocks of whales been catastrophically reduced? Is there logic in comparing catches of all foreign fleets over several centuries with that of the Soviet fleets in less than three decades?

In fact in the seventeenth century, all the fleets combined took about 1700 right whales. In the nineteenth century, they took up to 3000 right whales and 6000 sperm whales. In the twentieth century, in the 1930s alone, all countries combined (41 fleets) took 24,000–46,000 of all species annually. The average annual whale catch came to 13,670 blue whales, 12,704 fin whales, 1645 humpbacks, only 115 sei whales and 888 sperm whales. The 41 fleets and shore stations in the Southern Ocean caught 18,891 blue whales (including pygmy blue whales), 28,009 fin whales, 4477 humpbacks, 490 sei whales and 2585 sperm whales in the best season.

Let us compare these figures with Soviet whaling in the 1960s (in practice from 1959 to 1972). Remember that only two Soviet fleets, the *Slava* and the *Sovetskaya Ukraina*, took 22,564 humpback whales during the seasons 1959/1960 and 1960/1961 – proving that the species had not really been overexploited and that the population of humpbacks could have recovered! However, as a result of the Soviet's illegal catches, only two or three animals were encountered and taken by the end of the period.

The available stocks of fin whales allowed the USSR to take 1500–3500 and more whales annually without any particular effort. Over the entire period, the Soviet fleets took 23,320 fin whales. The stocks of sei whales were largely unexploited when the Soviet fleets started operations and the whalers annually took 2500–8000 animals. A total of 53,571 sei whales and 1486 Bryde's whales were taken in the 1960s. As a result, the population of sei whales close to the islands Tristan-da-Cunha was seriously over-exploited. Most of these were small animals, which were supposed to be protected. A total 72,000 sperm whales was taken during the Soviet period.

The stocks of blue and right whales had already been severely reduced by the period of Soviet whaling. However, the Soviet whalers took 100–200 blue whales and 500–3000 pygmy blue whales in disregard of the prohibition of catching them. In total, the Soviet fleets took 11,000 blue whales of both types. They also showed no mercy for right whales despite their protection. A total of 1443 Southern right whales were taken during the 1961/1962 season; by the time international inspection was introduced, the USSR had taken 3212 right whales in the Southern Hemisphere.

Let us now compare the intensity of active whaling in the 1930s with Soviet fleet whaling in 1960s. During the 1930s, the average annual catch was 330 blue whales, 310 fin whales, 40 humpbacks, 3 sei whales and 22 sperm whales – 700 whales total. In the 1960s, when many stocks had already been reduced, an average Soviet fleet annually took 242 blue whales (88 blue whales fewer), 512 fin whales (202 more), 1210 sei whales (1207 more), 853 humpbacks (813 more), 70 southern right whales (these were not caught in the 1930s, observing the restriction on whaling) and 1582 sperm whales (1560 more). One Soviet fleet took an average of 4470 whales per season (3770 more than in the 1930s). That means that the intensity of Soviet whaling was more than 6 times higher in terms of catches.

Let us compare the average catch of a Soviet fleet with the average catch made by an average fleet of the other countries in the 1960s (Table 1.4).

The number of whales caught by Soviet fleets in the 1960s compared with the average catch by the fleets of all other countries combined was 257% higher (i.e. by 2.5 times) and much more intensive than when whaling was at its height in the 1930s. The large-scale killing of whales had no economic basis. The profit of the whaling fleets in the 1960s from catches of 15,000 big whales per season was less than that of the 1980s from much smaller (in mass) 3000 minke whales whose meat was sold at sea to the Japanese.

Table 1.4 Average whaling catches per fleet in the 1960s

Whaling seasons	World countries (without USSR) Fleet numbers	World countries (without USSR) Average catch by fleet	USSR Fleet numbers	USSR Average catch by fleet	Difference (in %)
1959/1960	18	1770	2	7044	398
1960/1961	18	1895	3	5323	281
1961/1962	17	1742	4	4126	237
1962/1963	13	1619	4	3390	209
1963/1964	12	1694	4	3965	234
1964/1965	11	2190	4	4666	213
1965/1966	7	2497	3	3900	156
1966/1967	6	2186	3	5359	245
1967/1968	5	1908	3	5570	292
1968/1969	3	1786	3	4935	276
1969/1970	3	1789	3	5312	297
1970/1971	3	1976	3	4563	231
1971/1972	3	1885	3	5195	275

Whaling did not cover the expenses of building, repairing and maintaining whaling fleets and factory ships nor their crews. The money spent could have been used for the development of more cost-effective animal husbandry, aviculture and pisciculture. The management of whaling proved to be a disgrace for the Soviet leadership, and the same is true of the authors of the aforementioned publications (and those who ordered those publications), that continued to try to defend the predatory illegal whaling.

Summary

Historically whaling included several qualitatively different epochs:

1. primitive aboriginal;
2. expansion of the catch area during the Middle Ages;
3. active commercial whaling in the eighteenth to nineteenth centuries during the period of whaling fleet modernization and the growth of capitalism;
4. penetration of whalers into the Southern Hemisphere and the Southern Ocean from the beginning of the twentieth century;
5. post- Second World War period of the most intensive illegal whaling; and
6. introduction of international inspection and the decline of whaling.

In each succeeding epoch the intensity of whaling grew, while the length of the epoch dramatically declined. During the Middle Ages all countries together killed 1700 whales each year. During the nineteenth century about 7000 whales were taken. Prior to the Second World War up to 50,000 whales were killed in the Southern Hemisphere.

Whale poaching necessitated the passing of common laws for whaling and whaling regulations. As a result of the global political situation, the scant knowledge of whale biology, inadequate information about levels of whaling and biological information about the captured whales, these attempts proved to be ineffective. However, the main reason for this failure was the unwillingness to provide for international independent inspection.

In the post-war period, whalers killed up to 65,000 whales per season. The 1960s were characterized by an abrupt decrease in whale numbers and the cessation of whaling in many countries. Only Japan and the USSR continued commercial whaling in the Southern Hemisphere. The latter built new modernized whaling factory ships and catcher boats. The average annual catch by the Soviet fleet during that period was more than 2.5 times higher than that of any foreign fleet and more than 6.5 times higher than catches by foreign fleets in the 1930s.

The high Soviet catches were reached by large-scale, illegal whaling and a complete disregard of the whaling rules. Soviet whaling brought whale populations to near-extinction, was unprofitable, amoral and politically damaging for the country. The 1970s were characterized by the introduction of international inspectors and by a reduction in whaling and of the whaling fleet.

Beginning with the 1949–1950 season, Soviet whaling ships sent false whaling statistics to the IWC; these must be completely replaced by the true data, which is one of the objectives of this work.

Information about illegal catches by Japan needs to be stated more precisely.

Chapter 2
Analysis of the Correlation Between Whale Length and Weight

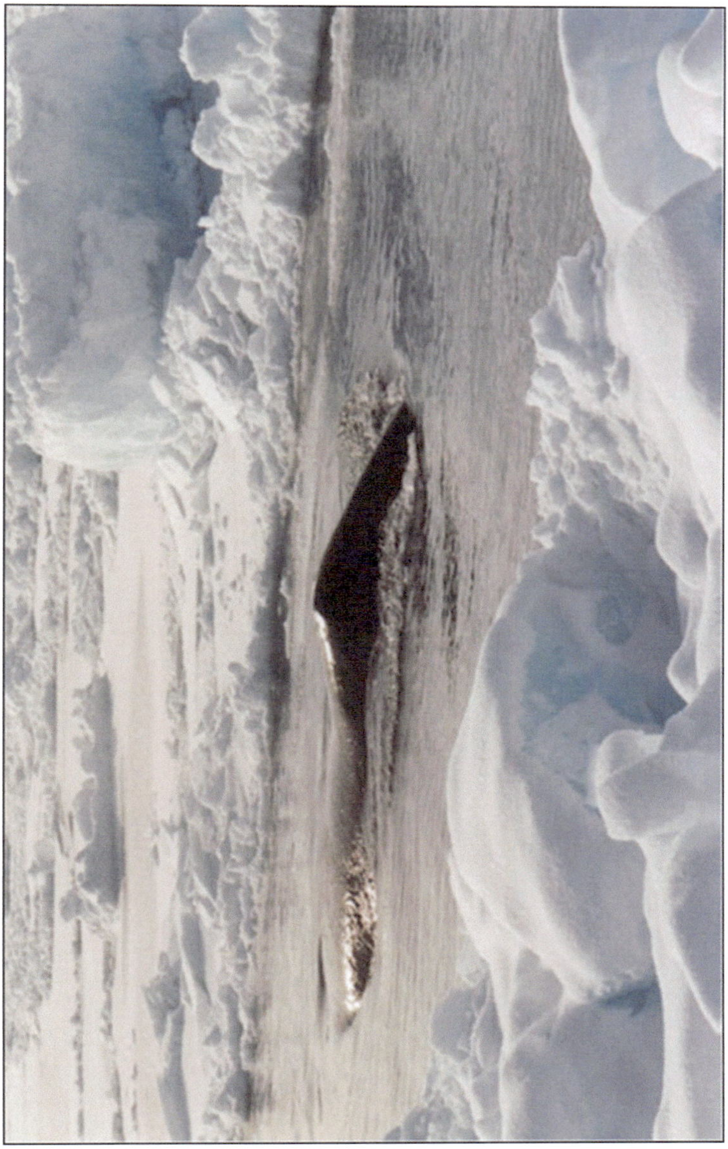

Y. Mikhalev, *Whales of the Southern Ocean*, Advances in Polar Ecology 5,
https://doi.org/10.1007/978-3-030-29252-2_2

Meeting of *Sovetskaya Ukraina* and *Slava* off Balleny Islands

Storm. Two unforgettable cruises on boat "Bodry 25"

Size is a fundamental characteristic and a species feature of animals. In addition, differences in size can be evidence of the ecological characteristics of species. For practical application, size is required to determine the biomass of different species, populations or herds, the amount of raw material that can be harvested from them, and the different types of produce, such as meat, fat and so on.

Studies of animal morphometry are necessary to determine the most effective measure by which size can best be characterized and compared. Biologists often prefer weight, as weighing is a less laborious procedure than measuring volume or determining the surface area of the body (or part of the body).

For whales, however, weighing often proves to be an extremely complicated procedure and is sometimes completely impossible. Hence, when characterizing the size of these leviathans, we use a more accessible, and sometimes more informative measure – the overall length. However, using length instead of weight is only justified if a constant relationship between these measures can be established. It is necessary to define whether this connection exists for whales.

2.1 Review of Previous Research Studies

In early research, on the basis of a small number of test weighings, several formulae were empirically determined and in addition to length (L), the girth or diameter (D) of the body was used, which was determined by the body transverse after the head was dissected. Let us examine some of these formulae.

Zenkovich (1937) suggested that the weight of fin whales, gray whales and beaked whales is determined by the formula:

$$P = \frac{LD^2}{3},$$

and the weight of humpback whales:

$$P = \frac{LD^2}{4}.$$

To determine the weight of blue whales, A.G. Tomilin (1937) suggested the formula:

$$P = \frac{5LD^2}{45}.$$

E. Eyerdal (Tomilin 1957) replaced the diameter of the body with 'extreme' girth (O) and empirically found the following dependence:

$$P = \frac{LO^2}{41}.$$

Zemsky (1958) believed that during the period of fetal growth, the relationship between length and weight of the embryo is not constant. He believed that the weight of a fin whale increased slowly until the embryo reached about 75–80 cm, after which it began to gain weight rapidly. He illustrated this conclusion by changes in the value of the coefficient that determined the relationship of embryo weight to length. Thus, when the length of a fin whale embryo was 50 cm and the weight was 1.2 kg, the coefficient was 24.2, whereas when the length was 565 cm and the weight was 1250 kg it was 2212.3.

Bodrov and Grigoriev (1963) suggested that the weight of sei whales could be determined by the formula:

$$P = LK,$$

where K -is an empirically-determined coefficient.

Rice and Wolman (1971) estimated the weight of gray whales using the formula:

$$P = aLD^2.$$

Similar relationships were offered by many other researchers (Omura 1950; Ash 1952, 1953, 1957; Fujino 1955; Laws 1959; Gihr and Pilleri 1968; Ohsumi et al. 1970).

Different opinions about the characterization of the relationship and numerous inaccuracies in the determination of whale weight according to the suggested formulae demanded further research and a more theoretically grounded method.

2.2 Theoretical Basis of the Method

Galileo, in his *Dialogue Concerning the Two Chief World Systems* (1632, in the 1914 edition), wrote that the weight of an animal increases in proportion to the cube of its length. Many researchers used the cubic function for a mathematical description of the correlation between whale weight and length in the prenatal and postnatal periods (e.g. Zasosov 1965; Hugget and Widdas 1951; Rice and Wolman 1971; Schmidt-Nielsen 1987). A small number of control weighings and empirically determined coefficients in the formula did not provide reliable results. The inaccuracy of their estimated whale weights was more than 10–15%; this was not satisfactory for either biologists or operatives.

Better prospects of obtaining a satisfactory relationship were created by considering a logarithmic dependence. The body shape of cetaceans has a certain similarity to that of some fish. For many fish that have an ellipsoid, drop-shaped form, it has been shown since the 1930s (Dunker 1924; Tyurin 1927; Yosipov 1929; Vinberg 1971; etc.) that that their weights lie on a straight line on a logarithmic plot.

It follows from this that the relationship between the logarithm of their body weight and the logarithm of their length shows a direct proportional dependence with a correlation factor close to one (this is very important) and can also be expressed by a power function of the type:

$$P = aL^b,$$

since, on the logarithmic scale we obtain the formula of a straight line:

$$\lg P = \lg a + b \lg L.$$

Since the whale's body, like many fish, has an elliptical shape, one might expect positive results for a logarithmic dependence for whale length and weight. A number of research works have shown such a dependence (e.g. Omura 1950; Ash 1952, 1953, 1957; Fujino 1955; Laws 1959; Mikhalev 1969, 1970a, b, c, 1975, 1978a, 1984; Lockyer 1993, 1995; etc.). This provides the basis, when studying whale growth, for relying on the length of the body, rather than their weight, as a characteristic of their size.

It is also very important that the coefficients in this relationship make biological sense: the constant of proportionality, *a,* characterizes the weight of the animal per unit of length, and the exponent *b* shows intensity of weight growth with increasing body length, which is inherently a regression characteristic. Another valuable quality of the coefficient *b* is that it is an abstract value. That is why it can be compared between different animal species as well as within one species, reflecting either population or ecological distinctions, as has been established for dolphins (Gihr and Pilleri 1968), or differences in the degree of fatness within the same groups of animals in different seasons.

2.3 Correlation Between Prenatal Length and Weight of Whales

We have processed the results of weighing 556 whales, from newborn to physically mature animals, and more than 1150 embryos, from the embryonic stage to prenatal fetuses, all of which were carried out on the factory ships *Slava*, *Sovetskaya Ukraina* and *Yuri Dolgoruky*. Data from the literature were also used (Tomilin 1957; Zenkovich 1937; Omura 1950; Laws 1959; Zemsky 1958; Govorokov 1934; Dorofeev and Arsenyev 1936; Drukker and Galichko 1936; Kiesewetter 1953; Kleinenberg et al. 1965; Omura et al. 1959; Budylenko 1970a; Vidal 1995; etc.)

Thus the results of the weighing of 888 whales and 1382 embryos have been analysed (Table 2.1).

Table 2.1 Types, quantity and sizes of weighed whales

Whale species	Prenatal period numbers, sizes					Postnatal period numbers, sizes				
		min		max			min		max	
	n	*L*, m	*P*, kg	*L*, m	*P*, kg	n	*L*, m	*P*, kg	*L*, m	*P*, kg
Blue whale	24	0.17	0.37	6.40	2050	57	7.0	2000	29.5	163.000
Fin whale	216	0.09	0.07	6.00	1550	34	6.0	1500	22.5	58.700
Sei whale	247	0.24	0.27	3.97	450	34	5.0	800	16.8	31.600
Bryde's	3	0.44	1.42	0.6	3	14	4.0	500	15.5	20.440
Minke	522	0.12	0.04	3.16	340	330	3.2	295	10.1	10.000
Humpback	150	0.10	0.04	2.69	300	17	5.0	2000	15.0	35.000
Southern Right	7	0.21	0.21	2.70	257	18	11.7	2865	17.4	10.650
Sperm whale	123	0.06	0.01	4.08	680	66	4.0	800	18.6	38.520
Killer whale	1	–	1.40	–	120	95	2.0	160	8.6	9900
Beluga	88	0.08	0.02	1.63	63	179	1.5	50	5.2	1373
Bottlenose	–	–	–	–	–	5	5.7	3800	7.0	5200
Common dolphin	1	0.72		6.7		38	0.64	4.3	1.48	44
TOTAL:	1382 embryos					888 whales				

So, in order to determine the correlation between body length and weight of whales in the prenatal period, we analyzed the results of the weighing of 1382 embryos from the embryonic to the prenatal fetal stage.

In all cases, for baleen and toothed whale species, the weights of embryos on a logarithmic plot lie on straight lines with correlation coefficients close to one. The exception is for very early embryos when their body still has a 'C'–shape and their zoological length cannot be measured precisely. Here we provide only the plots and regression coefficients we consider reliable (Figs. 2.1, 2.2, 2.3, 2.4, 2.5, 2.6, 2.7 and 2.8).

As we can see, the logarithmic correlation between embryo length and mass for minke whales (Balaenoptera), blue whales and fin whales is described by a straight line (high credibility value). The same is true for rorqual whales of medium and comparatively small size–sei whales and minke whales. The picture is the same for the baleen whales Megaptera (humpback whales) and Balaena (southern right whales) (Figs. 2.5 and 2.6).

It is possible to calculate approximate coefficients and exponents for every species within an accuracy of one-tenth with the help of the diagrams 2.1–2.8, determining the tangent of the angle between the regression line and the abscissa and defining the shift of this straight line from 0 (zero value). But we'd like to get the most precise coefficients (Table 2.2), as they are specific for each species and, besides, their changes show the conditions of growth of the animals (Farris 1956), which can also be used in ecological research.

Let's look at the specific character of these correlations for different whales. In the cases when the reliability of regression coefficient has been high, for practical

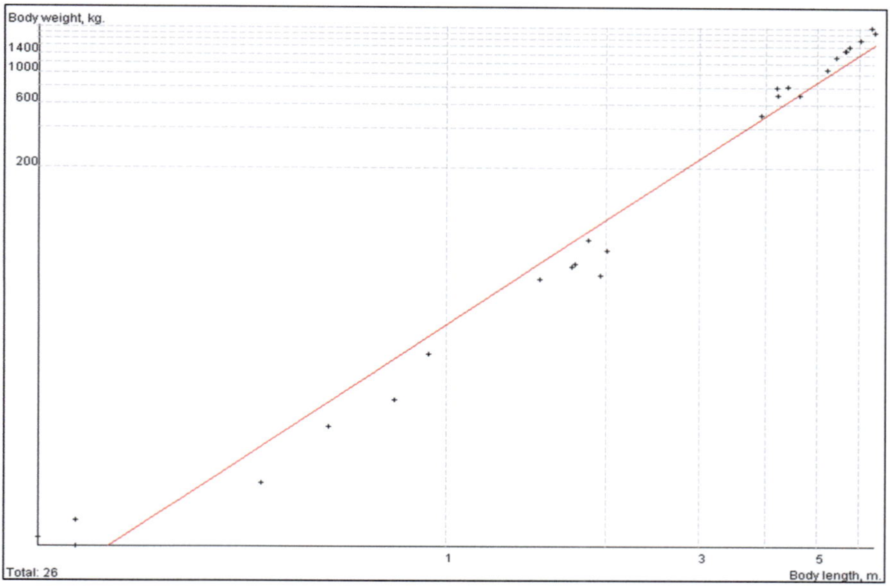

Fig. 2.1 Logarithmic correlation between blue whale embryo weight and length

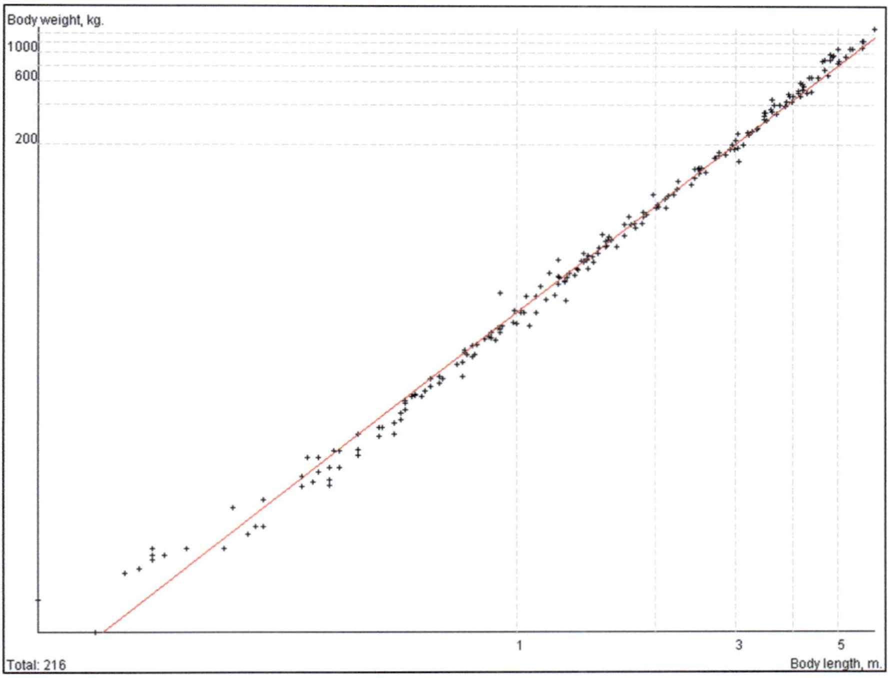

Fig. 2.2 Logarithmic correlation between fin whale embryo weight and length

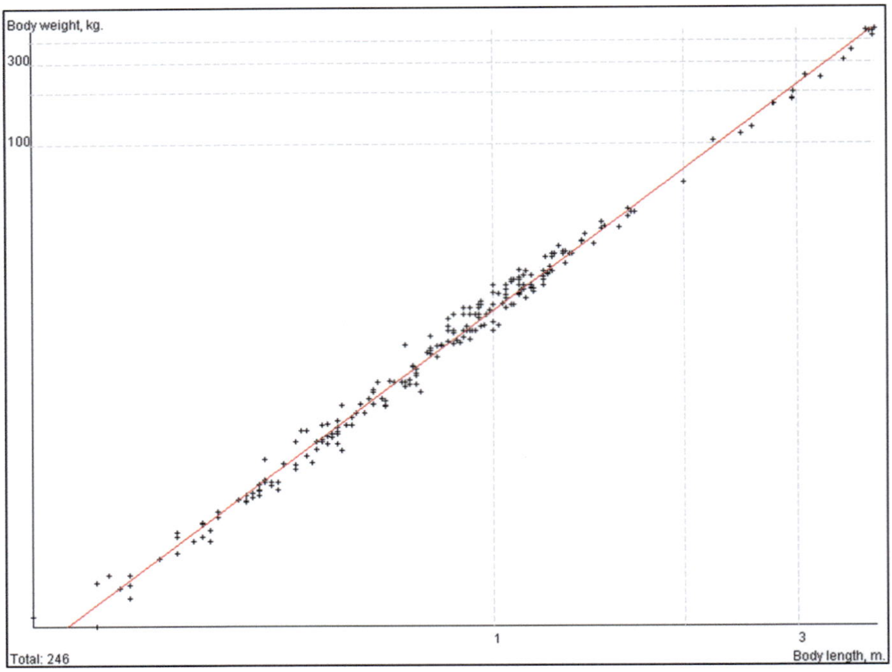

Fig. 2.3 Logarithmic correlation between sei whale embryo weight and length

reasons, we'll show the defining tables of weights, calculated on the basis of the formulae and the correction tables within 3σ.

2.3.1 Real Blue Whales and Pygmy Blue Whales

With regard to blue whales, we must point out that the embryos were weighed in the years when biologists and operatives had not yet distinguished two blue whale species. As they are rather close species, the correlation was high – 0.996. Although the difference between the blue whale species is not great, as the initial material is mixed, the power coefficient is a bit overstated. The embryo weight-length connection for the two blue whale species is described by the following formula:

$$P = 8.59L^{2.958+0.059}$$

Small sampling extracts and, judging by the margin of error (m + 0.059), a low credibility of regression coefficient makes the correlation between the length and the weight of the blue whales approximate and, using this formula, one cannot define reliably the weight of the embryos. To define the blue whale embryo weight the definition table for fin whales (Table 2.3) can be used.

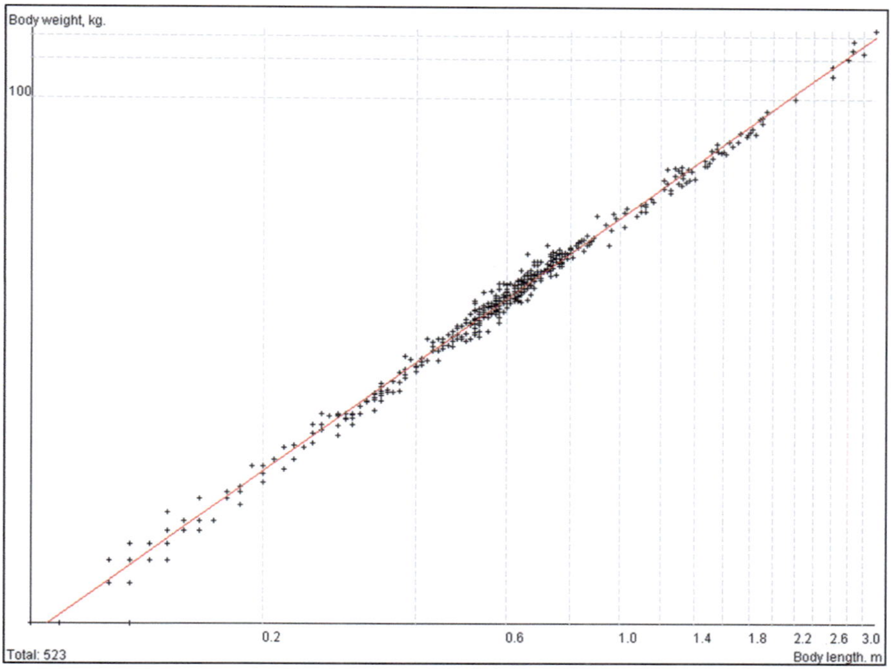

Fig. 2.4 Logarithmic correlation between minke whale embryo weight and length

Fin Whales The second biggest whale and the one most actively hunted in the 1960s and 1970s. The comparatively big sampling and the fact of weighing of embryos of practically all size groups (Tables 2.3 and 2.4) have determined high credibility of the "weight-length" correlation that we have researched:

$$P = 9.94L^{2.782+0.011}$$

Sei Whales and Bryde's Whales SAmong real rorqual whales is Bryde's whale, which by its size and appearance is very close to sei whales and similar to the two blue whale species. Operatives started to distinguish Bryde's whales from sei whales in the 1960s, before that time both were registered as sei wales. They were very rarely hunted. In this material three Bryde's whale embryos are included: a female 61 cm of length and 3.05 kg, a female 58 cm and 1.6 kg and a male 44 cm and 1.42 kg.

Of the remaining 247 embryos, the majority consisted of sei whales, while a minor number of Bryde's whale embryos (if any) in the samples could not have misrepresented the picture of weight-length correlation for sei whales. According to our calculations, the formula will be:

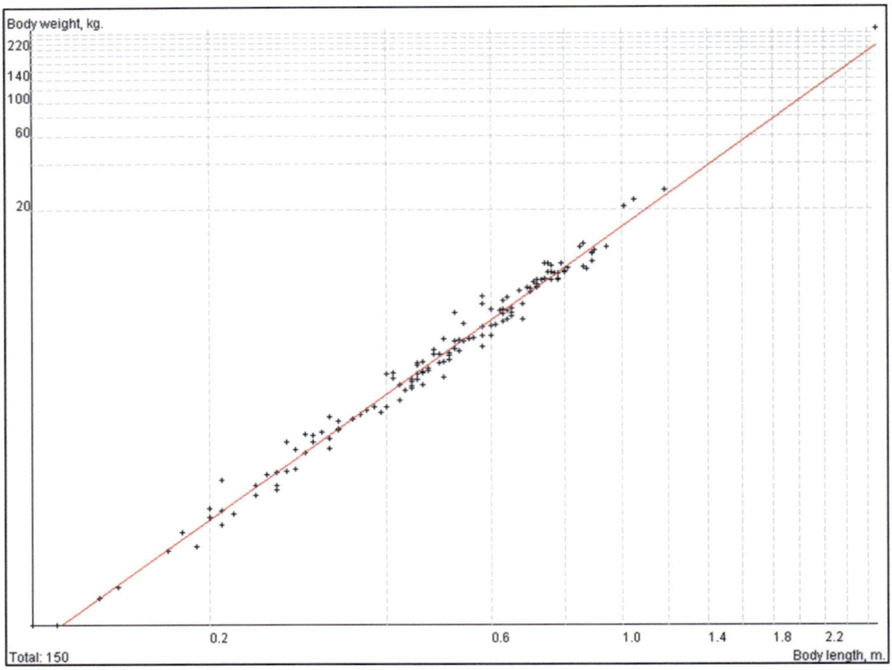

Fig. 2.5 Logarithmic correlation between humpback whale embryo weight and length

$$P = 10.5L^{2.772+0.016}$$

Lockyer C., (1993) researched the weight-length correlation for the South Atlantic sei whale embryos. According to her, the regression coefficient appeared to be overstated (b = 2.9198). To our minds, that happened because she underestimated the proportion coefficient (a = 9.68), which, as we already know, must be close to the weight of a one-meter embryo (Table 2.5).

Best (1977) received a power coefficient of b = 2.8368 for South African waters' Bryde's whale embryos. Because of the small sample numbers (six embryos), this index can only be accepted as approximate.

A lot of representative material, which we have used, allowed us to compose a rather precise definition table of sei whale embryo weights up to the newborn stage, excluding germ stage (Tables 2.5 and 2.6). The table can be used to determine the weight of Bryde's whale embryos as well.

Minke Whales Relying on weighing 191 minke whale embryos, Ivashin (1976) developed the formula:

$$P = 14.45L^{2.79}.$$

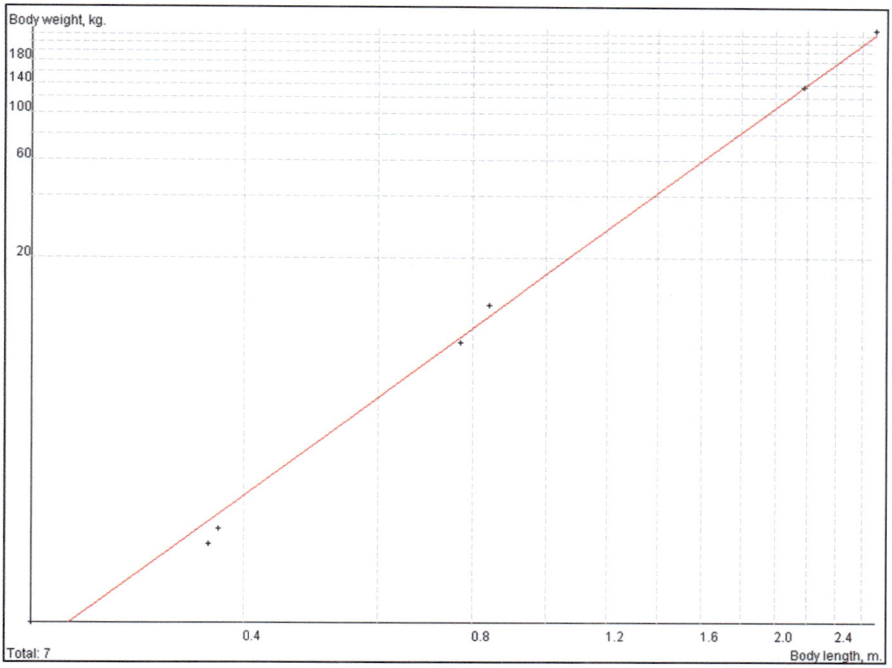

Fig. 2.6 Logarithmic correlation between right whale embryo weight and length

Active whaling of minkes in the 1970s allowed us to proceed with gathering material. We (Mikhalev 1978a, 1984; Ivashin and Mikhalev 1978) have processed the results of weighing 522 embryos, including sizes from the earliest development stages to prenatal fetuses, the biggest of which was 280 cm. A large quantity of samples has increased the credibility of our calculations (Table 2.7).

With a high correlation coefficient (r = 0.996) and a low standard regression coefficient error (m = +0.010), the following formula describes the connection:

$$P = 14.45L^{2.762+0.010}.$$

As the variability of minke whale embryo weights and the standard error are comparatively low, we have omitted the table of correction of weights. If necessary, the limits of embryo weight fluctuations can be calculated with the use of the formula given above.

Humpback Whales During our practice few humpback whales were caught, so we could only measure and weigh 150 humpback whales from 10 cm to 269 cm, which is not such a great number. Unfortunately, weight data for bigger embryos are absent, which has slightly reduced the authenticity of the correlation being analyzed (Table 2.8). As a result of regression analysis, the following formula has been set:

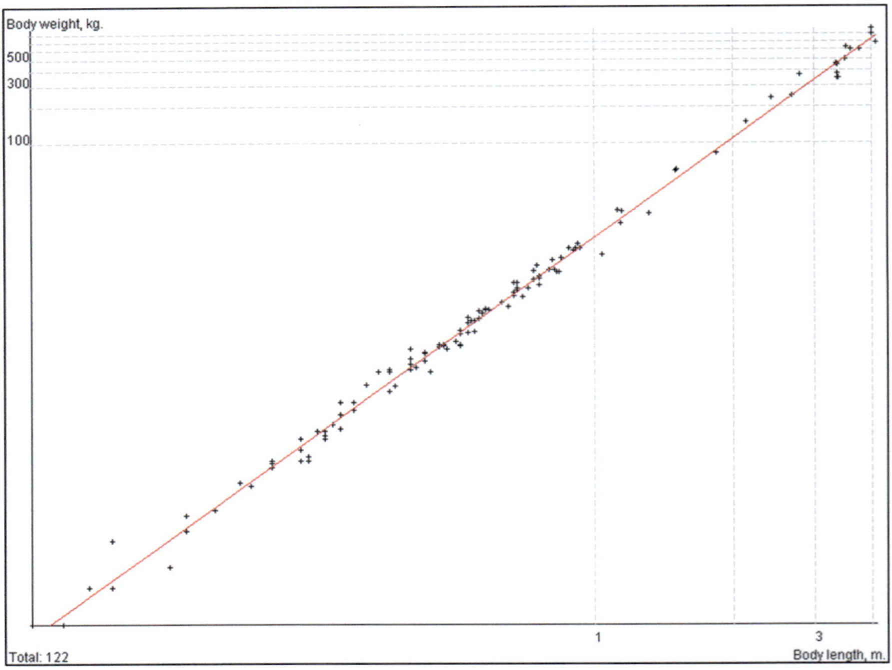

Fig. 2.7 Logarithmic correlation between sperm whale embryo weight and length

$$P = 15.6L^{2.741+0.028}.$$

Compared with the previous species, an abrupt increase in the constant of proportionality can be seen (a = 15.6), which is undoubtedly determined by the great difference between humpback whales' body shape, i.e. shorter and fatter, and that of minke whales. This difference can be also seen on the diagram by the regression line position (2.5).

Right Whales In the period of our practice not many right whales were hunted. We only have five reliable results for weighing Southern right whale embryos: four male embryos were 21 cm long and weighed 350 g, 36 cm and 840 g, 37 cm and 1.0 kg, 77 cm and 7.9 kg correspondingly; the female embryo was 84 cm and 12.0 kg. But even with such a small number, the connection between length and body mass proved to be rather close (r = 0.989). The error of regression coefficient is, naturally, high; consequently, the authenticity of the correlation, which we analyze, is low:

$$P = 15.7L^{2.608+0.222}.$$

Omura et al. (1959) have reported the weight of two embryos of Japanese right whales: a male 218 cm and 136 kg, and a female 270 cm – 257 kg. They were big

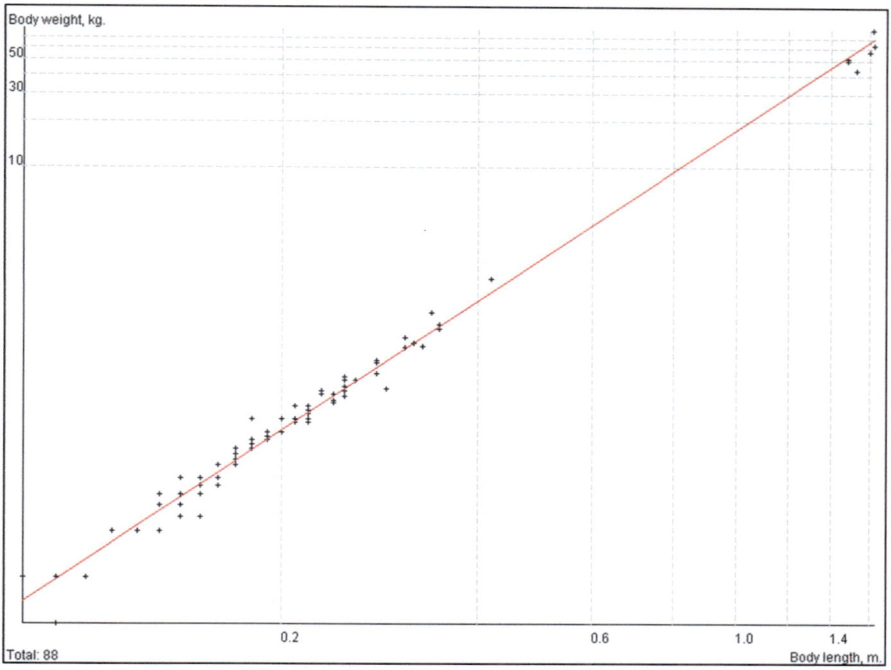

Fig. 2.8 Logarithmic correlation between beluga whale embryo weight and length

Table 2.2 Results of regression analysis of Whales' Embryo Weights

Whale species	Number, n	Coefficients of regression analysis				
		Correlation, R	Proportion, a	Regression, b	Sigma, $\sigma \pm m$	Reliability, s
Blue whale	24	0.996	8.59	2.958	0.059	69.7
Fin whale	216	0.998	9.94	2.782	0.011	36.4
Sei whale	247	0.995	10.5	2.772	0.016	6.95
Minke	522	0.996	12.3	2.762	0.010	3.89
Humpback	150	0.991	15.6	2.742	0.030	5.31
South	5	0.989	15.7	2.608	0.222	0.90
Sperm whale	123	0.997	16.8	2.726	0.019	29.5
Beluga	86	0.995	16.8	2.786	0.028	2.22

fetuses, which is important, and due to this the dimension line has become much longer, and the reliability of the connection, which we analyze, has increased. Taking into account these two embryos, the formula for the right whales becomes more reliable and looks like this:

$$P = 16.9L^{2.681+0.088}.$$

Table 2.3 Fin whale embryo weights definition table

Embryo length, cm	Embryo weight, kg				
	Length interval in 10 cm				
	0	**10**	**20**	**30**	**40**
0	–	0.018	0.119	0.365	0.809
100	10.14	13.15	16.75	20.93	25.64
200	68.71	78.52	90.99	101.2	113.5
300	210.2	229.8	251.2	272.9	296.3
400	465.2	498.7	531.8	566.6	590.0
500	862.0	906.7	960.2	1010	1063
600	1424	1489	1531	1599	1662
Continue embryo intervals (50–90 cm)					
–	**50**	**60**	**70**	**80**	**90**
0	1.463	2.415	3.787	5.476	7.571
100	31.05	36.56	43.75	51.29	59.70
200	127.3	141.8	157.0	173.8	191.1
300	321.8	347.3	374.9	404.6	433.8
400	643.4	685.7	709.4	768.8	813.9
500	1119	1177	1238	1294	1362
600	1778	–	–	–	–

Table 2.4 Value of corrections for fin whale embryos' weights for 3σ

Embryo length, cm	Embryo weight, kg				
	Length interval in 10 cm				
	0	**10**	**20**	**30**	**40**
0	–	–	–	0.055	0.119
100	1.402	1.811	2.288	2.837	3.462
200	9.030	10.29	11.67	13.15	14.74
300	26.85	29.32	31.94	34.67	37.59
400	57.17	62.17	66.33	70.66	75.16
500	106.0	111.8	117.8	123.9	130.3
600	173.0	180.0	188.9	197.2	205.7
Continue embryo intervals (50–90 cm)					
–	**50**	**60**	**70**	**80**	**90**
0	0.218	0.355	0.537	0.769	1.056
100	4.168	4.967	5.834	6.803	7.867
200	16.45	18.28	20.23	22.31	24.51
300	40.63	43.83	47.18	50.68	54.35
400	79.84	84.70	87.74	94.96	100.4
500	136.9	143.7	150.7	157.9	165.3
600	214.5	223.5	232.7	242.2	251.8

Table 2.5 Sei whale embryo weights definition table

Embryo length, cm	Embryo weight, kg				
	Length interval in 10 cm				
	0	**10**	**20**	**30**	**40**
0	–	0.028	0.173	0.507	1.083
100	12.18	19.69	24.34	29.58	35.50
200	75.90	86.24	95.17	110.0	122.7
300	221.3	240.8	262.3	283.9	307.2
400	473.0	505.9	537.4	571.1	606.8
Continue embryo intervals (50–90 cm)					
–	**50**	**60**	**70**	**80**	**90**
0	1.954	3.159	3.770	6.753	9.208
100	35.50	42.09	49.29	57.38	66.40
200	136.9	151.8	167.3	184.4	202.0
300	332.5	357.6	384.7	413.8	442.4
400	685.3	–	–	–	–

Table 2.6 Value of corrections for sei whale embryo weights for 3σ

Embryo length, cm	Embryo weight, kg				
	Length interval in 10 cm				
	0	**10**	**20**	**30**	**40**
0	–	–	–	0.036	0.081
100	1.077	1.409	1.800	2.256	2.780
200	7.593	8.712	9.932	11.26	12.69
300	23.80	26.10	28.54	31.13	33.86
400	53.52	57.38	61.41	65.62	70.01
Continue embryo intervals (50–90 cm)					
–	**50**	**60**	**70**	**80**	**90**
0	0.153	0.252	0.394	0.574	0.800
100	3.376	4.049	4.804	5.643	6.571
200	14.24	15.90	17.69	19.59	21.63
300	36.74	39.79	42.97	46.32	49.84
400	74.59	79.30	84.31	89.47	94.82

Sperm Whales After weighing 13 North Pacific sperm whale embryos, Omura (1950) calculated the regression coefficient = 3.18. The apparently high coefficient is certainly determined by the small number of embryos measured and weighed. Gambell (1966) showed a lower coefficient for South African sperm whales of the Durban area = 2.932 + 0.055.

Included in our regression analysis are 123 embryos of Southern hemisphere sperm whales with a length of between 6 and 408 cm, i.e. from prenatal to newborn. The following correlation was calculated, the correlation coefficient being close to 1 (r = 0.997), and the definition Tables 2.9 and 2.10 were created:

Table 2.7 Minke whale embryo weights definition table

Embryo length, cm	Embryo weight, kg				
	length interval in 10 cm				
	0	**10**	**20**	**30**	**40**
0	–	0.02	0.15	0.45	0.99
100	12.4	16.0	20.6	24.9	31.4
200	84.0	94.8	108	122	139
Continue embryo intervals (50–90 cm)					
–	**50**	**60**	**70**	**80**	**90**
0	1.84	3.06	4.62	6.68	9.25
100	37.9	44.2	53.5	64.7	73.5
200	158	174	191	217	–

Table 2.8 Humpback whale embryo weights definition table

Embryo length, cm	Embryo weight, kg				
	Length interval in 10 cm				
	0	**10**	**20**	**30**	**40**
0	–	0.034	0.225	0,677	1,480
100	17.97	23.15	29.37	36.57	44.68
200	118.0	134.5	152.5	172.8	193.5
300	355.1	387.8	423.3	459.2	498.0
400	776.9	832.6	943.6	1005	1069
Continue embryo intervals (50–90 cm)					
–	**50**	**60**	**70**	**80**	**90**
0	2.718	4.459	6.782	9.754	13.43
100	53.92	64.25	75.61	88.43	103.0
200	216.6	241.2	266.3	294.3	323.2
300	540.4	582.5	628.1	677.0	725.2
400	11.39	–	–	–	–

$$P = 16.7L^{2.726+0.018}.$$

Beluga Whales As we have already pointed out, the results of weighing 88 White Sea beluga whale embryos were given to us by G.A. Oganezov. In a logarithmic coordinate system, weights of the embryos lay on a straight line, beginning with 15–20 cm in length up to prenatal fetuses. The correlation coefficient for this length limit is 0.989. Regression analysis of this area resulted in the following:

$$P = 17.0L^{2.798+0.029}$$

Table 2.9 Sperm whale embryo weights definition table

Embryo length, cm	Embryo weight, kg				
	Length interval in 10 cm				
	0	**10**	**20**	**30**	**40**
0	–	0.034	0.223	0.669	1.460
100	17.49	22.59	28.64	34.81	43.50
200	114.4	130.4	147.8	167.4	187.4
300	343.2	374.5	408.6	443.2	480.6
400	748.5	801.7	853.3	908.2	966.8
Continue embryo intervals (50–90 cm)					
–	**50**	**60**	**70**	**80**	**90**
0	2.674	4.377	6.649	9.540	13.12
100	52.46	62.46	73.47	85.86	99.74
200	212.5	233.0	257.5	284.5	312.5
300	521.2	561.8	605.4	652.5	698.8
400	1029	1095	–	–	–

Table 2.10 Value of corrections for sperm whale embryo weights for 3σ

Embryo length, cm	Embryo weight, kg				
	Length interval in 10 cm				
	0	**10**	**20**	**30**	**40**
0	–	–	–	0.055	0.120
100	1.417	1.832	2.317	2.875	3.511
200	9.186	10.48	11.88	13.39	15.06
300	27.41	29.95	32.62	35.45	38.42
400	59.54	63.64	67.91	72.36	76.99
Continue embryo intervals (50–90 cm)					
–	**50**	**60**	**70**	**80**	**90**
0	0.219	0.357	0.542	0.776	1.067
100	4.229	5.033	5.927	6.914	7.999
200	16.77	18.64	20.63	22.76	25.02
300	41.54	44.82	48.25	51.85	55.61
400	81.80	86.80	91.98	97.35	102.9

We can point out that G.A. Ognetov (1987) received a much higher regression coefficient (b = 2.94). The difference can be explained by the fact that, while calculating, he mistakenly included minor embryos, whose zoological length cannot be measured precisely and whose weight does not lie on the straight line on the logarithmic graph. It is most likely, that, compared with our regression coefficient, the overstating of the latter in other authors' works about whales, as seen above, happened for the same reason.

Other Species of Toothed Whales We include several examples for minor species of toothed whales using data from the literature. For embryos from the North Atlantic grinds Globicephalus melas, the correlation between weight and length was calculated by Lockyer 1993; Bloich et al. 1993 … using the formula:

$$P = 13.6L^{2.677}$$

For porpoise embryos Phocoena phocoena from British waters, Lockyer, 1995 set the following correlation:

$$P = 15.3L^{2.633+0.059}$$

For Phocoena sinus of the Gulf of California. we have used the results of weighing 44 embryos published in Vidal's works (1995), and received the dependence with the correlation coefficient 0.979:

$$P = 18.4L^{2.550+0.080}$$

The examples given above show that in periods of prenatal development of two suborders of whales of both Northern and Southern hemispheres, there is high credibility value that the straight line describes the correlation between weight and length during the period of embryogenesis in the logarithmic scale.

In all cases the regression, or power, coefficient was less than three. It should be noted that although this difference of three in the regression coefficient seems to be moderate, it is really important, because, while calculating weights using lengths close to neonatal fetus size, the weight difference is about 20–30%. And more importantly, such a difference proves the thesis "that with growth of biological objects geometric similarity is not observed" (Vinberg 1966, p. 278). For embryos of fin whales, sei whales, sperm whales and Japanese right whales of the Kuril islands, this disproportionate growth was morphometrically shown by E.I. Ivanova (1961a, b, c, d). S. Ohsumi et al. (1958) have come to the same conclusion for fin whales of the Northern hemisphere on the basis of their own data and for those of the Southern Hemisphere on the basis of V. Zemsky (1958) data.

It is important to note that the functional dependence is species-specific. For species that are shorter-bodied but wide in the cross-section (humpback, smooth, sperm whales and others), the constant of proportionality is much higher than that for longer real rorqual whales. At the same time the intensity of weight growth with the increase in length is higher for the bigger species of real rorqual whales. Thus, the regression analysis has revealed a difference in correlation between fin whale and sei whale embryo length and weight, which R.M. Laws (1958) did not notice in his graphic analysis.

The constant of proportionality is higher with minor species, and in our opinion, the reason is because all calculations have been made in the m/kg system, therefore the constant of proportionality numerically equals the weight of a one-meter embryo. But the age of such an embryo of a minke whale, for example, is much higher than that of a one-meter blue whale embryo. Consequently, the body proportions of embryos of different ages are incommensurable. It appears from this, that the constant of proportionality is a function of age, whereas the power coefficient (index of regression) is a function of embryo length.

2.4 Correlation Between Postnatal Length and Weight of Whales

The procedure of weighing whales during a voyage on a whaling ship is rather laborious and complicated. The mast to which the dynamometer is fastened can only bear a weight of eight tons. Hence, whales heavier than eight tons must be weighed not as a whole, but in parts, with up to 5–7% losses of mass due to loss of blood and other liquids. During a voyage only several whales are usually weighed, which explains why fewer whales than embryos were weighed, and the higher inaccuracy. Naturally, the correlation coefficients turned out to be lower. Nevertheless, for all whale species we have achieved (Mikhalev 1970a, 1975, 1984) rather reliable indexes of correlation, and regression analysis of the correlation being researched in the meter/ton system. Based on the formulae, as for embryos, definition tables of weights and three sigma deviations were calculated, if the correlation authenticity was satisfactory.

It should be noted that because whales gain a significant amount of weight during the austral summer fattening period in the Southern Ocean waters, weight tables are compiled separately for each month of the whaling season. To illustrate this, we have included definition tables of weights in the meter/ton system for January and February (middle of the season) and for the whales of whaling sizes. Here are several examples.

Real Blue Whales and Pygmy Blue Whales It's difficult to separate the results of the weighing of the Southern Ocean blue whales and pygmy blue whales. That's why the data for both species have been processed together. Regression analysis has shown that the correlation between the blue whale weight and body length can be reliably described by the formula (Table 2.11)

$$P = 6.81L^{2.933}.$$

Fin Whales Functional correlation of fin whale weight and length is described by the formula, correlation coefficient being 0.989 (Table 2.12):

Table 2.11 Blue whale weights definition table

	Tenths of meter									
Length, *m*	0	1	2	3	4	5	6	7	8	9
19	38.3	38.9	39.5	40.1	40.7	41.4	42.0	42.6	43.3	43.9
20	44.6	45.2	45.9	46.5	47.2	47.9	48.6	49.3	50.0	50.7
21	51.4	52.1	52.9	53.6	54.1	55.4	55.9	56.6	57.4	58.2
22	58.9	59.7	60.6	61.4	62.1	63.0	63.6	64.6	65.5	66.2
23	67.0	68.1	68.9	69.7	70.6	71.5	72.4	73.3	74.2	75.2
24	76.1	76.9	78.2	78.9	79.8	80.7	81.9	82.8	83.8	84.8
25	85.7	86.7	87.7	88.7	89.8	90.8	91.9	93.0	94.0	95.1
26	96.4	97.3	98.4	99.5	100.6	101.8	102.8	104.0	105.2	106.3
27	107.5	108.7	109.8	111.0	112.2	113.4	114.6	115.9	117.1	118.3
28	119.6	120.8	122.1	123.4	124.7	125.9	127.2	128.6	129.8	131.2
29	132.5	133.8	135.2	136.6	137.9	139.3	140.7	142.1	143.5	144.9

Table 2.12 Fin whale weights definition table

	Tenths of meter									
Length, *m*	0	1	2	3	4	5	6	7	8	9
18	36.3	36.9	37.9	38.5	39.1	39.7	40.3	40.9	41.6	42.9
19	42.8	43.5	44.1	44.8	45.4	46.1	46.8	47.5	48.1	48.8
20	49.5	50.2	51.0	51.7	52.4	53.1	53.9	54.6	55.5	56.1
21	56.9	57.7	58.5	59.2	60.0	60.8	61.6	62.5	63.3	64.1
22	64.9	65.8	66.6	67.5	68.3	69.2	70.1	71.0	71.9	72.8
23	73.7	74.6	75.5	76.4	77.4	78.3	79.3	80.2	81.2	82.2
24	83.1	84.1	85.1	86.1	87.1	88.1	89.2	90.2	91.3	92.3

$$P = 9.89L^{2.840}.$$

Sei Whales Sei whales are much smaller than blue whales and fin whales, and that's why in calm weather some small species were weighed as a whole, which would seem to ensure higher accuracy in definition of body weight. Nevertheless the correlation coefficient, although rather high, proved to be a bit lower than that of the fin whales, with practically the same number of whales weighed. It is possible that among these weighed whales there was a small number of Bryde's whales. The mixture of two types could have reduced the correlation coefficient a little. Nevertheless, the following formula for sei whales is (Table 2.13):

$$P = 10.7L^{2.726}.$$

Bryde's Whales As has been already stated, processors usually mistake these whales for sei whales. However we have only processed data where there is no

Table 2.13 Sei whale weights definition table

	Tenths of meter									
Length, *m*	0	1	2	3	4	5	6	7	8	9
12	8.75	8.95	9.15	9.36	9.56	9.78	10.0	10.2	10.4	10.7
13	10.9	11.1	11.3	11.6	11.8	12.1	12.3	12.6	12.8	13.1
14	13.3	13.6	13.8	14.1	14.4	14.7	14.9	15.2	15.5	15.8
15	16.1	16.4	16.7	17.0	17.3	17.6	17.9	18.2	18.5	18.8
16	19.2	19.5	19.8	20.2	20.5	20.8	21.2	21.5	21.9	22.2

Table 2.14 Bryde's whale weights definition table

	Tenths of meter									
Length, *m*	0	1	2	3	4	5	6	7	8	9
11.0	9.68	9.92	10.2	10.4	10.7	10.9	11.2	11.5	11.7	12.0
12.0	12.3	12.5	12.8	13.1	13.3	13.6	13.9	14.2	14.5	14.9
13.0	15.2	15.5	15.8	16.1	16.5	16.8	17.2	17.5	17.8	18.2
14.0	18.6	18.9	19.3	19.7	20.0	20.4	20.8	21.2	21.6	22.0

doubt about their species. That may explain the fact that with the comparatively small number of weighings (14 Bryde's whales) the correlation coefficient is high −0.988. This high correlation can be explained by the fact that Bryde's whales are a bit smaller than sei whales and they were weighed whole. The correlation between the weight and length for Bryde's whales is expressed by the formula (Table 2.14):

$$P = 14.2L^{2.721}.$$

Minke Whales Active whaling of minkes started with the 1971/1972 season when the bigger whale types had been hunted to the extent that whaling targets could not have been fulfilled. Active whaling of a new species called for more detailed definition tables of weights. The correspondence between length and weight in this whale species was calculated separately for males and females, and for internal organs and different body parts.

Minke whales, which are comparatively small, were weighed whole. From 1961 to 1984, 316 minke whales were weighed by the Soviet whaling flotilla The results of weighing 12 minke whales published in other authors' works (Zenkovich 1955; Tomilin 1937; Ohsumi et al. 1970; Sleptsov 1961, 1965a, b). The data from the weight of two Northern hemisphere calves (Scheffer and Slipp 1948; Sleptsov 1952) have been used in graphic analysis only as approximations. The sample extract size is very big and, consequently, the coefficient authenticity rather high.

The analyzed males were 6.4–9.2 m; their weights were 2855 kg to 10 tons (r = 0.985 + 0.033). Females were from 6.5 m to 10.1 m long. Their weight varied from 3100 kg to 11,220 kg. Regression analysis for both sexes (r = 0.982 + 0.035) has resulted in the following (Tables 2.15, 2.16 and 2.17):

Table 2.15 Minke whale weights definition table (both sexes)

	Tenths of meter									
Length, *m*	0	1	2	3	4	5	6	7	8	9
6	–	2.62	2.75	2.82	3.03	3.13	3.33	3.49	3.65	3.81
7	3.98	4.16	4.34	4.53	4.72	4.92	5.12	5.33	5.54	5.76
8	5.99	6.22	6.46	6.70	6.95	7.21	7.47	7.74	8.01	8.30
9	8.59	8.83	9.18	9.49	9.60	10.1	10.5	10.8	11.1	11.5
10	11.8	12.2	12.6	13.0	13.4	13.7	14.1	14.6	15.0	15.4

Table 2.16 Male Minke whale weights definition table

	Tenths of meter									
Length, *m*	0	1	2	3	4	5	6	7	8	9
6	–	3.09	3.22	3.35	3.49	3.63	3.77	3.92	4.07	4.23
7	4.38	4.55	4.71	4.88	5.05	5.23	5.41	5.59	5.78	5.97
8	6.16	6.36	6.56	6.77	6.98	7.19	7.41	7.63	7.86	8.08
9	8.32	8.56	8.80	9.04	9.29	9.55	9.81	10.1	10.3	10.6
10	10.9	11.2	11.4	11.7	12.0	12.3	12.6	12.9	13.2	13.6

Table 2.17 Female Minke whale weights definition table

	Tenths of meter									
Length, *m*	0	1	2	3	4	5	6	7	8	9
6	2.31	2.43	2.56	2.69	2.83	2.97	3.12	3.27	3.42	3.58
7	3.75	3.92	4.10	4.28	4.47	4.66	4.86	5.06	5.27	5.49
8	5.71	5.94	6.12	6.41	6.66	6.91	7.17	7.44	7.71	7.99
9	8.28	8.57	8.87	9.18	9.49	9.81	10.1	10.5	10.8	11.2
10	11.5	11.9	12.3	12.7	13.1	13.5	13.9	14.3	14.7	15.1

$$P = 20.0L^{2.751+0.033}.$$

Comparing the correspondence between length and weight by region, we have only found reliable differences for minke whales to the south-west of both Heard and Prince Edwards islands, which, other conditions being the same, happened to be heavier than minke whales from other regions of the Southern Ocean (Tables 2.18 and 2.19).

It's important to note that the theoretical value of minke whale weight gives the summary an error within the limits of 0.7–1.0%, while it is higher than 10–20% when using the calculation under the formula of cubic dependence (Zasosov 1965).

Weight correlation of the Southern Ocean minke whale body parts and separate organs are given in Table 2.20.

Because of the small extracts and great variability of signs the correlation is low, as is the authenticity of regression coefficients. These can be only accepted as approximate. With some credibility value, it can be said that females gain integumentary fat more quickly ($b = 3.122 + 0.305$) than males ($b = 2.078 + 0.353$). The

Table 2.18 Minke whale weights definition table for region to the South-West of Heard and Prince Edward islands

	Tenths of meter									
Length, *m*	0	1	2	3	4	5	6	7	8	9
6	4.07	4.18	4.29	4.41	4.53	4.65	4.77	4.89	5.01	5.18
7	5.39	5.52	5.65	5.78	5.91	6.04	6.17	6.31	6.44	6.58
8	6.72	6.86	7.00	7.14	7.29	7.43	7.57	7.72	7.87	8.02
9	8.17	8.32	8.47	8.62	8.78	8.93	9.09	9.25	9.40	9.56
10	9.72	9.84	10.1	10.2	10.3	10.5	10.7	10.9	11.0	11.2

Table 2.19 Value of corrections for minke whale embryo weights for 3σ

	Tenths of meter									
Length, *m*	0	1	2	3	4	5	6	7	8	9
6	0.255	0.270	0.285	0.300	0.315	0.330	0.350	0.365	0.380	0.400
7	0.420	0.440	0.460	0.480	0.500	0.520	0.540	0.565	0.590	0.615
8	0.640	0.665	0.690	0.720	0.740	0.770	0.800	0.820	0.860	0.830
9	0.925	0.560	0.990	1.025	1.060	1.100	1.135	1.170	1.210	1.250
10	1.290	1.330	1.370	1.415	1.460	1.505	1.550	1.600	1.650	1.700

Table 2.20 Correlation between minke whale lengths and weights of body parts and internal organs

Body parts, internal organs	Number, n	Correlation index, $r \pm m$	Proportion index, $a \pm m$	Regression index, $b \pm m$
fat with peritoneum:	120	0.852 ± 0.067	2.495 ± 0.185	2.627 ± 0.349
muscle mass:	63	0.647 ± 0.097	19.104 ± 0.211	2.267 ± 0.347
head:	41	0.828 ± 0.089	3.170 ± 0.138	2.586 ± 0.279
skeleton without skull:	27	0.688 ± 0.145	0.651 ± 0.205	3.765 ± 0.793
heart:	26	0.718 ± 0.141	39.3 ± 4.3	7.998 ± 1.58
lungs:	24	0.644 ± 0.163	0.363 ± 0.002	4.718 ± 1.195
liver:	53	0.801 ± 0.145	0.247 ± 0.042	3.504 ± 0.524
kidney:	14	0.502 ± 0.249	0.900 ± 0.187	1.498 ± 0.709

weight of the liver also increases more quickly with females (b = 3.50 + 0.52) than males (b = 2.86 + 0.59). Biological sense of this phenomenon can be explained by the different biological functions of the sexes. Sexually mature females need additional energy storage for reproduction and feeding of the whale-calves.

Humpback Whales Humpback whales are the only representatives of the Megaptera family. Sexually mature humpback whales are as long as 15–16 m, but their body is less oblong than that of the Balaenoptera minke whales, and therefore their mass is much bigger. By the time of our voyage the hunting of humpback whales was coming to naught, and hence we weighed only 17 of them. With a correlation coefficient of 0.973, the connection between humpback whale length and weight is satisfactorily described by the formula (Table 2.21):

Table 2.21 Humpback whale weights definition table

	Tenths of meter									
Length, *m*	0	1	2	3	4	5	6	7	8	9
11	20.0	20.5	21.0	21.5	22.1	22.6	23.2	23.7	24.3	24.9
12	25.5	26.1	26.7	27.3	27.9	28.5	29.2	29.8	30.5	31.2
13	31.8	32.5	33.2	33.9	34.6	35.4	36.1	36.9	37.6	38.4
14	39.2	39.9	40.7	41.5	42.4	43.2	44.0	44.9	45.7	46.6
15	47.5	48.4	49.3	50.2	51.1	52.0	53.0	53.9	54.9	55.9

Table 2.22 Sperm whale weights definition table

	Tenths of meter									
Length, *m*	0	1	2	3	4	5	6	7	8	9
10	11.0	11.3	11.6	11.9	12.2	12.6	12.9	13.2	13.5	13.9
11	14.2	14.5	14.9	15.2	15.6	16.0	16.3	16.7	17.1	17.5
12	17.8	18.2	18.6	19.0	19.5	19.9	20.3	20.7	21.1	21.6
13	22.0	22.5	22.9	23.4	23.9	24.3	24.8	25.3	25.8	26.3
14	26.8	27.3	27.8	28.3	28.8	29.4	29.9	30.4	31.0	31.6

$$P = 24.8L^{2.794}.$$

Sperm Whales Out of the toothed whales, only sperm whales were actively hunted in the Southern Hemisphere. Their males are much bigger than their females, but no authentic differences in correlation between body length and weight due to sex have been found. Sixty-two sperm whales were weighed.

As for baleen whales, in logarithmic co-ordinates all weight values have satisfactorily lain on a straight line, the correlation coefficient being 0.978. Regression analysis set the following correspondence (Table 2.22):

$$P = 25.8L^{2.633}.$$

Killer Whales Only several of these animals of these were hunted during the season. We weighed 43 killer whales on board *Slava* and *Sovetskaya Ukraina* and used the results of the weighing of 12 killer whales from the works by Sleptsov (1952, 1961, 1965a, b). As these were not large whales, most of them were weighed as a whole. Seventeen males and 10 females were weighed in parts, in addition to two killer whales whose sex wasn't determined.

Weighing of a whole body ensured a close correlation between the whale length and weight, and between the length and weight of body parts and organs. Correlation coefficients happened to be higher than those of minke whales.

Male killer whales are much bigger than the females. But the regression analysis did not reveal differences in correlation between the whale length and weight because of their sex (the same was true for sperm whales).

Table 2.23 Results of correlation-regression analysis between length and weight of internal body parts of killer whales

Body parts, internal organs	Number, n	Correlation index, $r \pm m$	Proportion index, $a \pm m$	Regression index, $b \pm m$
fat:	25	0.95–0.07	3.25–0.03	2.95–0.28
muscle:	23	0.95–0.07	1.96–0.07	3.41–0.25
head:	17	0.87–0.13	3.50–0.04	2.53–0.38
skeleton:	20	0.94–0.08	4.58–0.05	3.00–0.25
heart:	16	0.84–0.09	0.36–0.3	2.14–0.37
lungs:	13	0.85–0.06	0.51–0.02	2.73–0.51
liver:	20	0.76–0.08	2.59–0.05	1.63–0.33

The correlation is expressed by the formula:

$$P = (16.97 + 0.18)L^{2.97+0.17}.$$

According to our material we can represent the difference between killer whales of the Balleny islands thus: $P = (18.94 + 0.03)L^{2.95 + 0.28}$ and the killer whales of Scotia sea: $P = (12.93 + 0.03)L^{3.145 + 0.52}$.

The results of the regression analysis can be illustrated by the following table (Table 2.23).

Compared with minke whales, the intensity of fat and muscle accumulation (loin and abdominal flesh) is higher for killer whales. Relative weight of the head is approximately the same. Relative weight of killer whale lungs is much higher than that of minke whales, not at the expense of the coefficient of the weight growth intensity, but at the expense of a higher correlation coefficient. By the neonatal period the killer whale lungs are already more developed, perhaps because they are better divers.

The liver of the killer whales also reaches a relatively big size, which is determined by differences in nutrition between minkes and killer whales. The relative mass of the skeleton is higher than that of the muscles and fat. A relatively heavy skeleton doesn't burden the whales, as they are hydrobionts and practically live in a zero gravity state.

As to the comparison of killer and sperm whales, the coefficient of accumulation of integumentary fat intensity of the former is lower than that of the sperm whales, and that of the muscles is higher. The reason for this may be the fact that killer whales are more active predators.

Beluga Whales A total of 179 weighing results have been included in the correlation analysis of connection between length and body weight. This comparatively big sample has ensured the reliable authenticity of the coefficient according to the formula (Table 2.24):

$$P = 21.2L^{2.647}.$$

Table 2.24 Beluga whale weights definition table

	Tenths of meter									
Length, *m*	0	1	2	3	4	5	6	7	8	9
2	133	151	171	192	215	240	266	294	323	355
3	388	423	460	499	541	584	623	676	726	777
4	831	887	946	1006	1070	1135	1203	1274	1347	1422
5	1500	1581	1664	1750	1839	1931	2025	2122	2224	2325

Table 2.25 Correlation between length and weight of whales in the postnatal period

		Regression analysis coefficient					Connection
Whale Species	Number, n	correlation, r	proportion, a	regression, b	sigma, $\sigma \pm m$	reliability, s	formula, $P = aL^b$
Blue whale	63	0.987	7.47	2.933	0.059	5414	$P = 6.81L^{2.933}$
Fin whale	34	0.989	9.89	2.840	0.074	4080	$P = 9.89L^{2.840}$
Sei whale	38	0.985	10.7	2.726	0.096	1807	$P = 10.7L^{2.726}$
Bryde's	14	0.988	14.2	2.721	0.118	1940	$P = 14.2L^{2.721}$
Minke	328	0.998	20.0	2.751	0.017	871	$P = 20.0L^{2.751}$
Humpback	17	0.973	24.8	2.794	0.168	4874	$P = 24.8L^{2.794}$
Sperm whale	66	0.978	525.8	2.633	0.068	4261	$P = 25.8L^{2.633}$
Killer whale	95	0.970	518.3	2.938	0.075	689	$P = 18.3L^{2.938}$
Beluga	179	0.967	21.2	2.647	0.052	94.8	$P = 21.2L^{2.647}$

All examples of correlation between weight and length of the cetaceans in the postnatal period have shown the same pattern as for the embryos: with growth of whale size their weight grows in power function, the power index always being less than 3. But as some whales were weighed in parts and because of much fewer samples for every species, the authenticity of the index has been reduced a little in comparison with that for the embryos (Table 2.25).

Nevertheless, one can see that in general the coefficients of proportionality and the coefficient of regression for whales in the postnatal period have increased. This might have been caused by intensive fat accumulation in the given period. To a certain extent the mass has increased because of the females' gaining weight late in pregnancy. The following graphic charts (Fig. 2.9) illustrate the directly proportional correlation between the logarithm of weights and the length of the whales researched.

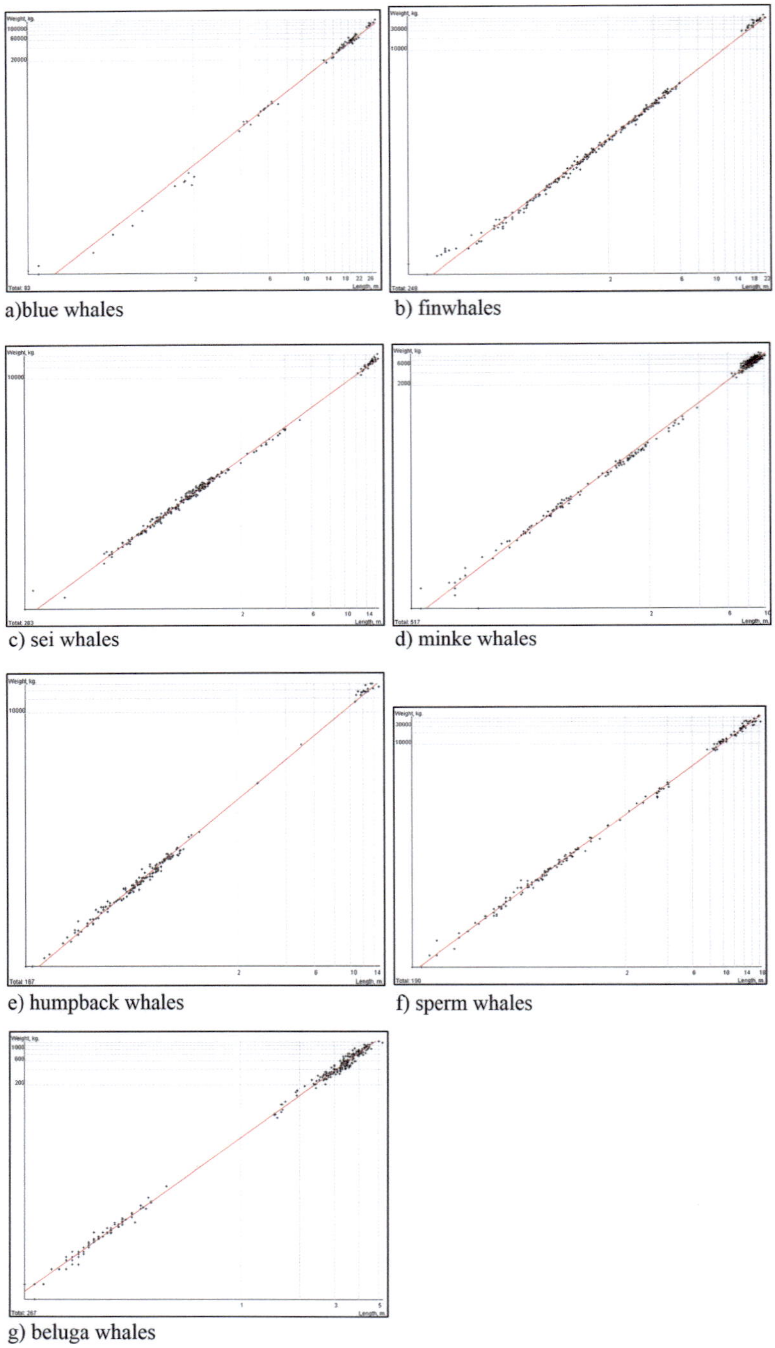

Fig. 2.9 Correlation between whale weight and length. (**a**) blue whales (**b**) finwhales (**c**) sei whales (**d**) minke whales (**e**) humpback whales (**f**) sperm whales (**g**) beluga whales

2.5 Summary

Size is a fundamental characteristic and a species feature of all animals. Cetaceans, like other animals, are most precisely characterized by weight, but the length of the whale is more accessible and often more informative. Hence the necessity to determine the existence of constant correlation between these measures.

Having weighed 888 whales and 1382 embryos, through graphic and correlation-regression analysis of the results it has been found that there is a close functional connection between whale embryo length (L) and weight (P), with the exception of the earliest stages when the fetus body still has a C-shape form:

$$P = aL^b.$$

This connection is species-specific: the constant of proportionality of shorter-bodied but more massive whales (humpback, right, sperm whales) is much higher than of that of real rorqual whales. The coefficients have not only mathematical, but also biological sense, viz.: the constant of proportionality *a* characterizes the weight of the animal with length for the unit of measure, and the power index *b* the intensity of mass growth with the increase in embryo length. The regression coefficient *b* is an absolute value, so it can be compared with different animal species and with taxons within a species, depicting either population or ecological differences.

Within the real rorqual whale genus, smaller species have a higher constant of proportionality (*a*), while the intensity of weight growth with body length increase (*b*) is higher in bigger species. In all cases of baleen whales Balaenoptera, Megaptera, Eubalaena and toothed whales Physeter, Delphinapterus, Phocoena, Globicephalus researched, the power coefficient proved to be less than three in the formulae of correlation between embryo length and weight.

The connection between baleen and toothed whale length and weight in the post-natal period happened to be, per se, the same as for the embryos. Compared with the prenatal period, the coefficients of proportionality and regression have notably grown, caused probably by intensive fat accumulation and by fetus development in the case of pregnant females.

For production purposes, the definition tables of whale weights and tables of correlation within standard error have been created on the basis of functional dependence. For minke whales, the definition tables have been differentiated according to sex, as well as for two regions – to the west of Heard and Prince Edward islands.

The rate of integumentary fat accumulation is higher for female minkes ($b = 3.122 + 0.305$) than for males ($b = 2.078 + 0.353$). The same applies to the weight of the liver: $b = 3.50 + 0.52$ for females, $b = 2.86 + 0.59$ for males. The intensity of killer whale integumentary fat accumulation is higher than that of minke whales, but lower than that of sperm whales. As to the muscles, it is higher for killer whales than for minke and sperm whales. The relative mass of the skeleton is also higher. The relative lung weight of killer whales is higher than that of minke whales, not at the expense of its mass growth density coefficient, but at the expense of a higher constant of proportionality.

With distinctive sexual dimorphism of sperm and killer whales, no essential differences in functional dependence between length and weight have been found. As to the regions, differences have been found for killer whales of the Balleny islands and the Scotia Sea.

Chapter 3
General Growth Laws

Y. Mikhalev, *Whales of the Southern Ocean*, Advances in Polar Ecology 5,
https://doi.org/10.1007/978-3-030-29252-2_3

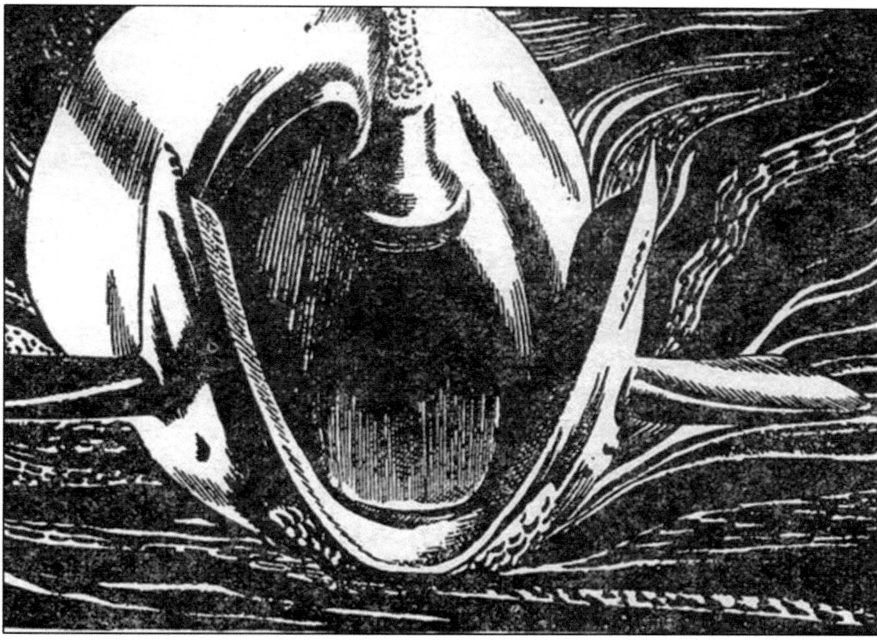

The following topics are considered in this chapter:

- Linear growth of whales during the period of total ontogenesis
- Sexual dimorphism and distinctions in growth rate between males and females
- Variational variability and animal growth features
- Relative growth of body parts (allometry) and morphostructural changes of animals

3.1 A Brief Review of Theories of Growth of Superior Vertebrates

The question of animal growth has always interested biologists. Of particular interest during the period 1900–1940, was the question of quantitative laws of total growth. Later on, as G.G. Vinberg (1966) fairly notes, this problem was not given "enough attention, in spite of the fact that when discussing many actual problems of theoretical and applied biology, biologists come across the necessity of organising knowledge in this area" (p. 276).

In our research we are more interested in the applied side of the question, i.e. the mathematical expression of quantitative laws, animal growth in formulae, which have, however, a deep theoretical basis. The review and analysis of such formulae and theories of total animal growth are not new (Hoffmann 1938; Mikhalev 1969, 1970a; Vinberg 1971; Mina and Klevezal 1976; Schmidt-Nielsen 1987; etc.). Several directions are already known.

3.1.1 Variational-Statistical Study of Total Growth

At its core, the study of the total growth of animals is reduced to the calculation of the statistical indicators characterizing the size of the organism at a certain age. In the phenomena of growth, deviations (decrease or increase) of the subjects under investigation are constantly encountered.

Therefore, to identify the optimal growth process the following methods are applied:

(a) the individualizing method (Camerer 1910; Dulitsky and Moshkevich 1926; Mikhalev 1970a; et al.), the essence of which is that only individuals which develop in optimal environmental conditions are selected for study;
(b) the regular method, a generalizing research method, whose growth techniques have been discussed in detail in many textbooks on biometrics and variation statistics (Dulitsky and Moshkevich 1926; Nemchinov 1945; Rokitsky 1964; Ventsel 1969; Plokhinsky 1978; et al.).

3.1.2 Investigation of Growth Using Empirical Formulae

The result of statistical processing of the material is a series of numbers which can be used to construct empirical formulae of growth, which roughly reflect the relation between the size of animals and their age as observed in experiments. The task is only to find the most simple formula. The oldest attempt to express growth dependence mathematically was undertaken by B. Gompertz (1825). The exponent he proposed had a wide application in the past (Loird 1966; Loird et al. 1968; Siliman 1969; et al.).

$$P_m = P_0 l^{at}$$

Parabolas of higher orders as empirical formulae were widely used as growth equations. For example, J. Donaldson (1924) studied the postnatal growth of rats using a second order parabola (P = at2 + bt + c). R. Pearl and L. Reed (1925) simplified this formula by a selection of constants, which, however, do not have any biological meaning:

$$P = at + b.$$

where

P – weight of animal;
t – age,
a and b – constant

Other researchers (Backmann 1931; Janisch 1927) offered similar growth formulae. R.P. Spenser and others (1966), comparing the growth of the nuclei of three populations of monkeys and pigs, found that in the second half of fetal development, embryo size is in linear dependency on the duration of pregnancy. For their formulae

$$\text{size growth} : L = at + b.$$

or

$$\text{weight growth} : \sqrt[3]{P} = at + b$$

factors also chosen empirically. A modification of this dependency is the formula by Hugget S. and B. Widdas (1951) for the entire period of prenatal growth in mammals, including whales, discussed in detail below.

As was already mentioned, the main drawback of these formulae is that their constants do not have biological meaning. However, V.V. Bunak's (1946) investigation into the formulae of higher order parabolas are a subject for special discussion.

3.1.3 Mathematical Analysis of Growth Phenomena

The mathematical analysis of growth phenomena means finding dependence between the size of animals and their age using an analysis of the relationship between growth and the processes determining this growth. The main difficulty of this analysis is to establish the fundamental causes of the change of growth rate with age. Hence several hypotheses arose based on various analogies:

(a) with the chemical processes of growth (Ostwald 1908; Robertson 1925; Crosier 1927; Brody 1927; Naydenov 1927);
(b) with the physical processes of growth (Przibram 1922);
(c) with increasing population (Lotke 1925);
(d) with physiological processes (Vinberg 1966; Payne and Wheeler 1967).

The first attempt in this direction was made by M. Rubner (1908), who investigated animals' energy consumption in the growth process. A. Putter (1920) created a dynamic theory of growth, which was later endorsed by L. Bertalanffy (1934) and E. Fischer (1928). G.G. Vinberg (1966) tried to explain the growth rate of animals by the intensity of energy metabolism, and P. Payne and E. Wheeler (1967) by the degree to which the growing body was supplied with nutrients. Interestingly, the authors of these studies, which were based on other theories than academician I.I. Schmalhausen's (1927–1935, see below), came to the same mathematical function. Their conclusions will be discussed in more detail when the theory of parabolic growth is considered.

C. Taylor (1958, 1962) and St. C.S. Taylor (Taylor 1968) developed a growth theory using an assumption that weight gain represents the resultant of two oppositely directed metabolic processes – anabolism and catabolism.

3.1.4 The Hypothesis of Growth and Organism Differentiation

The fact that the growth of an organism depends on the organism's cellular differentiation in the process of growth was first shown by K. Minot (Minot 1907, 1908). Some combining of ideas related to growth differentiation and intensity of metabolism were outlined by T.B. Robertson (Robertson 1925). He believed that the process of age-related changes in metabolic rate and energy also decreases the growth rate. Graphically, this growth is represented by a logistic curve. This dependence characterizes well growth throughout ontogeny, although in reality the inflection point of the S-shaped curve is attributed closer to its end. I.I. Schmalhausen (Schmalhausen 1927, 1935, 1935a) developed the ideas further and more deeply, and proposed putting a power function on the dependency of the growth of differentiation, which will be discussed in more detail later.

Thus, there are four basic types of growth formulae reflecting linear growth, exponential, S-shaped and parabolic (mono-parabola and higher-order parabolas). Due to the fact that in the first type of formulae the constants have no biological meaning, and linear growth can essentially be expressed by a power function, there is no sense in analyzing them, especially since, to all intents and purposes, they reflect a relationship between age and size less accurately than, say, a parabola of higher orders. This is evident from studies by V.V. Bunak (1946). Logistic formulae reflect growth throughout ontogeny more accurately than linear formulae, but they, like the first, can be successfully replaced by parabolas of the higher orders, as I.I. Schmalhausen (Schmalhausen 1935) originally suggested.

Clean, simple growth at a constant rate is expressed by an exponential formula:

$$P = P_0 e^{at}.$$

where

P – total size of the growing organism,
P_0 – the size at the initial moment,
a – constant growth rate,
t – time from the beginning of the observation,
e – base of natural logarithms.

This growth occurs exponentially. It is observed in bacteria and in multicellular organisms during cell division. When the egg divides the number of blastomeres also increases exponentially (Schmalhausen 1935). The growth of the developing organism is best reflected not by an exponential formula, but by the so-called "one-term parabola":

$$P = at^k$$

where

P – the weight of the animal
a – coefficient of proportionality
k – the power factor

Empirically, for embryo growth in higher vertebrates, this formula was devised a long time ago and found practical application (Jackson 1909; Murray 1925). E. McDowell and E. Allen (McDowell and Allen 1927) clarified that the right time to apply this formula is starting from the moment when a primary streak appears on a dividing egg, i.e. at the time t_0 of conception of the embryo.

Theoretically, researchers used the one-term formula parabola differently. Based on the bulk of factual material on the vertebrate's growth at embryo and postnatal developmental periods, I.I. Schmalhausen (1927, 1935, 1935a) showed that relative growth rate is inversely proportional to age or, in other words, the product of the relative growth rate (C) by age (t) is a constant (K): Ct = K. I.I. Schmalhausen (1935) called this "the law of growth." However, the constant in his opinion "is not constant throughout the growth period, but only within a certain biological time" and "in the transition from one period to another the constant changes sharply (p. 39). By integrating the expression Ct = K, I.I. Schmalhausen (1935) established a relationship between growing body weight (W) and time (t):

where m – parameter, which he called the "mass factor", since at t = 1, W = mt^k.

According to I.I. Schmalhausen (1935), excellent concordance of the empirical and theoretical data proves the truth of the formula. Since the expression is a power function, or a parabola (in the broadest sense of the word), I.I. Schmalhausen (1935) called this type of growth parabolic.

Assuming that the growth of the embryo depends on its supply of nutrients through the surface of the embryonic-vascular epithelium, P. Payne and E. Wheeler (Payne and Wheeler 1967) came to parabolic dependence using the following considerations. They assumed that the surface area of absorption during embryogenesis is proportional to the surface area of the embryo which may be expressed as $\sqrt[3]{s^2}$ or $s^{\frac{2}{3}}$ that is $\frac{ds}{dt} \sim s^{\frac{2}{3}}$,

$$\text{Thus,} \frac{ds}{s^{\frac{2}{3}}} \sim dt$$

which after integration gives:

$$s^{\frac{1}{3}} \sim t \text{ or } s = at^3$$

where the constant a – number of nutrient penetrates into the embryo per unit of its surface.

The coincidence of these experiments conducted on two classes of higher vertebrates (birds and mammals) with theoretical data, in the opinion of Payne and Wheeler, is proof of the truth of "the cubic law of growth."

G.G. Vinberg (1966) found a parabolic relationship between energy exchange and the weight of the animal within the same growth phase, and this again shows the possibility of finding a biological foundation for the parabolic dependence of growth over time. A study by V. Bunak (1946) is purely mathematical proof that only formulae expressing parabolic dependence can truly reflect the total growth of animals. The author pointed out that the absolute and relative gains during the whole growth period or in one of its phases, expressing growth dynamics in the most specific form, have very diverse trends of changes, cannot be expressed by different curves, and require a specific formula. These are formulae of parabolas of higher order, and because the absolute growth rate is the first derivative of growth, the formula that reflects total growth must also be a parabola, but one order of magnitude higher.

So, V.V. Bunak (1946) found that "the basic type of growth curve for one phase is a parabola of the third order or a parabola of the second order when the rates are aligned in a straight line" (p. 23). Further, and this is very important to emphasize, "this basic type characterizes the growth for both volume and linear dimensions, but the latter is often followed by the second option" (p. 23). For practical purposes the value of these formulae is that they are very flexible and may relatively easily describe a wide variety of dependencies. The constants can be set with sufficiently high accuracy on the basis of experiments by different methods (the least square method, the method of finding the weighted average, the graphic-analytical method) described in the manuals for the preparation of the empirical formula and methods of approximate calculations (Bailey 1970; Lakhtin 1922; Bezikovich 1931; Semendyaev 1933; Bradis 1935; Plokhinsky 1978; et al.).

Note the remark by V.V. Bunak (1946) that the same results for a single growth phase can be obtained by some monomials formulae, including the one-term parabola.

Moreover, the fact that the empirical data in the embryonic stage of growth for higher vertebrates in a logarithmic system of coordinates lies on a straight line, according to I. I. Schmalhausen (1935), is evidence that prenatal growth is reflected more accurately by a one-term parabola (for mass, or surface or for linear dimensions), when the relationship between the animal size measurements are constant. It is important that its coefficients have biological meaning. Thus, m – size of the growing organism in a first unit time, and k – intensity indicator growth.

The latter indicator is unnamed, and therefore it becomes possible, by comparing the power coefficients, to compare different periods of growth in the same species, or the same period of growth in different species.

All of the above can be summarized by the opinion of I.I. Schmalhausen: "The growth process is basically a process of exponential character, and under permanent external and internal conditions a pure exponential growth can be realized (e.g. bacteria and certain periods of "pure" animal growth). Yet we must admit that as a result of non-compliance with one of the basic premises of exponential growth, i.e. constant internal growth conditions of the developing organism, in fact, this period in complex organisms is not an exponential process. In the embryonic period of higher vertebrates, when both size increase and cellular differentiation is very intense, and conditions are almost constant – such as inside the mother's body – growth is closer to parabolic than to the exponential line. On the contrary, in this case, the curve has a parabolic shape and is most accurately expressed in a very simple formula: $W = mt^{K_w}$. The most common form of the growth of the higher animals is obviously parabolic growth "(Schmalhausen 1935, p. 25).

3.2 Prenatal Growth of Cetaceans

Let's define the curve type and specific growth rate of cetaceans in the prenatal period of development. The difficulty of studying the prenatal growth of whales is primarily because the seasonal nature of catching them means authentic embryonal material can only be obtained for five to 6 months of the year. In other months, the material is either lacking altogether or is too insignificant. Therefore, the actual 6 months' worth of material has to be extrapolated to the whole embryonal period. The conclusions made depend on the degree of authenticity of this extrapolation. Following are the major researches of this question.

N. Hinton (Hinton 1925) and S. Risting (Risting 1928) believed that during the first two to 3 months, the growth of baleen whale embryos is slow, and in the subsequent 9 months is characterized by constant daily gains.

N. Mackintosh and Z. Wheeler (Mackintosh and Wheeler 1929), guided by the dot diagram and average monthly large whale species embryo sizes, have drawn a concave arch and also considered that embryo growth rate increases gradually. S. Ohsumi, M. Nishiwaki and T. Hibiya (Ohsumi et al. 1958) have drawn an approximate growth curve on the dot diagram for fin whales of the Pacific and, without any calculations or theoretical substantiations, have accepted it to be a parabola. Corresponding coefficients were fitted empirically.

N. Peters (Peters 1938) used an analogy with terrestrial mammals. However he compared the sizes of different aged embryos not by weight but by the length of fetuses, which should not be done when speaking about four-legged animals, because there is no continuous communication between these measures. Comparison of whale embryo growth with that of agricultural animals not by body length but by weight have been made by S. Hugget and U. Widdas (Hugget and Widdas 1951). Their study is frequently referred to, therefore let's discuss it more in detail. They believed that the increase in weight of whale embryos is a function of age, and that their growth can be described by cubic dependence:

$$P = at^3$$

Where

P-is embryo weight,
t- age,
a – coefficient of proportionality.

They suggested describing the linear growth of embryos by the formula:

$$L = b(t \quad t_0)$$

Where $t_0$represents the short period of time of accelerated embryo growth as curvilinear from conception up to placentary circulation, which depends on the duration of pregnancy for different whale types;

t – the period of rectilinear growth,
b – a coefficient expressing the inclination of the line of regress during the period of uniform growth, named by them the "specific growth rate of embryos".

It is necessary to note that the calculation of coefficients made by S. Hugget and U. Widdas (Hugget and Widdas 1951) on length, has given an underestimated result, compared with the calculation on weight. This circumstance specified that there is no cubic dependence between whale weight and length. Hence, while the conclusion of the authors about the cubic formula of growth is not quite correct, they have not noticed this contradiction.

R. Law's work (Laws 1959) presents a point of interest. He agreed that in first half of pregnancy whale embryo growth corresponds to the law assumed by S. Hugget and U. Widdas (Hugget and Widdas 1951). However in the second half of pregnancy, in his opinion, the growth rate increases and comes nearer to exponential.

The main reason for disagreements between the researchers of the prenatal growth of whales was their failure to observe the rules of statistical processing of data.

3.2.1 *Prenatal Growth of Baleen Whales*

Most fin whale embryonal material was gathered on coastal stations in subtropical and tropical zones in course of the whole year, so the technique applied for drawing the curve of baleen whale growth will be demonstrated in more detail than the example for fin whales.

Fin Whales Data of BIWS including 58,605 embryos for the period 1923–1956, published by Naaktgeboren, Slijper, Utrecht (Naaktgeboren et al. 1960) and our own material, 2539 embryos, were processed. All the lengths of embryos have been

Table 3.1 Distribution of fin whale embryo lengths in monthly variation sets

	Months											
Size classes, cm	V	VI	VII	VIII	IX	X	XI	XII	I	II	III	IV
0–29	7	3	6	1	5	26	33	66	109	34	9	3
30–59	13	27	12	5	6	87	88	242	575	184	23	5
60–89	25	43	20	18	6	104	145	457	1188	530	118	3
90–119	23	54	64	19	7	84	167	722	1785	843	165	5
120–149	14	73	59	24	9	53	153	969	2296	1129	303	5
150–179	10	59	73	31	19	12	98	931	2848	1546	421	6
180–209	4	42	75	42	10	12	47	737	2977	1871	501	10
210–239	8	39	71	39	20	5	14	517	2683	1771	511	7
240–269	2	5	55	57	27	9	4	343	2674	2135	728	6
270–299	1	4	31	48	31	2	1	174	1961	1865	630	15
300–329	–	–	21	61	36	8	2	86	1724	2065	859	20
330–359	1	1	13	26	41	4	3	21	879	1599	600	8
360–388	1	–	3	21	40	7	3	8	757	1931	994	17
390–419	1	1	1	19	29	8	1	1	255	1194	652	15
420–449	1	–	1	5	22	11	–	6	143	1016	715	19
450–479	1	–	–	5	17	23	1	2	85	652	599	15
480–509	1	–	1	1	6	7	–	–	18	316	389	15
510–539	–	–	–	–	4	4	1	–	10	118	237	10
540–569	–	–	–	1	–	1	–	5	7	80	203	12
570–599	–	–	–	–	–	2	1	1	3	24	54	5
600–629	–	–	–	–	–	–	–	1	2	12	57	5
630–659	–	–	–	–	–	–	–	–	–	2	16	1
660–689	–	–	–	–	–	–	–	–	1	1	4	–
690–719	–	–	–	–	–	–	–	–	–	1	3	–
Total	113	351	506	423	335	469	762	5288	22,980	20,919	8791	207

shown in monthly variational sets with class intervals of 30 cm, with the percentage of their occurrence in each class calculated (Table 3.1).

The percentage of fin whale embryo length occurrence in classes of monthly variation sets is shown in the diagram (Fig. 3.1).

In the diagram (Fig. 3.1) attention is paid to the presence of not one but two groups of embryos: the basic A (October – April) and an additional B (May–September) that testifies that presence of the second, less significant season of fin whales with a shift of approximately 6 months.

For further processing of variational sets transgression of these two groups in October, November and September we had to divide them artificially. For example, in October the basic set of embryo group is limited by dimensional lengths of 120–210 cm, and in November by 210–240 cm Only after the division of groups A and B into average embryo sizes were quadratic deviations and standards mistakes calculated (Table 3.2).

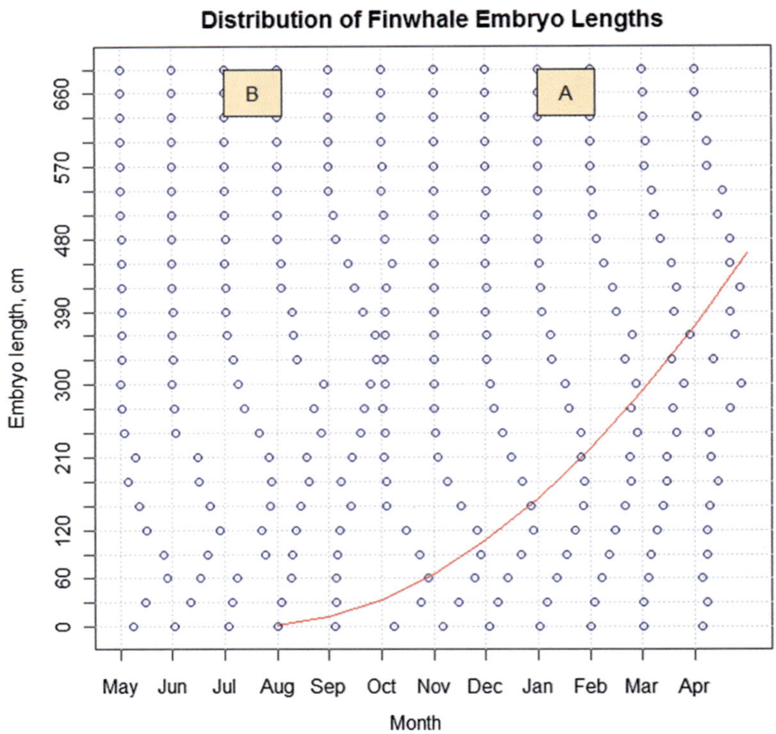

Fig. 3.1 Distribution of finwhale embryo length

Table 3.2 Fin whale embryo length statistics for group A

Month	Arithmetic median, *M, cm*	Deviation, σ, cm	Variation, *CV*, %	Standard error, *m*, cm
October	85.24	42.00	49.20	2.16
November	109.57	48.60	44.30	1.79
December	157.54	60.81	38.70	0.83
January	207.70	84.00	40.04	0.56
February	278.42	103.20	37.06	0.71
March	333.07	111.84	33.60	1.11
April	385.39	121.20	31.40	8.08

Incidentally, previous researchers did not use quadratic deviation to judge variability within different sets of minke whale embryo lengths, yet the application of limits of deviation only for this purpose does not give serious results, as "extremes in the sets are not very steady and are easily shifted when the quantity of analyzed

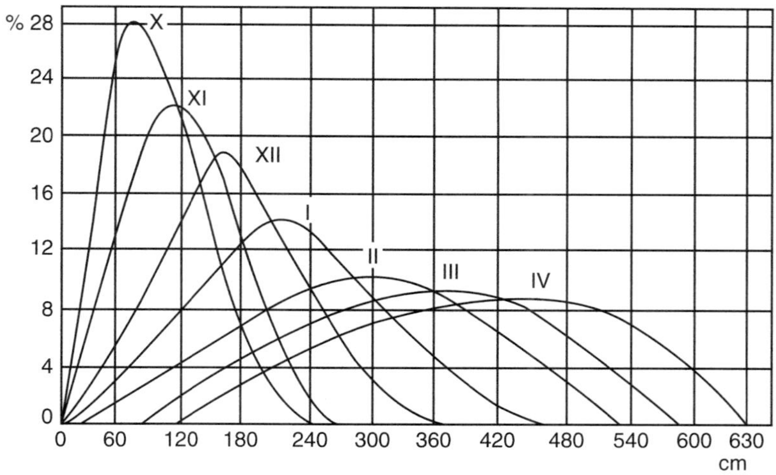

Fig. 3.2 Curves of fin whale embryo lengths distribution per month

animals changes" (Rokitsky 1961, with. 35). It is absolutely impossible to judge variability of embryo lengths in any particular month if, as in case of Southern Hemisphere fin whales, transgression of monthly variational sets of two groups of embryos is observed. Average sizes and extreme limits will remain constant for a long time. A table by A.G. Tomilin (1957, p. 177) with the data on 21,450 the Southern Ocean fin whale embryo lengths well illustrates this point. It shows that for seven out of 8 months, minimal sizes remain the same, as if the smallest embryos do not grow at all within the six-month period.

Again we will address the percentage diagram of the incorporated material (Fig. 3.1) to consider in more detail the group including the bulk of embryo lengths (group A) from October to April.

In October and November the distribution of embryo length is a little asymmetric, with the right side of the distribution curve less steep than the left side. In December and January the distribution is more normal, but several peaks appear which will be better seen in the following months. Therefore it is necessary to smooth these sets. Smoothing (Fig. 3.2) is executed by a graphic method and by simple sliding average (Rokitsky 1961; Plokhinsky 1978).

The smoothed monthly variational sets illustrate and confirm that the symmetry of distribution of embryo lengths in October and November is slightly negative, in December, January and February is practically normal, and in March and April is positive. Thus, on the IWC data the same law in distribution of lengths can be found, to which R. Laws (Laws 1959) paid attention on a rather small amount of data, but which was more accurately collected by biologists.

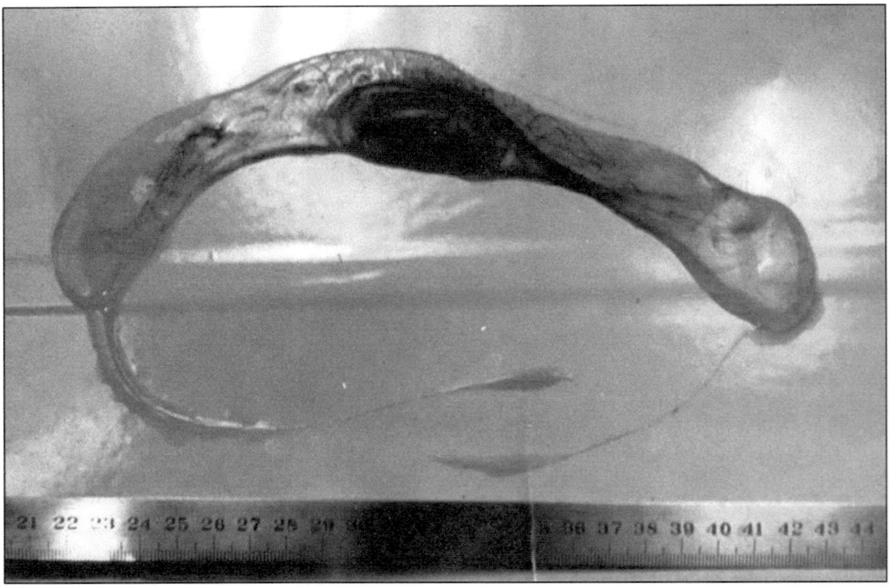

Whale embryo in membrane

What is the reason for the abnormal distribution of frequencies of embryo length in October and November? The main reason is that the sets are composed of different aged embryos whose growth rate is different. The character of distribution is also influenced when part of the embryos are for some reason not included in samples. While embryos at the embryogenesis stage are not always found (Brinkmann 1948; Salnikov 1956; Laws 1959; Mikhalev 1970a, 1978a, b, 1984; Ivashin and Mikhalev 1978; etc.), in October and November, there must be a lot of such embryos. Remember that embryos at the embryogenesis stage can be aborted before the whale body is cut and the female whale's uterus is examined. Remember that before villous appear on the germinal bubble, the embryo is not fixed in the uterus horn and moves rather freely within it (Sleptsov 1955a). This means the miscarriage of small fetuses can occur when the air-inflated whale body is being drawn up the slip to the stern of the whaling vessel.

The miscarriage of small sized fetuses sometimes occurs on deck when a whale is turned over, especially when the additional rope around its belly creates great pressure on it. After this procedure we sometimes found embryos up to 10 cm long with intact membranes on the deck near a whale. In our practice and in the practice of our colleagues in the scientific group, it happened that despite the presence of a functioning corpus luteum of pregnancy, the embryo was not found, though the uterus was thoroughly examined. We can say with confidence that the loss of fetus is a frequent phenomenon.

Fin whale embryos are rare in the embryogenesis stage in December and January. By this stage of pregnancy the placenta has developed significantly and loss of fetus

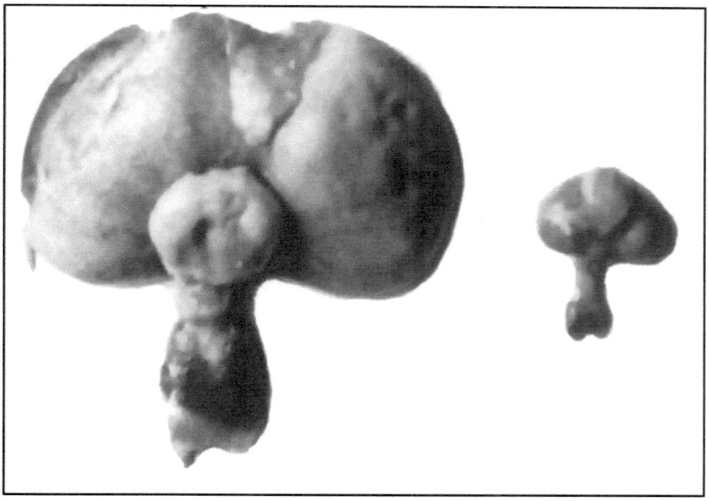

Fig. 3.3 Hypophysis of pregnant female fin whale (left) and hypophysis of embryo (295 cm long) found in the female

becomes impossible. Therefore it is not surprising that the distribution of frequencies of embryo length occurrence in variational sets at these months are normal. It is only during the late stages of pregnancy, when the female organism is almost ready for natural delivery, that we again come across cases of embryo loss. The corpus luteum of pregnancy in the ovaries proves the abortion case. Moreover, in such cases the uterus is big, friable and bleeds strongly. The phenomenon is sometimes observed for fin whales at the end of March and in April.

When for some reason it is impossible to find the ovaries, a direct confirmation of pregnancy can be an enlarged, wrinkled, reddish and knobby on the edge hypophysis. Adenohypophysis (Fig. 3.3) (Mikhalev 1966, 1970a).

Harpooners recounted losses of large embryos at the moment when female whales were harpooned and towed. Sometimes fetus loss occured at the moment when the whale was taken up the slip on to the deck (Risting 1928; Zenkovich 1935; Mikhalev 1970a, 1978a, b, 1984; Ivashin and Mikhalev 1978).

The differentiated females of various biological group migrations from one feeding area to another influence the nature of the distribution of embryo lengths (Jonsgard 1951; Laws 1959; Berzin 1971). As R. Laws (Laws 1959) specifies, at the end of the whaling season, first of all female fin whales pregnant with large embryos (the senior age groups), move from the area where the catch is conducted (and, hence, where scientific material is being gathered) to subtropical waters. By this time the breeding season starts, which reduces the average size of embryos found. A.V. Yablokov (Kleinenberg and others, 1964 section "Reproduction") points out the influence of the season of procreation on the low monthly average of embryos for beluga whales in April.

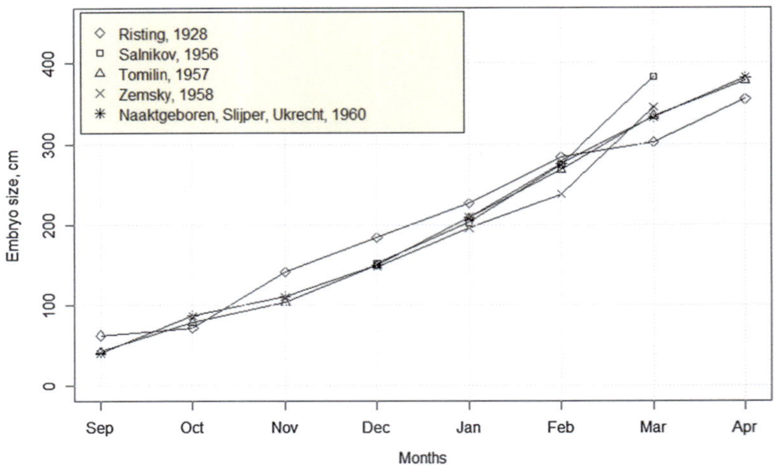

Fig. 3.4 Growth of fin whale embryos according to different authors

There are probably other reasons also, but in this case the most important fact is the establishment of abnormal distribution of embryo lengths in variational sets, because in this case "the main characteristics of the random value of a set" is not the average, but modes, which are "the most probable values" of embryo lengths in the sets researched (Rokitsky 1961; Ventsel 1969; etc.). So, in our case the average length of fin whale embryos in October is 85 cm, and the mode within 55–60 cm. The March average is 333 cm, the mode being within 375–385 cm.

Unfortunately, this rule was not observed by previous researchers of prenatal whale growth, which distorted real average sizes, and the growth curves drawn on their basis (Fig. 3.4) presented broken lines and essentially differed.

However, both average and mode make sense only in cases in which the set is qualitatively homogeneous. In variational sets in any month, fluctuations of embryo lengths are averaged for the same-age and also for embryos of different ages (those conceived at different times), and with different growth rates. While studying growing objects it is more correct to use compound values (Rokitsky 1961, 1964; Plokhinsky 1961, 1967, 1978; etc.), as logarithms of embryo lengths in variational sets give distributions closer to normal. Unfortunately, previous researchers of whale growth did not observe that rule either.

But even observing those rules of biometry, because of the artificial division into two groups of variational sets and the necessity of smoothing them due to their having several peaks (Fig. 3.2), average geometrical fin whale embryo lengths with a sufficient degree of reliability are recorded for 7 months only: for October, November, December, January, February, March and April. It is significant that these values on the schedule make the concave arch like a parabola, i.e.. a curve with a constantly accelerating growth rate.

Still, though the graphic method of the analysis has "great heuristic value, helping to open law, which by other methods are opened quite often with difficulty" (Schmalhausen 1935), mathematical analysis of the recorded data is necessary.

Table 3.3 Fin whale embryo length, geometric means and monthly increments

Median and increments, cm	Whaling season months						
	October	November	December	January	February	March	April
Geometric mean	≈ 35	≈70	≈115	≈170	≈235	≈310	≈395
Increment of 1st order	–	≈35	≈45	≈55	≈65	≈75	≈85
Increment of 2nd order	–	–	≈10	≈10	≈10	≈10	≈10

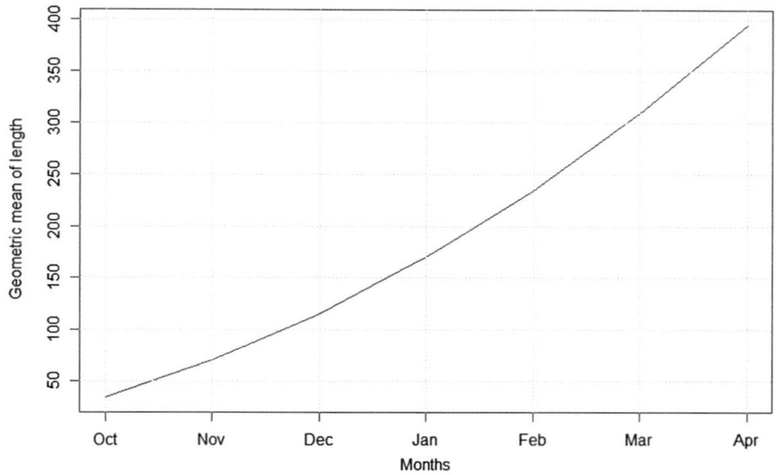

Fig. 3.5 Growth of finwhale embryos, based on geometric means

On the basis of the average geometric lengths of fin whale embryos rounded off to the nearest integer, we have determined increments of these average values for every month of embryo growth.

The following average lengths and increments (Table 3.3) have been calculated.

As we see, the second-order increments (an increment of increments of the first order) are approximately identical and, hence, there are reasons to consider that the curve we are studying (Fig. 3.5) is close to the parabola of the second order.

To define the curve type it is necessary to know the average age of embryos in every month of the researched set, however, it is only possible to determine that age approximately. Therefore in a number of trial variants, a range of age values was taken, so that guided by the smallest quadratic mistake on the regression curve we could come closer to the true age.

In trial schedules of the logarithmic system of coordinates, the average geometric lengths of fin whale embryos lie on a straight line, when the average age of embryos in October has been accepted to equal 3.0 months (a quadratic mistake = 0.0079), and, hence, the average age of embryos in November – four, December – five, etc. months. A similar trial and error method to determine age was used by H. Clark, B. Florio and R. Hurowitz (Clark, Florio, Hurowitz, 1955) when studying the growth of ribbon snake embryos.

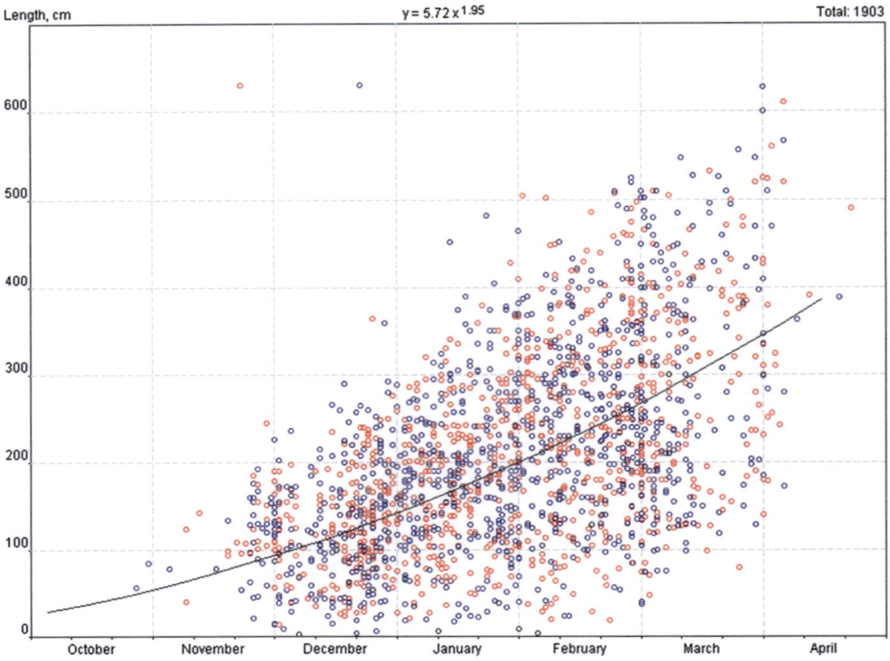

Fig. 3.6 Fin whale embryo growth

Regression analysis of the recorded average values has shown the dependence of fin whale embryo lengths (L) on age (t) is well described by the formula:

$$\mathrm{Lg}(\mathrm{L}) = \lg(5.72) + 1.95\lg(\mathrm{t})$$

After potentiation the logarithmic expression the formula of the researched curve is:

$$\mathrm{L} = 5.72\mathrm{t}^{1.95}$$

Where

L – embryo length, cm;
t – age, month.

The growth curve (Fig. 3.6) constructed with this formula on the basis of theoretical values of monthly average fin whale embryo lengths shows the density of location of points on the dot diagram of embryo lengths specified by biologists. They practically coincide with the regression line drawn in approximation, which illustrates that sometimes the approximate method of drawing of the growth curve gives rather exact results.

Using this growth curve formula, it follows that the size of the animal in the prenatal period is determined by the coefficient of embryo growth intensity (b), parameter of age (t), and also depends on "the mass factor", on the size (p – weight growth; l – linear), which the animal reaches during parabolic growth and differentiation per the first time-unit. This unit can vary (day, week, month)…and is selected depending on the size of the animal and the duration of the investigated period of growth (Schmalhausen 1935).

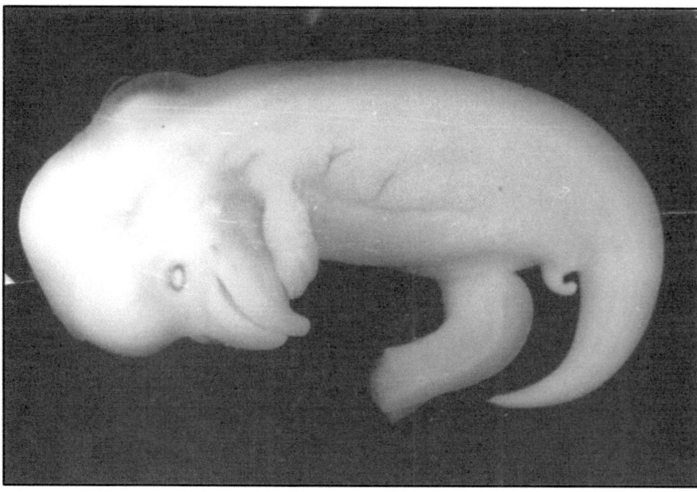

Fin whale embryo, 60 mm long, 1.5–2 months old

In its turn "the mass factor" depends on fetus size at the blastula stage, because further growth of the embryo remains, in general, proportional to this size. Up to the blastula stage the growth is not subject to parabolic, but exponential, correlation. This period of "pure" growth takes the period of time $(t-t_{o)}$, which was taken into account and included in the formula of monomial parabola by E. McDowell and E. Allen (McDowell and Allen 1927). The length of the period for whales is difficult to assume, but, probably, its duration does not differ from those of large terrestrial mammals and, together with the latent period, is about 10–15 days.

Some words about specific growth rate. Usually, when speaking about growth rate, we mean the speed, of growth, which is necessarily connected with absolute or relative gain. Previous researchers of whales operated by pure gain, and on this basis calculated absolute growth rate as:

$$(l_2 \quad l_1)/(t_2 \quad t_1)$$

which is true only when this value does not change with time, that is when growth goes with constant pure gain. Such growth is rather rare in the animal world, and even if it is observed, it is impossible to compare the growth of various animals, as "absolute length gain of approximately 1 cm for the mouse and for the elephant – are absolutely incomparable values" (Schmalhausen 1935, p. 13). If the gain is expressed as a percentage of the size of an organism some time between conception and birth, it is possible to measure and compare the relative speed of growing animals. Yet both relative and absolute gains only indirectly characterize growth rate at a given period of time and do not fix changes in growth rate.

Similarly, growth rate during a given period does not fix growth rate changes occurring during each subsequent moment. Therefore it is necessary to apply specific growth rate (C) which reflects changes in the growth intensity and truly quantitatively characterizes the growth process as such.

The concept of specific, or true, growth rate was introduced independently by both S. Brody (1927) and I.I. Schmalhausen (1927). It is determined by the following formula:

$$C = b / t,$$

where

b – is a constant of parabolic growth intensity,
t – is the age of the animal at the moment of research.

Empirical values of linear growth at a specific speed are determined by the formula:

$$\frac{\lg(l_2) - \lg(l_1)}{(t_2 - t_1)0.4343},$$

where 0.4343 is the basis of the natural logarithm.

Since specific growth rate is comparable only in biologically equivalent age, it cannot be compared in the same month of intro-uterine development of different species, if pregnancy duration of the animals being compared is not the same. But it is possible to make such a comparison using relative age, for example, to compare specific growth rates in the middle of the intro-uterine development period.

We have calculated the specific speed of fin whale prenatal growth, and the schedule of its theoretical value changes (Fig. 3.7) has been created.

As is proper for the parabolic growth, the specific growth rate (C) of fin whale embryos falls with age, and the schedule looks like a hyperbole. Notice that specific growth rate is very sensitive to fluctuations of growth conditions and its empirical values usually deviate from theoretical values, revealing non-uniformity of growth – periodization. However, as a great interval of time is taken (a month), the periodization is not appreciable.

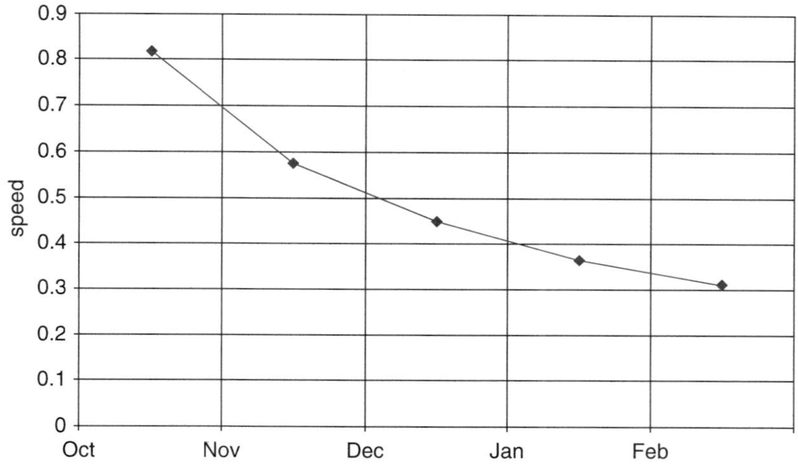

Fig. 3.7 Specific speed of fin whale embryo growth

On the subject of specific growth rate, it is interesting to note that if the smoothed monthly variational lines of the original data on the graph (Fig. 3.2) are connected with the top points of the lines for each month, the curve will remain a hyperbole. Is it coincidental? It is possible that, this fact is mathematically connected to specific growth rate, and such a curve should have led, from the beginning of the research, to the assumption that fin whale prenatal growth is parabolic.

As to the research of embryo growth using another group of original fin whale data (Fig. 3.1, group B), processing these data has produced less reliable results, due to the small sample, as smoothing of monthly variational lines of embryo lengths and determining the average sizes could not be exact. But these values are also fairly well described by a parabola of the second order with the growth intensity parameter close to that found for the basic group of the Southern Ocean fin whale embryos.

Pygmy Blue Whales From October to May, 637 pygmy blue whales embryos were measured. Apart from the basic group of embryos, a group of larger embryos and a group of embryos of earlier developmental stages were also included, which is especially visible in November and December (Fig. 3.8).

According to a technique similar to that applied for fin whales, the basic groups of pygmy blue whale embryos can be determined and the digital material processed. With the least error the compound values of pygmy blue whale embryos received in the logarithmic system of coordinates have lain on a straight line (the variant when

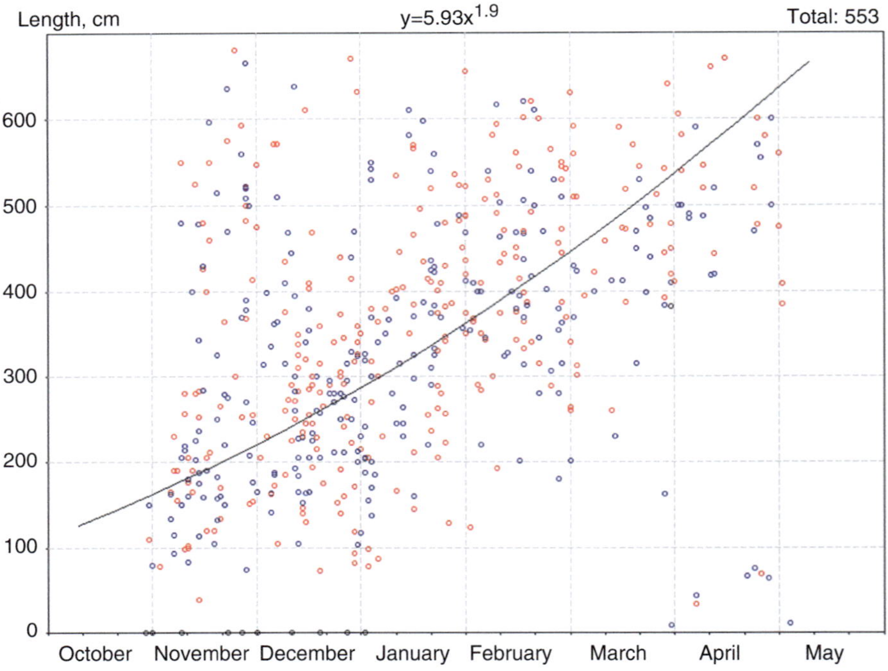

Fig. 3.8 Pygmy blue whale embryo growth

the age of November embryos has been accepted to equal 6 months, hence, December – seven; January – eight, etc.). The regression analysis shows that the curve for pygmy blue whale embryo growth in a rectangular system of coordinates is well described by the formula:

$$L = 5.39\, t^{1.90}$$

Sei Whales Measurements of 2416 sei whale embryos cover the period December to April. Preliminary data from Britain's Natural History Museum have been used.

It is important that in contrast to pygmy blue whales and fin whales, only one set of sei whale embryos is included in the "basic" group (Fig. 3.9). This has allowed us to calculate compound values of embryo lengths more precisely. With the least error, logarithms of these values have lain on a straight line (variant, when the November embryo age has been accepted to equal 4.5 months).

By means of regression analysis the following linear growth formula has been determined for sei whales in the prenatal period of development:

$$l = 4.52\, t^{1.98}$$

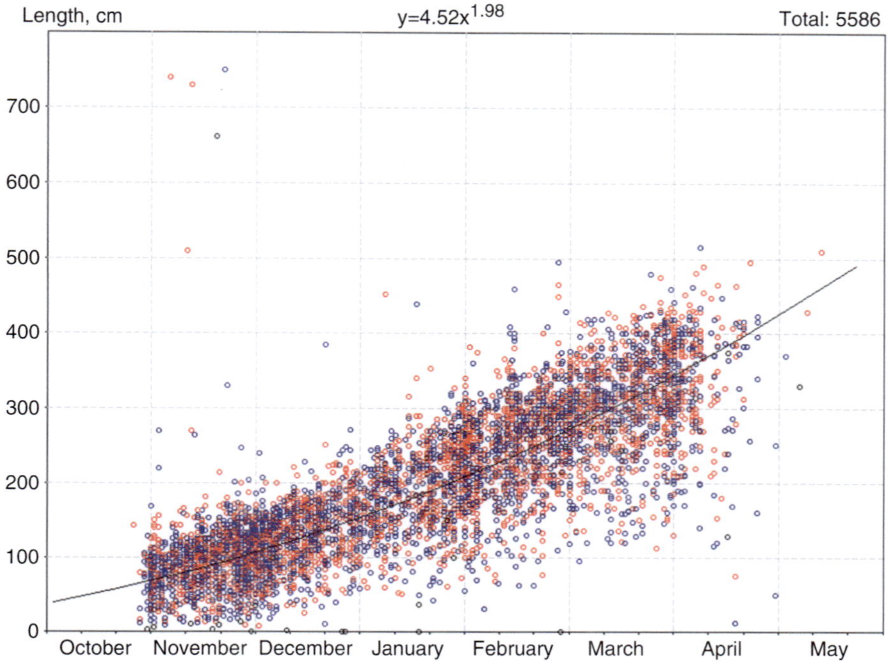

Fig. 3.9 Sei whale embryo growth

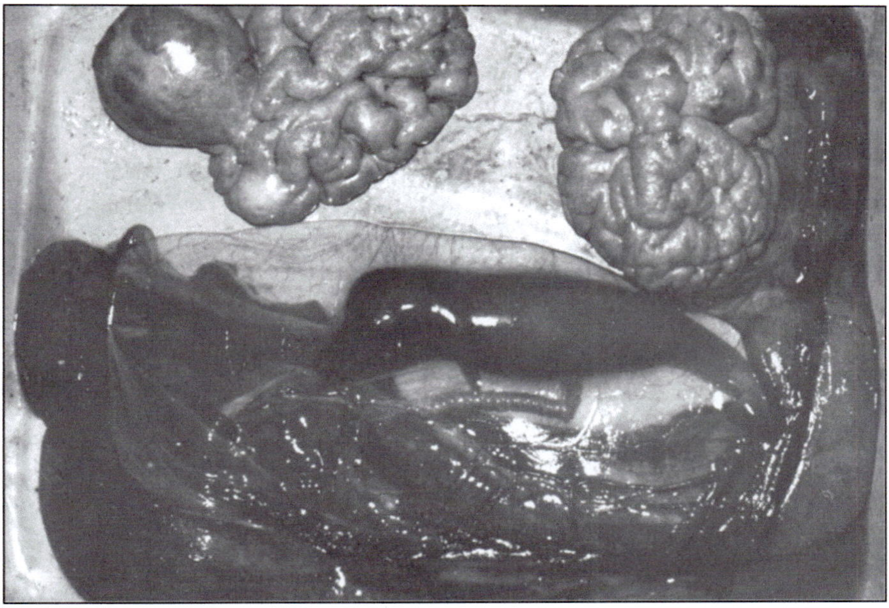

Sei whale embryo in membrane in left ovary

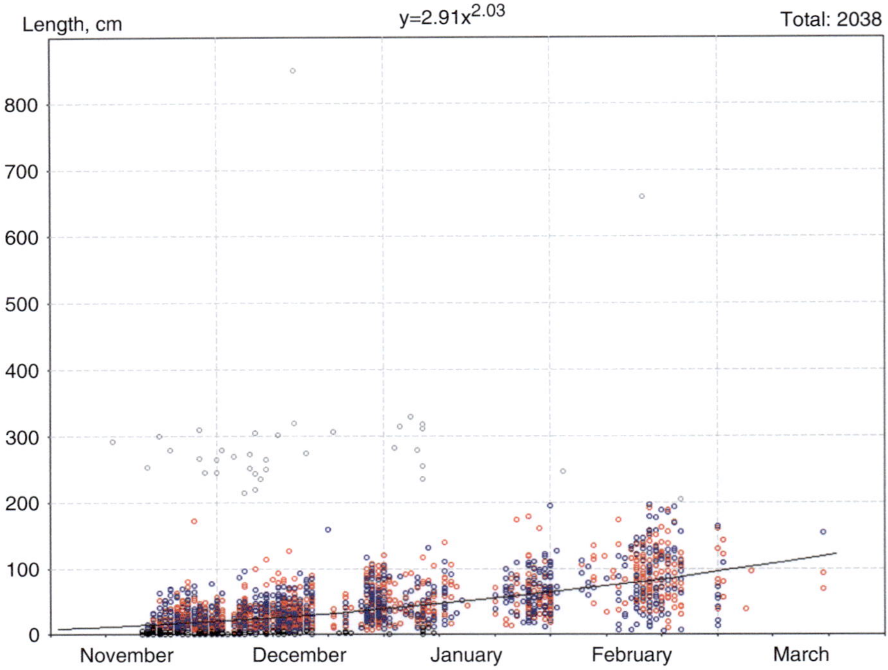

Fig. 3.10 Minke whale embryo growth

Minke Whales More than 1500 minke whale embryos were measured on the whaling ships *Slava*, *Sovetskaya Ukraina* and *Yuri Dolgoruky*. Additional data for the sizes of about 500 more fetuses have been handed to us by VNIRO employee M.V. Ivashin, who measured them on Japanese whaling flotillas. In total 2222 minke whale embryos from the embryogenesis stage to prenatal sizes have been processed (Fig. 3.10).

During the researched period of time (from November till March), as for fin and pygmy blue whales, there were two groups of minke whale embryos. The second group, with embryo lengths of more than 200 cm, was small and easily separated from the basic set.

The compound values of embryo lengths of the basic group, calculated by using a logarithmic system of coordinates, lie on a straight line, assuming that the average

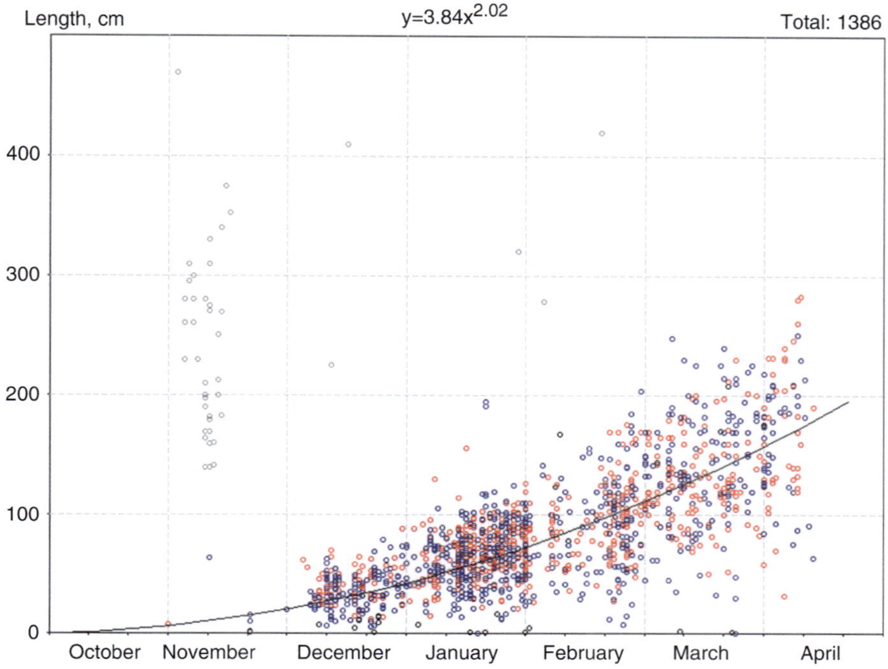

Fig. 3.11 Humbpack whale embryo growth

age of December embryos is 3.5 months. By means of regression analysis the following formula for minke whale growth in the prenatal period has been calculated:

$$l = 2.91\,t^{2.03}.$$

Humpbacks During the period from November till April, 1435 humpback embryos from the embryogenesis stage up to prenatal fetuses were measured and included for processing (Fig. 3.11).

On the dot diagram the additional group of embryos is well defined in November and is less appreciable in December and January. This group is easily separated from the basic set.

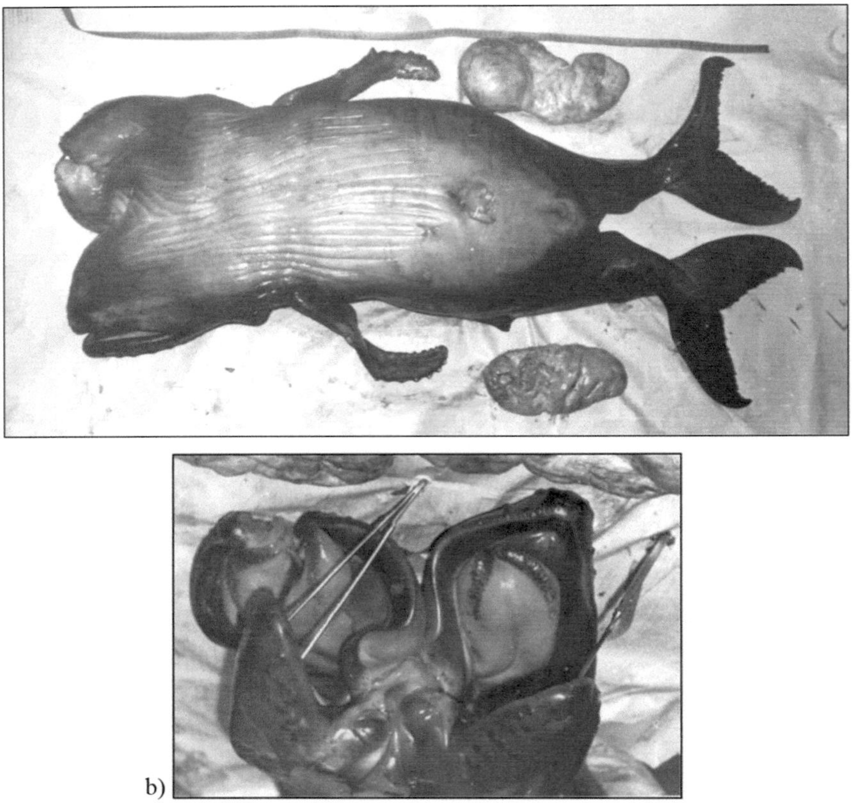

a) b)

Conjoined twins of humpback whale: (**a**) ventral view (**b**) heads only

The calculated compound values of embryo lengths in a logarithmic system of coordinates lie on a straight line and caused the least deviation error, assuming that an average age for the basic group of embryos equaled 2 months in December. As a result of regression analysis of these values the following formula of humpback embryo growth has been calculated:

$$l = 3.84\, t^{2.02}$$

3.2.2 *Prenatal Growth of Toothed Whales*

In considering the prenatal growth of toothed whales, a representative of the Southern hemisphere – sperm whales – and of the Northern hemisphere – beluga whales have been taken as examples.

Sperm Whales The prenatal growth of sperm whales is the most uncertain among the cetaceans. We shall specify only the most fundamental research: R. Laws (1959) considers that the nature of sperm whale embryo growth is slower than that of the baleen whales and is represented by a straight line on the graph. S. Ohsumi (1965), having processed many thousands of IWC data, has confirmed this opinion. A.A. Berzin (1971), with practically the same data, defines three periods of sperm whale embryonal growth: slow in the first months, then more intensive, and the third period – slow again.

To preserve the integrity of the original data, we did not include the IWC figures in processing and used only the material collected by biologists on the Soviet whaling flotillas: 6233 sperm whale embryos from embryogenesis stage up to prenatal fetuses (Fig. 3.12) in total.

As can be seen on the graph, in addition to the basic group of embryos, there is a group of prenatal fetuses and a group of embryogenesis stage embryos. Unfortunately, and in this case very importantly, there are no data from May till October, which does not allow the curve of sperm whale embryo growth to be plotted with a sufficient degree of reliability. However, the basic group of embryos is easily separated from additional groups, which facilitates processing of the material. As a result, the regression line on the schedule looks like a concave arch of the parabola type.

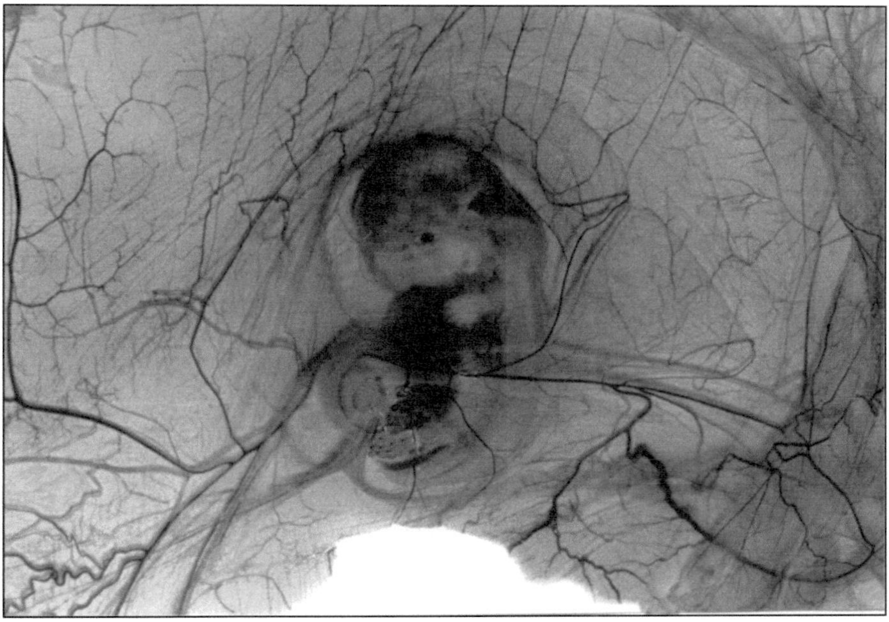

Spermwhale embryo in membrane

The graphic analysis shows that the general law of sperm whale prenatal growth looks like a parabola and does not essentially differ from the prenatal growth of

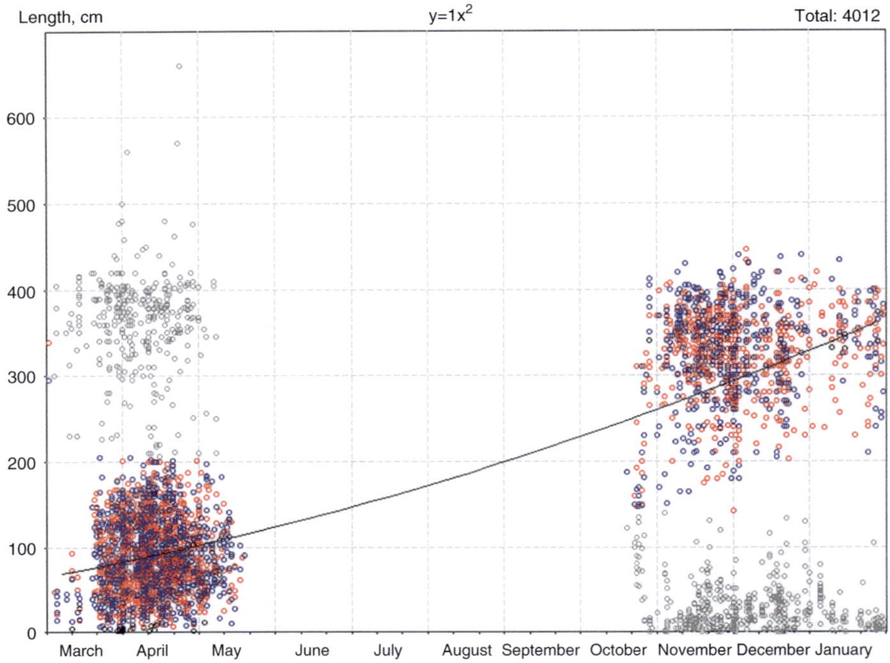

Fig. 3.12 Sperm whale embryo growth

other whales. But even if sperm whale embryonal growth were described by a straight line as it was in works by R. Laws (1959) and S. Ohsumi (1965), there would not have been a contradiction, because function with an exponent equal to one degenerates in a direct line:

$$L = at^b,$$

when b = 1 it gives the formula of a straight line: L = at.

Beluga Whales In the in section titled "Reproduction" of the "Beluga Whale" monograph (Kleinenberg etc. 1964) A.V. Yablokov analyzed the sizes of 231 beluga whales embryos. Having defined monthly arithmetical lengths, the author could not find any definitive law of embryo growth. At the same time (his Fig. 99, p. 328) he drew a curve on the cyclogram representing the general tendency of embryo growth based on the areas where the data was most densely concentrated.

On this curve A.V. Yablokov calculated the average "linear gain of embryos during pregnancy", and for each subsequent month the following were produced: 4 cm; 7 cm; 11 cm; 14 cm; 18 cm; 21 cm; 25 cm; 28 cm; 31 cm; 35 cm; ("Beluga whale", p. 329).

As we see, monthly gains grow smoothly. However, what is wonderful is that gains of the second order/sequence (the difference between monthly gains) are practically identical. So, the difference between the third and the second months' gains equals 3 cm, between that of the fourth and the third – 4 cm, the fifth and the fourth – 3 cm, the sixth and the fifth – 4 cm, the seventh and the sixth – 3 cm, the eighth and the seventh – 4 cm, the ninth and the eighth – 3 cm, the tenth and the ninth – 3 cm, and, at last, a difference between the eleventh and tenth months – 4 cm. That is, the gain of the second order is practically constant and, on average, is equal to 3.5 cm. This implies that the beluga whales' prenatal growth is described by a monomial parabola with the intensity of growth close to two:

$$L = at^2.$$

This is another vivid example of the hopelessness of using arithmetic-average values of embryo lengths to determine the prenatal whales' growth curve. The smoothed curve growth based on the most dense concentration of data appears to characterize the general law of embryonal whale growth more precisely. The example given is more proof of the paradoxical fact that "the graphic analysis also has a big heuristic value, helping to open a law, which is opened by other methods with quite a big work" (Schmalhausen 1935, p. 35).

* * *

The examples given are proof enough that cetaceans are not exceptional among mammals and the general law of their prenatal growth is parabolic, with the exception of the small period of exponential growth at the earliest stages of development and some deceleration of growth during the prenatal period. It is very important, that in the formula of a monomial parabola, all constants make biological sense. Power coefficient (the index of growth intensity) is an abstract value and, consequently, can be compared with similar coefficients of other types, within one type and even at different stages of an animal's growth.. The factor of proportionality, or "the mass factor" determines the size an embryo reaches in the process of parabolic growth per the first time unit (Schmalhausen 1935). Both these factors, together with the duration of pregnancy, determine the size of the newborn.

"The mass factor" in turn depends on the fetus size developed during the period of "pure", exponential growth, which comes to an end when the differentiation of tissue (McDowell and Allen 1927) begins. It is difficult to say how long this period lasts for fin whales and other whale types, as cetaceans belong to a group of highly organized mammals and probably, its duration does not differ much from that for large terrestrial mammal – approximately 10–15 days.

In similar works on terrestrial animals weight, rather than linear growth is more often investigated, so for comparability we shall determine the coefficient of whale embryo increase with the years by weight, instead of length (Cw).

The task is not difficult as the connection between embryo length and weight is also expressed by a power function (see. Sect. 3.2). Therefore it is enough just to multiply the factors of regress. Blue whale Cw = 5.62; fin whale Cw = 5.42; sei whale Cw = 5.49; minke whale Cw = 5.61; humpback Cw = 5.54; i.e. for minke whales, on average, the factor of weight growth intensity in the prenatal period approximately equals 5.5.

In comparison with the mass growth intensity of other kinds of animals: leghorn hens – 3.20; human beings – 3.90 (Schmalhausen 1935); Beijing ducks – 4.03; Muscovy ducks – 3.988; white Plymouth Rock hens – 3.711 (Magakyan 1967); mice – 3.4 (MacDowell 1928); Northern seals – 3.82 for males and 3.18 for females (Kuzin 1970). As can be seen, the intensity of prenatal growth is considerably more in cetaceans than in birds and other mammals, with which the comparison has been made.

3.3 Features of Individual Growth of Embryos

The question of whether all cetacean embryos grow with the same speed is disputable. However, this question has not only theoretical, but also practical value, as the size of embryos from any given month are used in attempts to define their age and thus to establish the periods of mating and procreation of whales. For example, in January the average embryo measures from one up to four meters. If we assume that they had grown with identical speed, the periods of mating and procreation will be determined by the amount of time embryo requires to grow up to three meters. If the growth of individual embryos occurs at different speeds, it is necessary to subtract individual fluctuations of their growth rate in order to calculate the duration of reproductive periods.

By analogy to other animals and the human being (Harrison 1969) it seems there should not be doubts about differing growth speed in individual whale embryos growth. Nevertheless, researchers have different opinion on this question.

R. Laws (1959) has drawn a dot diagram of fin whale embryo sizes. The dots made a cone-shaped graph, from which he has come to the conclusion that the growth rate of individual fin whale embryos is not the same. Naaktgeboren, Slijper, Utrecht, 1960, on the contrary, considered that fin whale embryos grow with identical speed, in proof of which they have cited the extremely unpersuasive fact that the line connecting maximal embryo length values on the dot diagrams parallels the line of monthly average values. Kimura, 1957 has shown that, irrespective of sex, growth of one of the twin embryos passes ahead of the other, and by the moment of birth the difference between them reaches 2 feet (approximately 60 cm). On the basis of this fact the author considers that approximately the same distinction in growth rate is true for non-twin embryos. We shall consider this problem in detail.

3.3.1 Variational Variability of Embryo Growth Rate

Distinctions in the growth rate of individual embryos are specified by the difference between minimal and maximal sizes (limits), variability parameter, and monthly dimensional variational lines coefficient of relative variability (CV). This is illustrated by the example of fin whale embryos (Table 3.4).

Table 3.4 Variability in fin whale embryo length

Month	Number of embryos, n	Median, M	*Min-Max*	Sigma, σ	Relative variability, CV
December	334	138.3 ± 3,0	EE-359.0	54.7	39.8
January	969	194.4 ± 2,5	22,0–468.0	78.0	40.2
February	795	257.5 ± 3,3	24,0–532.0	92.8	36.1
March	436	326.3 ± 4,4	4,0–620.0	93.0	28.4

The data shows that with embryo growth, size variability grows with simultaneous reduction of relative variability (CV). That is., fin whale embryos of younger age show more relative variability in sizes, which was recorded as a general law for fauna by K.M. Baer (1950) in 1882.

Undoubtedly, above all, variability is determined by distinctions in the growth rate of embryos of the same age, but it is also determined by the fact that, for every month, there are embryos of different ages whose growth rate varies.

Other factors, some of which will be considered, influence the variability of whale embryo sizes.

3.3.2 Distinctions in Male and Female Embryo Growth Rate

Sexual dimorphism of cetaceans, in particular, is expressed by the fact that female baleen whale are larger than males, however male toothed whale are larger than females. According to Mackintosh and Wheeler (1929), and also N. Peters (1938), the difference in length of physically mature males and females does not just occur at the moment of approaching sexual maturity. That is, their difference in size increases with age. This fact allows the assumption that distinctions are visible in the embryonal period.

V.A. Zemsky (1950) on the basis of analyzing the length of 66 female and 67 male the Southern Ocean fin whale embryos noted that female embryos are on the average larger (247 cm) than male embryos (221.1 cm). However in his subsequent work, V.A. Zemsky (1960) did not find any authentic distinctions.

R. Laws 1959 did not find any statistical authenticity of distinctions.

We have analyzed the average sizes of minke whale, sei whale and fin whale male and female embryos during the whole catch season.

The average length of the 439 male minke whale embryos of measured equalled 36.27 cm, female embryos – 41.5 cm, the difference in length being 5.18 cm. The average size of 438 male sei whales embryos was 229.5 cm, and of 491 females – 237.9 cm, the difference being 8.4 cm). The average length of 1260 male fin whale embryos equalled 223.8 cm, and females – 231.0 cm, the difference being 7.2 cm. That is, in all cases without exception, female embryos appeared longer than male embryos with a reliable distinction within the limits of 93–96%.

There is no doubt, that without exception female baleen whales are larger than males. Looking at a group of sei whale embryos and a group of fin whale embryos of the same average size, the difference between the females and males is greater for sei whales. Where sei whales were smaller than fin whales, their embryos in the group above were older (to have reached the same size). Hence the conclusion that distinctions in whale embryo length become more evident during the prenatal period and, probably, by the new-born period, become even more significant.

3.3.3 Dependence of the Embryo Size on the Size of Females

We have measured pregnant females and their embryos of 439 minke whales, 889 sei whales and 1719 fin whales to set the dependence shown in the next table (Table 3.5).

From this adequate sample data it is evident as the female size increases, the average length of their embryo grows a little. A similar dependence has been noted by S. Ohsumi (1965), V.A. Zemsky (1950, 1960) and R. Laws (1961), who considered that this correlation connection is determined by earlier terms of mating of older females. However, in this case of female sizes, minimal and maximal sizes of embryos should also have grown, which has not been observed. Therefore, in our opinion, it is necessary to recognize, that larger females have larger embryos.

Graphically, in a logarithmic system of coordinates, the correlation is well described by direct lines and can be quantitatively expressed by the following formulae:

Minke whales: $l = 0.00187\ L^{2.45}$
Sei whales $= 0.00139\ L^{2.70}$
Fin whales $= 0.00175\ L^{2.33}$,
Where l – length of an embryo; L – length of a female.

Within the limits of the embryo sizes investigated, the difference in the length of sei whale and fin whale fetuses was about 85 cm. Compared average embryo lengths are close to the average sizes of fetuses in February, and, obviously, this difference should increase in the following months. According to Zemsky's data (1960), it reaches 120 cm in March. At the moment of birth the difference will probably have increased by a few centimeters and will have become twice as big as S. Kimura's calculations (1957) on the basis of comparison of the sizes of twin embryos.

Thus, the analysis allows the following conclusion to be made: the growth rate of the same aged cetacean embryos is not the same and depends on the embryo sex, on the female size, and individual, genetically inherent features of whale growth. It is necessary to take these differences in embryo growth rate into account when calculating the length of mating and calving of cetaceans.

Table 3.5 Correlation between female size and the size of their embryos

Female length,	Number,	Embryo length, *cm*.		
m	n	Minimum	Maximum	Median, M
Minke				
8,5–9,0	244	–	–	37,46
9,1-9,5	195	–	–	40,61
Sei whale				
13,6–14,0	4	110	340	196
14,1–16,5	36	73	296	181
14,6–15,0	125	63	356	199
15,1–15,5	228	57	475	224
15,6–16,0	233	63	480	244
16,1–16,5	206	65	455	257
16,6–17,0	52	98	410	265
17,1–17,5	5	210	356	265
Fin whale				
18,6–19,0	8	83	249	159
9,1–19,5	36	30	317	165
19,6–20,0	100	39	545	192
20,1–20,5	250	25	563	204
20,6–21,0	242	10	510	215
21,1–21,6	275	5	547	211
21,6–22,0	230	16	512	227
22,1–22,5	231	24	520	227
22,6–23,0	166	15	533	240
23,1–23,5	124	17	620	244
23,6–24,0	41	52	465	213
24,1–24,5	13	39	411	207

While saying that the differences in length of the same aged whale embryos can mainly be explained by the differences in individual growth rate, we do not deny that the fact that same sized females start thereproduction season at different times has some influence on embryo size. It is difficult to say what the quantitative ratio of these two factors is. However, it is evident that differences in individual embryo growth rate are significant and, according to our calculations, are more than 20% at the moment of birth

3.4 Relative Growth of Body Parts and Changes in Morphological Structure of Cetacean Embryos

A detailed study of the prenatal development of whales, being the subject of an independent research, is not the task of the following chapter. Instead, some examples of the relative growth of body parts and changes in the process of cetacean embryo form and colour during the process of growth are examined.

3.4.1 Allometric Growth of Cetaceans in Prenatal Period

The study of the laws of relative growth of animal body parts in a logarithmic system of coordinates, was developed by Huxley (1932), and received the name allometry (Huxley and Teisier 1936), from the Greek word alloios – various (Schmidt-Nielsen 1987). Allometric growth of body parts or organs is considered as a function of the general sizes or sizes of other body part and is expressed by the general formula:

$$Y = ax^{b},$$

Where y is the size of a body part (either an organ or the measurement of a body part,

x is the size of an organism as a whole,
a and b are coefficients.

If the power index in the formula is more than one, the allometry is called positive, if it is less – negative, if it equals one – isometric. On the logarithmic schedule such correlations are described by direct lines.

Initial research showed that the intensity of the growth of body parts of the highest order terrestrial vertebrates occurs explosively. Mathematically it is expressed in the change of the power index of allometric correlation (Huxley 1932; Schmalhausen 1935).

Let's consider allometric correlation using the example of whale embryo growth in fin and minke whales.

Fin Whales Fifty-five embryos of various lengths were measured, and 125 fin whale embryo measurements, published by S. Ohsumi (1960) are included in the mathematical processing.

The analysis showed that during fin whale embryo growth the distance from the end of the snout to the center of an eye increases rather quickly (b = 1.16) in the beginning, later growth becomes isometric. Distance from the end of the snout to the ear aperture during the whole growth shows slightly negative allometry (b = 0, 92). The relative growth of the head in width (the distance between the eyes centers) is rather intensive (b = 1, 23) during all embryogenesis.

The tail part of the body (from the anus and from the navel to the fork in the tail blades) grows proportionally to the zoological length of the embryo. The back fin practically does not change its relative position, neither does the border of belly strips endings after they have been completely formed.

The chest fin grows lengthwise isometrically at first, and when the fetus reaches 110 cm in length, intensity of its growth increases (b = 1, 19). The growth of a fin in width in relation to the length of the body is a little slow in the first phase, and then becomes close to isometric. Fin width changes in correlation to its length with constant but lower speed (b = 0, 91).

The relative growth in the height of the dorsal fin is very intensive at first (b = 1, 46). When the fetus reaches 120 cm in length, its growth intensity is reduced (b = 1, 19). Relative growth in length of the tail fin blades is initially also very high (b = 2, 0), then reduced (b = 1, 19), and in the pre-fetus period of development becomes isometric. Growth of tail fin blades in width shows negative allometry at first (b = 0, 69), and then also becomes isometric.

Minke Whales Minke whale embryos from 5 to 300 cm measured, the most precisely fixed measurements will be considered. The analysis has shown that in the process of embryogenesis the distance from the end of the snout to the eye grows quickly (b = 1.17) at first, then the growth is reduced (b = 1.10) and when the embryo is 110–115 cm long the growth rate becomes almost isometric. The distance from the end of the snout to the ear aperture in the first half of embryogenesis shows positive allometry (b = 1, 05–1, 10), and in the second half allometry becomes isometric.

The distance from the tail fork blades to the anus shows slightly positive allometry at first, then growth dramatically reduces (b = 0, 90) and at the last stage of fetus development becomes isometric. The measurement from the tail fork blades to the navel were different according to sex. Male growth rate was isometric practically all the time, while females at first showed negative allometry (b = 0, 93), by the middle of embryogenesis grew a little (b = 0, 95), and then became isometric.

The borders of the ventral grooves are completely formed by the second half of embryogenesis. From this moment, the growth of the measurement from the tail fork to the ventral grooves shows dramatically negative allometry (b = 0, 67), and in the last months of uterine development becomes practically isometric. By the middle of embryogenesis, the distance from the tail fork blades to the dorsal fin increases, showing slightly positive allometry (b = 1.06), and in the second half of uterine development its growth rate is sharply reduced (b = 0.90).

Flipper length in the pre-fetus period shows slightly positive allometry (b = 1, 05). By the fetus period of development flipper growth rate it is reduced (b = 0, 96), but during the last stage of embryogenesis it grows dramatically (b = 1, 16). The tendency of flipper growth in width is similar.

Growth of the dorsal fin in height in the beginning goes very quickly (b = 2, 07), then it is sharply reduced to negative allometry (b = 0, 91), and in the last stage of intro-uterine development it grows again (b = 1, 08). Tail fin blades actively grow lengthwise in the first half of uterine development (b = 1, 40). At the beginning of the second half, growth rate is reduced to b = 1.05, and at the last stage of allometry even becomes negative.

Hence, in the process of fin whale and minke whale embryo growth, geometrical similarity of body form is not observed, and growth intensity of separate body parts during certain moments changes in leaps. Except for the embryogenesis stage with minke whales, these "leaps" often occur when their body length is 10–15 cm and about 55–60 cm. With fin whales the change of allometry speed growth occurs when their body length is about 25–30 cm and 110–115 cm.

3.4.2 Change in the Body form and Coloring of Whale Embryos

This question is considered using fin and minke whales as examples.

Fin Whales The smallest embryos collected were 1–2 cm long. Fetuses up to 10–15 cm have well-formed visceral arches, which are the early stages of hind legs. In the eye layings, where the eyes begin to form, an external pigmentary layer can be seen. By the end of the germinal period, the hind legs are reduced, and eye lens and eyelids become visible. Blood vessels are slightly formed on the body surface. Hemispheres of cerebrum can be seen through ectoderm. Gender cannot be determined by external genitals.

When embryos are 35–40 cm, pigmentation starts to appear on the vertex, tail fins and on the left side of the lower jaw. An appreciable white strip appears at the ear apertures. The dorsal fin at this time looks like a rounded fold. It will become triangular later on. Vibrissa can be seen on the jaws. The ventral grooves are visible in the center of the belly and on the bottom jaw. Pili appear on the extraembryonic membrane (chorion).

At 100–120 cm, all of the embryo's back is pigmented, and a V-mark and asymmetric coloring of the jaws are well-seen. All ventral grooves and the connections between them become apparent.

The chorion of fetuses longer than 500 cm easily come off the uterus walls. Baleen plates are 3–4 cm in height, and asymmetry in their coloring can be seen. Tail blades swivel freely from the "cam". The dorsal fin is tucked and pressed to the body, but can be easily straightened to a vertical position. The umbilical cord parts easily from the umbilicus.

It is important to note that the formation of various morphological structures parallels those revealed during the allometrrical growth study. The moments of qualitative changes correspond to the ending of germinal, pre-fetus and to the beginning of the fetal periods. Embryos larger than 500 cm have all the features of newborns. So, as in the case of terrestrial mammals, cetacean embryos have different periods of development.

Minke Whales The body form of embryos up to 9.5 cm long remains C-shaped. The frontal lobe looks like a bubble with cerebral hemispheres visible inside. Rudimentary eyes are visible through the transparent skin. The chord, hepatic outgrowth and ribs can be seen. The rudiment of a dorsal fin is also visible. In the tail department there are 11–18 vertebral segments. The tail fin looks like a rounded fold.

After embryo growth reaches 10–15 cm, the head and tail start to straighten quickly. Vibrissa dorsal fin appear and anus appear, the gender of the embryo becomes distinguishable and the ear aperture and blowhole are visible.. "Pair formations" become noticeable as white color swellings at the end of the upper jaw on the ventral side (Mikhalev and Zelenaya 1986; Mikhalev 1991) (see Chap. 4). Blood vessels begin to appear.

When the embryo's body is longer than 50–60 cm, it is almost straight, and the frontal bubble becomes flatter and loses transparency. The crest of the upper jaw becomes visible and the eyelids are divided by a crack through which the eyeballs can be seen. The blowhole looks like two cracks and its skin is pigmented. "Pair formations" become appreciable on both upper and lower jaws. The notch in the posterior margin of the dorsal fin and the ventral grooves are visible.

In relation to the general duration of embryogenesis of fin and minke whales, the phases of the germinal period are approximately: germinal – 17%, pre-fetus – 11%, fetus – 72%. It is important to notice that these percentages correspond to those of agricultural animals (Bogolyubsky 1967, etc.).

3.5 Postnatal Growth of Whales

To establish the growth rate of cetaceans–terms of the onset of sexual maturity, reproductive period, life expectancy, age variability, age structure of populations–and to solve many other questions of their biology, it is necessary to be able to define the age of an animal.

3.5.1 Problems of Defining Whale Age

Flaky structures in which age and other physiological conditions of animals are fixed are called "recording structures" (Mina and Klevezal 1976). The best structure for defining the age of baleen whales is the ear plug, and in toothed whales, the teeth. The age of baleen and toothed whales can also be judged by the amount of traces of corpus luteum on the ovary of pregnant females. However, it is necessary to know age of approach of a sexual maturity, periods of repetition of reproductive cycles; and to be able to distinguish traces of ovulation from traces of bodies of pregnancy.

Fig. 3.13 Whale recording structures: (**a**) fin whale ear plug, (**b**) minke ear plug, (**c**) spermwhale male tooth, (**d**) spermwhale female tooth

The ear plug of baleen whales is formed on a finger-shaped shoot, which projects inside the external auditory canal from the ear-drum bone cavity. During the life of a whale, the ear plug is formed layer by layer from degrading cells of epithelium (Fig. 3.13).

P. Purves (Purves 1955) suggested using these stratifications to define the age of an animal To make a graphic record of stratifications and to estimate them more objectively, T. Ichichara (Ichichara 1966) used a photometric plant. The author and his colleagues recorded graphic records of lamination (recording structure graphs) of ear plugs with a registering microphotometer IFO-451 directly from colored histologic preparations and from photographic copies (photographic plates) of these preparations, recorded by passing them through light (Mikhalev 1971a, 1972, 1973, 1975, 1977, 1978a, b, 1982, 1986, 1990a, b, c, 1991, 2000, 2002a; Lockyer 1972; Dubin etc. 1982; Mikhalev and Ivashin 1982).

Recording structure graphs of the ear plugs of baleen whales reveal a rather complex character of layering. Layers of all kinds of physically mature baleen whales become more dense and frequent from the top to the bottom. A change in the nature of the stratifications is observed. Such change has been named "trasphase" (a transitive phase) by K. Lockyer (1972). Section 3.4.1 provides more detail about the onset of sexual maturity.

It is more difficult "to read" the ear plugs of baleen whales than the teeth of sperm whales. Ear wax of young animals is especially difficult to read. While about 30% of ear plugs of physically mature minke whales cannot be read, that of non-mature whales is up to 50%. According to Sigurjonson (1980), about 42% of ear plugs of Northern hemisphere minke whales, and up to 80% of blue whales and fin whales cannot be read

The ability "to read" ear plugs is very different for animals from different areas, and insignificant distinctions of sex are appreciable. According to our data, 4.7% of female minke whale ear plugs can be better can read, in contrast to fin whales, where 74% of male ones were better read. According to K. Lockyer (1979), the difference between male and female fin whales is even higher than that −8% to 10%.

Layers in the toothed whale teeth are more visible on painted or polished sections, which have been submerged in organic acids, usually formic acid from ants. The first researchers considered the chromaticity and optical density of layers in teeth to be determined by their degree of calcifications, though their opinions differed on which layer is hypo- and which one is hypercalcified.

Some considered lighter, but optically less transparent strips, to be hypercalcified (McLaren 1958; Sergeant 1959, 1962; Berzin 1961, 1963). Others believed narrow, more optically transparent strips to be hypercalcified, as they are more intensively colored by silver nitrate or hematoxylin (Klevezal 1963, 1966; Klevezal and Kleinenberg 1967; Ohsumi et al. 1963). The latter conclusion has been confirmed by X-ray spectrum analysis (Sukhovskaya and Klevezal 1982).

We have analysed the Ca content in various stratifications and parts of the teeth of nine sperm whales: eight males 9.2–14.0 m long and one female (10.3 m). Tests were undertaken on cuts from various points of each tooth: the cement zone; the tooth axis; the average dentin line; and the dentin edge where it borders with the cement. The samples from dark and light layers were scratched out from the sections with wider and more contrasting layers using a low magnifying lens. In the tests, Ca content was defined by the atomic-absorption method (Cudgels etc. 1982; Mikhalev 1990a).

The analysis has shown insignificant distinctions in Ca content, the amount of Ca in 100 g of dentin being on average less (2.347 g) than in cement (2.741 g). In more transparent layers there appeared more Ca, but only by 0.085 g. And the reliability of this difference is low. Still, the presence of such differences is shown by the fact that the Ca content on more transparent tooth axes (2.452 g) is higher than on the optically less transparent dentin periphery (2.195 g). In the teeth of females the average Ca content was smaller than that in the males, being 1.915 g (1.395 g in dentin; double that in cement – 2.695 g).

Small differences in light and dark stratifications suggest that the optical density of stratifications of the teeth of sperm and other toothed whales is determined not so much by the content of carbonic salts but by the different organization of collagen fibrils and features of structure of albuminous stromas. This conclusion is proved by the fact that in a completely decalcified microsection of a tooth, lamination remains appreciable. L.I. Sukhovskaya and G.A. Klevezal (1982), who investigated teeth of a bottle-nose dolphin (Tursiops truncatus), also came to similar conclusions on the basis of X-ray spectrum analysis.

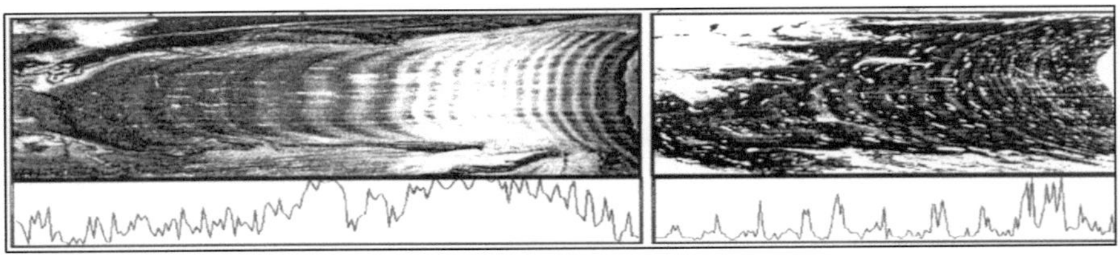

Fig. 3.14 Ear plug of fin whale (left) and minke (right)

3.5.2 Expert, or Subjective, Estimation of Stratifications

When defining the age of a whale, the question arises as to how many and what layers are formed within 1 year. Opinions differ very much. For example, some (Berzin 1961) consider that two transparent and two opaque strips are formed in sperm whale tooth dentin; others (Ohsumi et al. 1963) that one opaque with additional transparent and one more transparent strip are formed; the third opinion (Lockyer 1980) is that one opaque plus one more, often double, transparent strip is formed. G.A. Klevezal and S.E. Kleinenberg (1967) tried to explain such divergences in opinion by the fact that "additional" layers can be so clear that sometimes they are mistaken for "basic" ones (Fig. 3.14).

The complex nature of stratifications in recording structures of toothed whales, uncertainty in qualitative estimation of layers (narrow, wide, dark, light, transparent, opaque, etc.) and a lack of standardised terminology were considered at a special International Whaling Commission conference (IWC, Spec. Issue N 3., 1980). In view of this discussion, under the initiative of the International Whaling Commission Scientific committee, it was suggested that six experts from different countries (the author was the expert from the Soviet Union) define sperm whale age using sections from 50 teeth (Table 3.6).

Experimentation has shown (Donovan et al. 1981) that the experts' estimations are considerably varied. The subsequent increase in the number of highly skilled experts A.A. Berzin and G.A. Klevezal have joined the experiment) has only increased lack of order in these estimations (Fig. 3.15).

In our analysis (Mikhalev 1982, 1986) Wilcoxin's ranking method was used to estimate the reliability of data differences, which has a great capacity for reliability (Urbakht 1964). Paired comparisons have shown there is a real difference in the estimations of all eight experts. This lack of order does not mean that the experts' estimations can be averaged and that the average will be closer to the truth. Averaging acts only as an expression of a common, but subjective, opinion. It's important that the fact of using qualitative experts' estimations assumes the absence of reliable enough ways to determine annual stratifications. What makes it more complicated is the fact that from the top to the pulp cavity of the tooth (Fig. 3.15) the nature of stratifications varies (Mikhalev 1990a).

Table 3.6 Age estimation by decalcified sections of sperm whale teeth

#	# of whale	Length, m	Sex	Sperm whale age by expert estimation:							
				# 1	# 2	# 3	# 4	# 5	# 6	# 7	# 8
1	2	3	4	5	6	7	8	8	9	10	11
1	655	10.0	Female	18	19	16	6	14	16	18	15
2	4520	10.6	Female	32	32	30	29	30	32	29	28
3	3863	10.7	Female	27	24	26	25	23	27	27	26
4	4125	10.7	Female	30	30	31	31	30	33	34	30
5	3810	10.8	Female	17	18	17	17	16	17	17	17
6	4132	10.8	Female	30	26	28	26	24	27	30	27
7	3683	10.9	Female	21	22	21	22	20	18	22	19
8	3944	10.9	Female	34	34	31	34	26	35	34	29
9	3885	11.0	Female	31	28	29	29	29	32	31	28
10	3645	11.0	Female	26	23	22	22	18	23	24	21
11	4905	11.1	Female	33	30	31	31	33	30	33	30
12	3884	11.5	Male	18	16	16	16	11	17	18	16
13	3284	12.0	Male	18	20	19	20	17	20	20	18
14	682	12.3	Male	21	28	24	25	23	23	22	23
15	3902	12.3	Male	20	19	18	18	17	21	20	18
16	3259	12.4	Male	26	24	23	25	23	24	26	23
17	4009	12.4	Male	25	29	25	25	22	30	25	22
18	4083	12.4	Male	25	22	23	22	18	23	25	20
19	3301	12.5	Male	22	27	24	25	18	23	21	20
20	3883	12.5	Male	20	18	21	21	19	20	20	19
21	4622	12.6	Male	25	23	23	23	22	25	25	22
22	3885	12.6	Male	25	22	22	22	18	22	23	22
23	678	12.7	Male	23	28	23	26	22	26	23	22
24	3771	12.7	Male	27	25	25	25	14	24	27	23
25	4528	12.8	Male	28	29	29	29	29	28	28	28
26	3758	12.8	Male	28	25	25	25	21	25	28	24
27	3772	12.9	Male	22	24	22	22	22	21	23	21
28	4127	12.9	Male	30	32	30	31	28	32	31	29
29	4526	12.9	Male	29	23	25	25	23	24	26	24
30	4403	13.0	Male	23	24	22	22	25	21	23	21
31	4052	13.0	Male	25	24	21	24	24	25	25	24
32	4028	13.1	Male	30	26	23	24	25	27	26	24
33	3928	13.2	Male	24	25	24	25	19	24	24	22
34	4036	13.3	Male	27	25	26	25	26	26	27	26
35	4048	13.3	Male	22	22	20	22	20	23	23	20
36	3485	13.4	Male	22	20	22	22	19	22	22	22
37	3856	13.4	Male	32	27	27	27	27	27	31	27
38	3926	13.4	Male	30	25	27	25	26	26	28	25
39	4140	13.4	Male	27	20	23	21	20	20	21	21
40	3515	13.5	Male	32	30	33	30	29	33	32	23

(continued)

Table 3.6 (continued)

#	# of whale	Length, m	Sex	Sperm whale age by expert estimation:							
				# 1	# 2	# 3	# 4	# 5	# 6	# 7	# 8
41	3472	13.6	Male	30	32	26	26	27	26	28	26
42	3984	13.6	Male	31	33	27	31	26	26	32	29
43	805	13.9	Male	40	40	39	40	40	38	38	37
44	5178	14.0	Male	33	33	32	32	35	37	33	32
45	3529	14.0	Male	30	33	32	32	31	32	34	30
46	802	14.1	Male	41	37	38	39	44	34	40	34
47	2641	14.2	Male	40	36	35	35	34	38	35	35
48	1734	14.3	Male	45	41	39	40	37	37	43	37
49	4891	14.4	Male	31	27	29	28	21	28	29	27
50	4004	14.7	Male	29	21	22	21	23	27	24	28

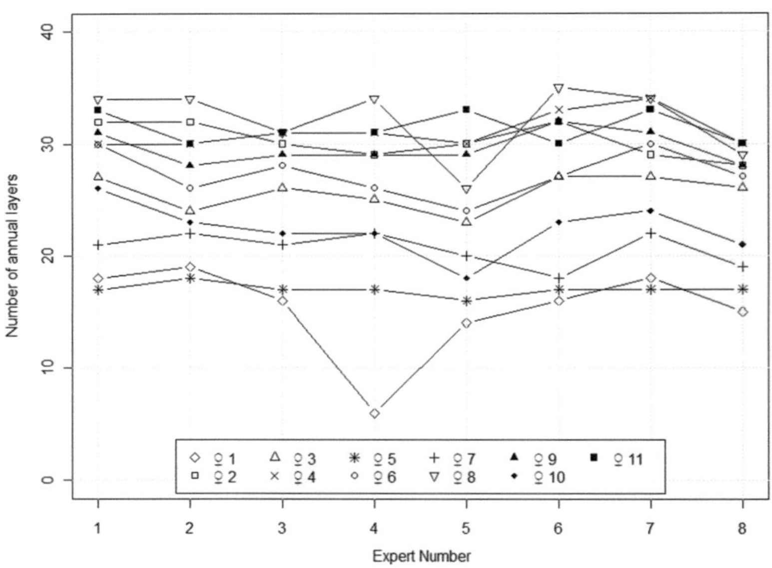

Fig. 3.15 Differences in estimates of spermwhale female age using teeth by different experts

3.5.3 *Quantitative Characteristic of Stratifications*

Graphs have in themselves, as academician I.I. Schmalhausen (1935) said, "heuristic value". Therefore we believe that recording structure graphs have some perspective. At a certain scale, recording structure graphs allow the measurement of quantitative parameters of layers and the mathematical processing of figures.

The first attempt to produce recording structure graphs for baleen whales were J. Ruud (1940), A.G. Tomilin (1945) and S. Ohsumi (1966). Then followed (Van

Utrecht 1965; Van Utrecht and van Utrecht-Cock 1966; Mikhalev 1975, 1982, 1986, 1990a, 2002a. Having portrayed graphically the relief of the surface of baleen whale plates,the authors tried to reveal the annual cyclic recurrence of stratifications by comparing the quantitative characteristics of stratifications with the animals' physiological condition. Unfortunately, baleen plates wear down with time, and record information from only about the last five or six last years of the animal's.

In our practice (Mikhalev 1972, 1973, 1975, 1977, 1978a, b, 1982, 1984, 1986, 1990a, 1991, 2000, 2002a; Ivashin and Mikhalev 1978) we used plethysmographs, profilographs and microphotometers to produce recording structure graphs. The best results have been obtained by using profilograph N-201, an English device for measuring surface curvilinearities Talyline-4, and with the help of a recording microphotometer IFO-451. Baleen plates, ear plugs, and the teeth of sperm whales, killer whales, pilot whales, bottlenose whales and dolphins were examined.

Profilograms were recorded from decalcified polished sections of teeth, while photograms were recorded from photoplates. To produce the photogram, objects were photographed both through reflection–baleen plates, decalcified polished sections of teeth–and by passing light through them–microsections of teeth and cuts of ear plugs.

The tooth lamination charts thus produced (Fig. 3.16) have allowed the measurement of the quantitative characteristic of stratifications. Experiment has shown that on a sperm whale tooth decalcified with a 5% solution of formic acid, the height of "rollers" fluctuates within the limits from 0.001 mm up to 0.02–0.03 mm. The width of the layers is about 1.5–3.0 mm.

On profilograms (Fig. 3.16) wavy repeating zones of groups of layers of various configuration and height are appreciable more precisely than on sperm whale polished sections (Fig. 3.13). In mature females it is less apparent, as it is deformed by strongly prominent "labels" (Fig. 3.17) – the clearest layers correlating with the periods of reproduction (Pike 1966; Klevezal and Myrick 1984; Mikhalev 1973, 1975).

Labels on the polished sections of male sperm whale teeth (Fig. 3.13) are expressed more feebly. Labels ("ovulation peaks") have also been revealed on the baleen plates of baleen whales (Van Utrecht 1965).

3.5.4 *Speed of Stratification Formation*

Formerly the width of annual stratification of sperm whales was defined by the gain of pre-pulp dentin (Berzin 1961, 1971; Yablokov etc. 1972; Gambell and Grzegorzewska 1967). Profilograms allow the measurement to be made more precisely. We have processed records of a surface relief of the first or the second pairs of sperm whales of different sizes and with teeth of differing biological condition (Mikhalev 1986).

It appeared that from November till January the gain of dentin was 0.25 mm, and from January till March 0.56 mm, i.e., double the amount. During the next months (April–May) the data was unreliable because there were too few samples.

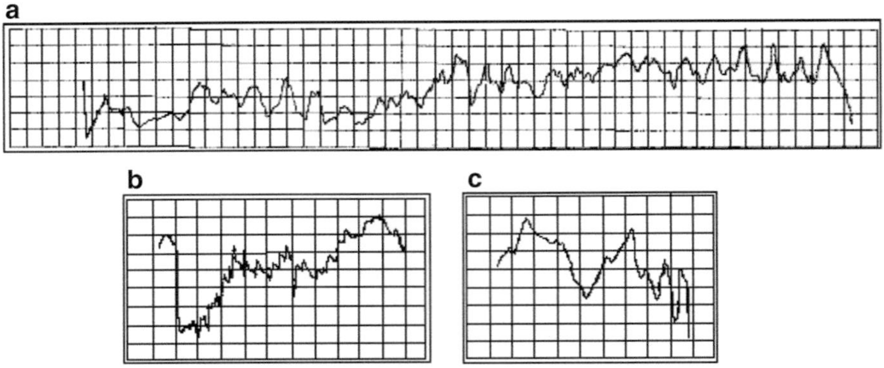

Fig. 3.16 Profilograms of teeth of sperwhale (**a**), bottlenose whale (**b**) and piloc whale (**c**)

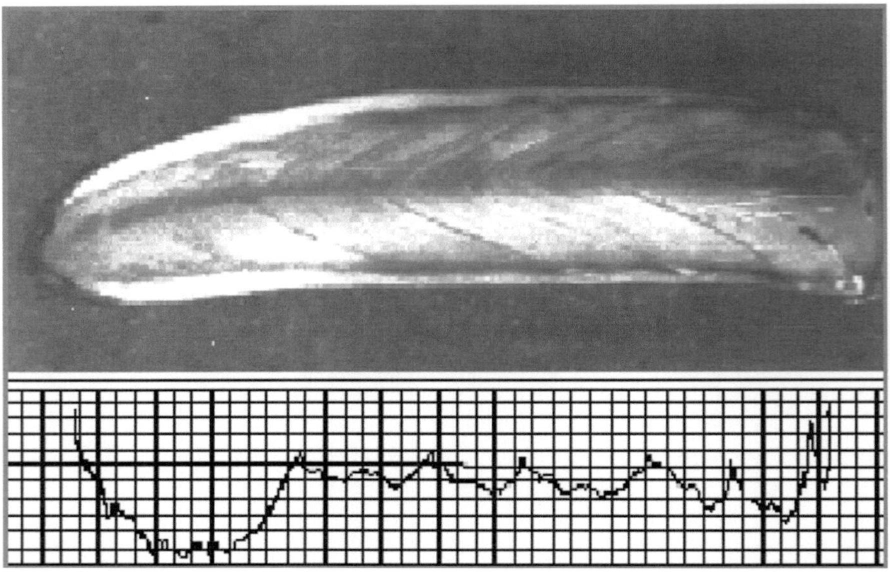

Fig. 3.17 Photo and profilogram of female spermwhale tooth

Yet it can be noticed that in the second half of the year dentin accumulates with the same speed, i.e. the width of an annual layer on average equals approximately 2.4 mm. It has also been established that in sperm whales the annual dentin gain reduces with age: usually the width of the light (less transparent) layers on the top of the tooth equals 0.7 mm, and at the cavity of the pulp layer, the width decreases up to 0.4–0.3 mm.

Defining the position of the neonatal line in sperm whale teeth, one proceeds from the fact that prenatal dentin is homogeneous and has no lamination. However,

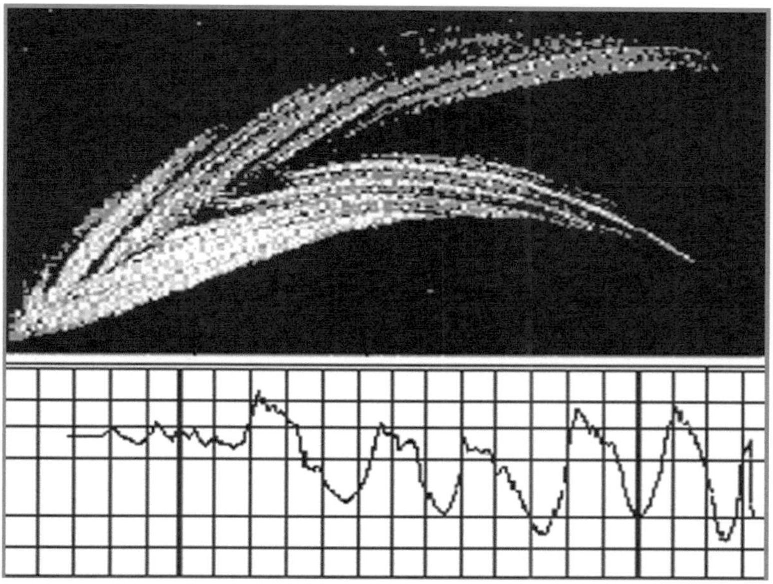

Fig. 3.18 Tooth of young spermwhale female (7 m long)

lamination is also poorly appreciable in the first annual layers, as evidently illustrated by the photo of a young sperm whale tooth profilogram (Fig. 3.18).

As a result, the first or even the second layers five or more mm from the top of the tooth are frequently taken for the neonatal line. As a matter of fact the prenatal dentin width is no more than 1.5–2.5 mm. A newborn sperm whale taken by the whaling vessel *Bodry-25* had prenatal dentin of about 2 mm (Mikhalev and Shevchenko 1973, 1986). The teeth of prenatal embryos had about the same prenatal dentin width.

3.5.5 Decoding of Recording Structure Graphs

Research into decoding the recording structures of cetaceans has given certain results. With some degree of reliability the so-called "Lunar layers" have been revealed; the labels connected with female reproductive cycle are described, and the rhythm of annual dentin gain has been determined (Van Utrecht 1965; Pike 1966; Cock 1966; Klevezal and Tormosov 1971; Klevezal 1973, 1975; Kasuya 1977; Mikhalev 1977, 1978a, b; Kasuya and Brownell 1979; Hohn 1980; Myrick 1980, 1984; Klevezal and Myrick 1984). Yet the definition of whale age remains problematic.

Homoiothermy of mammal assumes their rather constant physiological homeostasis. Therefore it is necessary to recognize that lamination of recording structures is basically genetically determined (Klevezal 1988). Nevertheless it is impossible to exclude the influence of exogenous factors, especially one as powerful as scarcity of food in for some years (starvation), the pan-epizootic, cyclic recurrence of solar activity and the change of seasons. This is, for example, testified by differences in the lamination of recording structures of dolphins living in natural conditions and those in captivity (Grue-Nielsen 972; Brodie 1982). A similar orientation in the recording structure graphs of different individuals also specifies the influence of common external factors (Mikhalev 1990a).

We have attempted to use this fact for decoding events fixed in recording structure graphs, and, primarily, for defining whale age. With this purpose, a comparison of the profilograms of the teeth of male sperm whales taken during the same period (December) in the same area (Atlantic sector of the Southern Ocean), but in various trade seasons from 1966 to 1975, has been made. For comparison we have selected only the profilograms in which a feature inherent to all is more clearly appreciable (Fig. 3.18). A similar technique is widely used in dendrochronology (Schulman 1958; Kolchin 1963; Huber et al. 1965; Fergson 1968; Kolchin and Rybina 1982; etc.).

One of the general features of the compared profilograms that was easily seen was the narrow and deepest label, or " dropping out ring", to use dendrochronological terminology, which, in teeth sampled in 1966, was formed a year earlier, in 1965 (by the way, 1964 was a peak of low solar activity). In our p rofilograms these characteristic labels were on one vertical line (in Fig. 3.19 a vertical line is designated by a dotted line).

As a result it appeared that in each subsequent profilogram, the number of peaks from the right side of the initial layer was growing and, hence, their number increased by 1 year. Due to this, the opportunity to define the configuration of the annual stratification approach arose. It is typical that every new layer at the pulp cavity edge is always well defined, but with aging of the animal, the layer structure deforms a little. It appears that it is not always easy to define the exact configuration of the annual stratification, though it is clear how many years there are in the interval between the initial label and pulp cavity. The sperm whale age determined thus practically coincided with the age calculated by pre-pulp layer gain.

Understanding there is insufficient validity in the demonstrative basis of the research, we still adduce it as a promising method for further study.

First of all the collection of sperm whale teeth for a much longer period is necessary for this purpose and also the introduction of mathematical computer processing profilograms. The first step in this direction, both for plants and animals (Fritts, Mosimann and Bottorff 1969; Eckstein and Bauch 1969; Goncharov and Smetanin 1974; Smetanin and Kasyanov 1979), gave encouraging results.

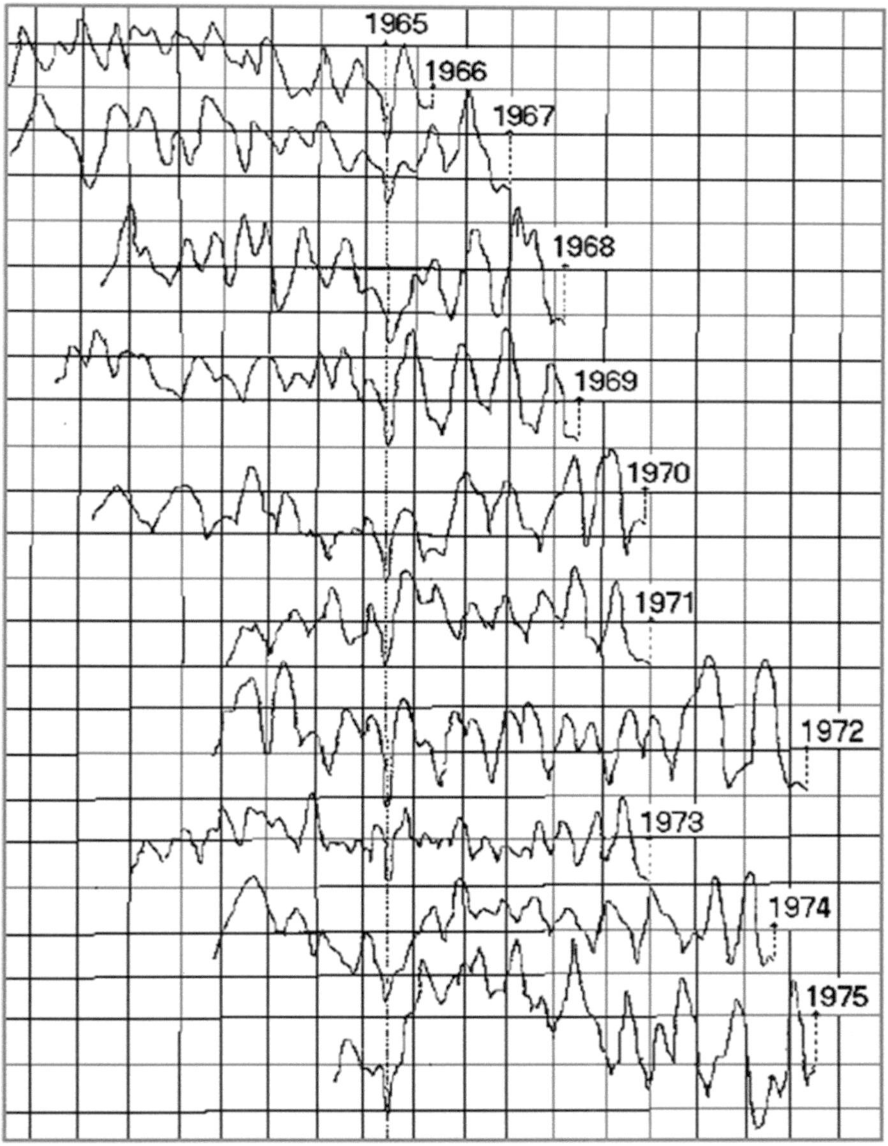

Fig. 3.19 Profilograms of spermwhale teeth taken in different years (1966–1975)

3.5.6 *Plotting of Postnatal Cetacean Growth Curves*

Quite a lot of experiments tracing postnatal whale growth curves are known: fin whales (Ohsumi et al. 1958), pilot whales (Sergeant 1962), sperm whales (Berzin 1963), gray whales (Rice and Wolman 1971; Zimushko 1968), pygmy blue whales (Sazhinov 1970), and other minor cetacean species (Pirrin and Myrik 1980), etc.

When drawing growth curves, ear plugs were used to determine the age of baleen whales and teeth to determine that of toothed whales. Only whole ear plugs and the first or the second pairs of teeth with the top and the blank pulp cavity were used. No reconstruction of any missing tops of the teeth was allowed. Only clear laminated recording structures were selected. The age was determined by taking into account the growth rate of the pre-pulp layer and the analysis of recording structure graphs.

The growth curves of whales and terrestrial mammals can be well described by logistical function, Bertalanffy's equation, Gompertz's equation, polynomial parabolas (Bunak 1946) and other formulae. Postnatal growth curves drawn by us are also well described by monomial parabolas. We shall give some examples.

Pygmy Blue Whale The age of 444 males and 351 female pygmy blue whales 12.5–23.5 m in length was determined,. The age varied from one to 58 years (Fig. 3.20).

A rather big data sample has produced rather satisfactory results. The post natal growth curve (regression coefficient mistake m = 0.003) is described by the formula:

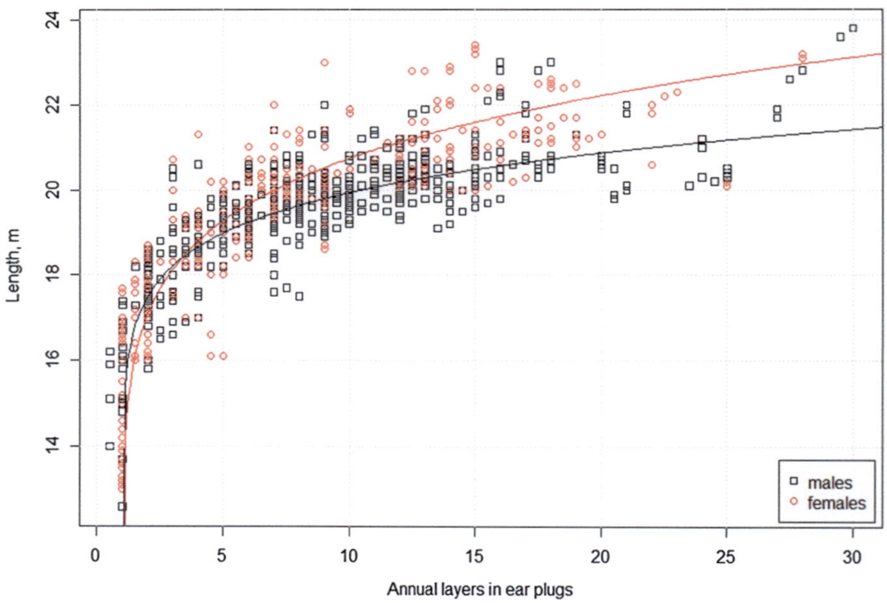

Fig. 3.20 Post-natal growth of females (red) and males (black) pygmy blue whales

For males: $L = 12.5(t - 1.0)^{0.081 \pm 0.003} + 5.0$,
Foe females: $L = 12.09(t - 1.0)^{0.12 \pm 0.003} + 5.0$,
Where L – length of a whale, m;
t – age of a whale, years.

From the growth curve, sexual maturity of pygmy blue whales comes at the age of 5–6 years when the male body length is 18.0–18.5 m, and the female is 19.0–19.5 m. Physical maturity of both comes at 10–12 years, the body length being 20–21 m. Pygmy Blue whales live, probably for 55–60 years.

Minke Whales The age of 544 males from 6.5 to 10.0 m long and 630 females from 7.0 to 10.5 m long was defined. The larger amount of sample material (in comparison with fin whales) has also increased the degree of reliability of correlation – the mistake of regression factor equaled 0.001. The age of whales varied from four to 59 years for males and three to 49 years for females (Fig. 3.21).

The median age of females was 21.5 years, and males, 19.9 years. The growth curve of minke whales is described by the formula

for males: $L = 8.09(t - 5.0)^{0.02 \pm 0.001}$
for females: $L = 8.54(t - 5.0)^{0.018 \pm 0.001}$

According to the diagrams, sexual maturity of minke whales comes at the age of 5–6 years, when the body length is about 8 m. Males reach physical maturity at the age of 18–20 years, with a body length of about 8.5 m, and females, 9.0 m, which is

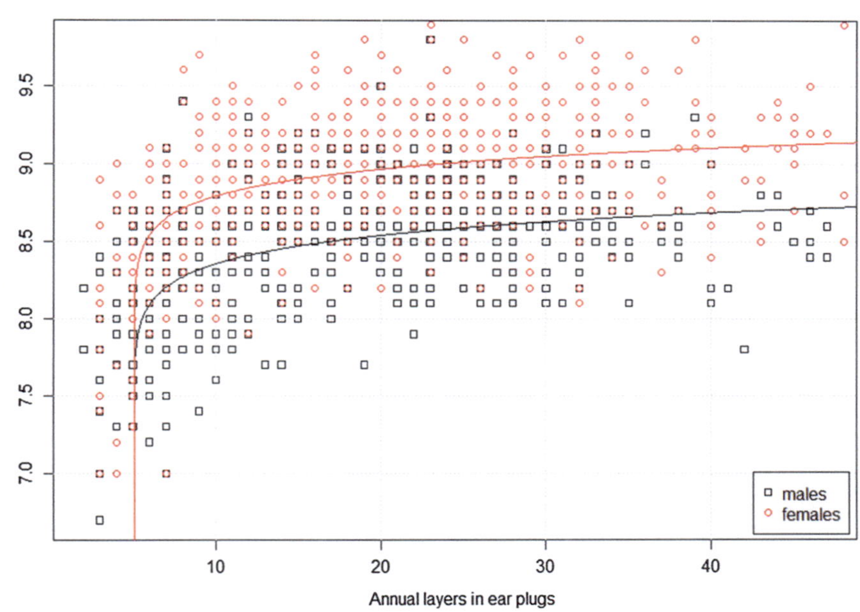

Fig. 3.21 Growth of females and males of Minke whales

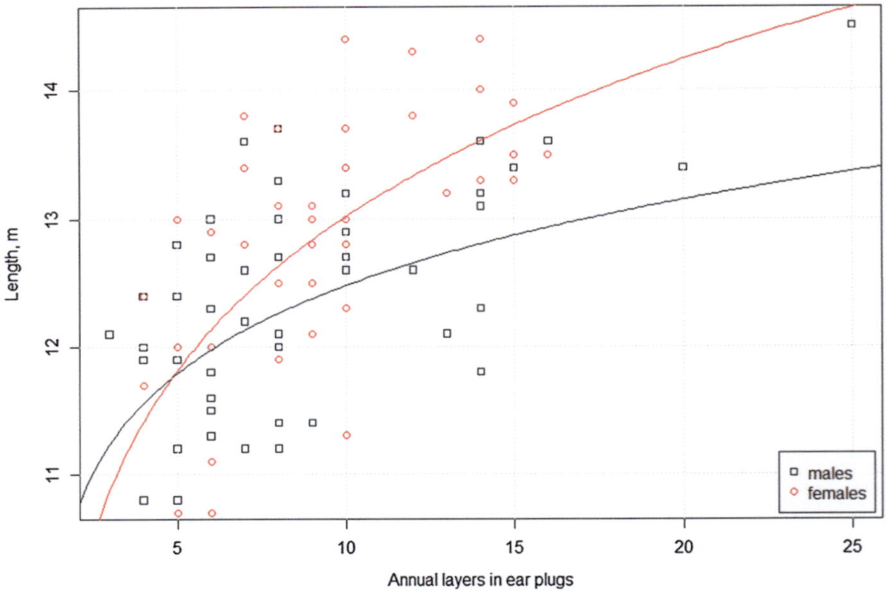

Fig. 3.22 Growth of females and males of humpback whales

20–30 cm more than that recorded by S. Ohsumi, J. Masaki and A. Kawamura calculations (Ohsumi et al. 1970).

Humpbacks During our practice in the 1960s, this whale species was rarely taken, therefore there is much less material than for the two previous species. The age of 45 males and 41 females was determined (Fig. 3.22).

Naturally, the reliability of the power parameter means that mistakes in the correlation factor were lower. The following formulae describe growth curves drawn on the basis of our material

for males: $L = 10.7(t - 1.0)^{0.07 \pm 0.02}$
for females: $L = 10.0(t - 1.0)^{0.12 \pm 0.02}$

These data do not allow a judgement with a sufficient degree of reliability regarding the time of sexual maturity, physical maturity or life expectancy.

A large amount of actual data on the age of Northern Pacific humpbacks derived from 561 animals has been published by N.V. Doroshenko (2000). And what is extremely valuable is that these data are from young animals. However, the technique of age definition causes doubts, as, judging by the growth curves published by Doroshenko, the life expectancy of humpbacks does not exceed 27 years. The age is obviously doubly underestimated, which led to a distortion of the growth curves. Intensity of the growth of humpbacks is doubly overestimated.

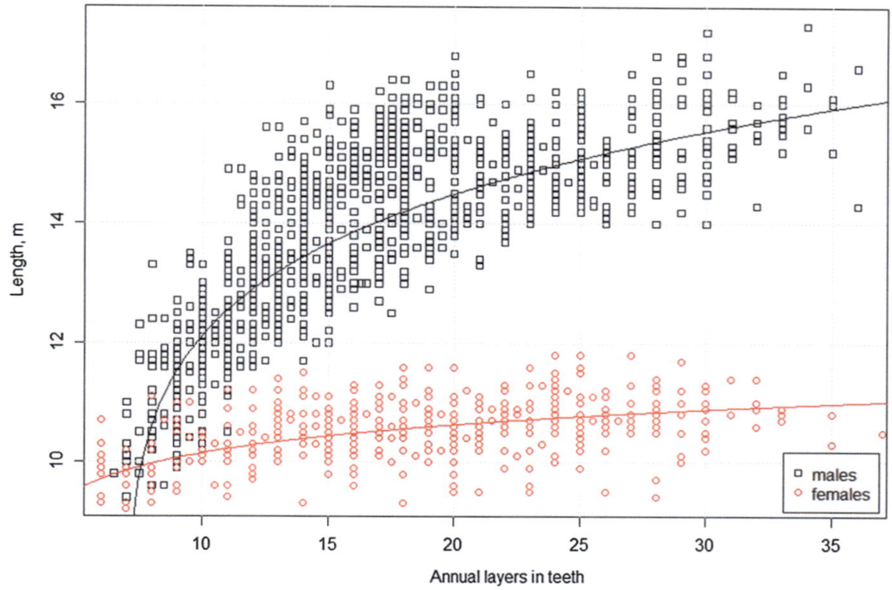

Fig. 3.23 Growth of females and males of sperm whales

Sperm Whales The age of male sperm whale was determined from 1368 teeth, and from 453 teeth for females (Fig. 3.23).

The large amount of sample material lowered the error of the regression coefficient: t = 0.002 for males, t = 0.003 for females. The growth curve for males is described by the formula:

$$L = 10.6(t - 7.0)^{0.122 \pm 0.002}$$

The growth rate of females is much lower:

$$L = 9.17(t - 3.0)^{0.052 \pm 0.003}$$

Judging from the growth curves, sexual maturity of females comes at the age of five-six years, when the body length is 8.0–8.5 m. Physical maturity is reached at 15–20 years, with a body length of 10–10.5 m, which is a little bit less than that determined by A.A. Berzin (1971). Specificity of the male growth curve (absence of a well defined excess of growth curve by puberty) does not allow the estimation of even an approximate age for the onset of sexual or physical maturity, which can be successfully determined by other methods (see below: Sect. 4.1).

Summary The graphic, correlation and regression analysis of embryo sizes have proved that cetaceans are not an exception to the class of mammals, so the common rule for their prenatal growth is also parabolic:

Pygmy blue whales- $l = 5.39\ t^{1.90}$
Fin whales – $l = 5.72\ t^{1.95}$
Sei whales- $l = 4.52\ t^{1.98}$
Minke whales – $l = 2.91\ t^{2.03}$
Humpbacks – $l = 3.84\ t^{2.02}$

Growth among the representatives of baleen whales is similar. The intensity of cetacean weight growth coefficient is much higher (Cw = 5.5), than that of other mammals and birds.

Differences in individual rates of growth of cetacean embryos are significant, and by birth the difference in sizes is more than 20%. A large comparative variability (CV) of sizes is shown in younger embryos. The size of the same-aged embryos depends on the size of their mothers.

Fin whales – $l = 0.00175\ L^{2.33}$
Sei whales- $l = l = 0.00139\ L^{2.70}$
Minke whales – $l = l = 0.00187\ L^{2.45}$

Thus the same-aged female embryos overtake male embryos in their rate of growth.

While the embryo is developing, both whales and terrestrial mammals have uterine development periods: the established moments of qualitative changes correspond to the end of germinal, pre-fetus and the beginning of fetus periods. Tables for the definition of whale embryo age are composed on the basis of the occurrence of one or other morphological attributes.

An insignificant difference in the amount of Ca in dark and light layers of sperm whale teeth has been found by means of atomic and adsorption methods.

The amount of Ca in the teeth of females is less than that of males. In cement it is twice as much as that found in the dentine. Optical density of layers is determined not so much by the content of carbonic salts but by a different organization of collagen fibrils and by features of the structure of albuminous stromas.

A subjective (qualitative) estimation of the number of annual stratifications in recording structures does not give reliable results. Quantitative estimation of stratifications, with the help of graphs that record tooth structure, has enabled the annual amount of gain to be determined more precisely.

An original method has been developed for the comparison of profilograms, enabling the amount of annual stratifications to be determined, that is to determine the age of whales.

The post-natal growth curves of whales are described thus:

Whale types	Males	Females
Pygmy blue whales	$L = 12.5\ (t-1.0)^{0.081 + 0.003} + 5.0$	$L = 12.09\ (t-1.0)^{0.012 + 0.003} + 5.0$
Minke whales	$L = 8.09\ (t-5.0)^{0.02 + 0.001}$	$L = 8.54\ (t-5.0)^{0.018 + 0.001}$
Humpbacks	$L = 10.7\ (t-1.0)^{0.07 + 0.002}$	$L = 10.0\ (t-1.0)^{0.12 + 0.002}$
Sperm whales	$L = 10, 6\ (t-7.0)^{0.122 + 0.002}$	$L = 9.17\ (t-3.0)^{0.052 + 0.003}$

Growth curves enable judgements about the life expectancy of whales, and the time of sexual and physical maturity to be determined.

Chapter 4
Biology of Reproduction of Cetaceans

Y. Mikhalev, *Whales of the Southern Ocean*, Advances in Polar Ecology 5,
https://doi.org/10.1007/978-3-030-29252-2_4

This chapter examines the following aspects of cetacean reproduction:

- age of sexual maturity
- defining age of sexual maturity by "transphase" in the aural plugs
- determination of average size of newborns
- correlation between the dimensions of females and their newborns
- period of pregnancy, peaks and phase of mating and calving
- lactation period
- reproductive ability of whales

The most important questions in the biology of reproduction are the duration of the reproductive cycle, time between pregnancies and number of such cycles during their life. In order to answer these questions, it is necessary to address the aspects outlined above.

4.1 Age of Sexual Maturity

In considering this question, it is necessary to notice that the term 'sexual maturity' is not quite correct. If we take the first maturing of the graafian follicle on ovaries as the moment of approach of sexual maturity in female whales, it still remains unknown how often ovicell maturing results in fertilization and leads to subsequent pregnancy.

We defined the approach of sexual maturity by comparing the results of several methods (Mikhalev 1984, 1990, 1991). As a rough approximation, we judged the age of sexual maturity by retardation of growth rate. More exact results are yielded by the comparison of female age with the number of ovulation traces on both ovaries. Good results, especially if all animal age groups are represented in the analysis, show the graphic analysis (cumulative distribution percentage curve) of the percentage ratio of mature and non-mature animals. This method can be corrected by the analysis of primigravida females in the early stage of development. Only females with a functioning yellow body of pregnancy on the ovaries have been analyzed.

Our method was also based on the analysis of the rate of accumulation of traces of ovulation on the ovaries. The regression line specifies exactly the moment of occurrence of the first trace of ovulation, the variation line being sufficiently expanded. The advantage of this method is that comparable results are yielded. Its objectivity is significantly high compared to other methods, and consequently its application is preferable.

Minke whales. We will consider the use of this method taking minke whales as an example.

In the Northern Atlantic, minke whales are believed to become mature at the age of two, the length of the female body being 7.3 m and the male 6.7–7.0 m (Jonsgard 1951; Stephenson 1951; Sergeant 1963). The same has been recorded for the Northern Pacific area (Mattsuura 1936; Omura and Sakiura 1956).

Having used data from only 24 females for the Southern Ocean waters, the Japanese researchers S. Ohsumi, J. Masaki, A. Kawamura (Ohsumi et al. 1970) estimated the age by lamination of aural plugs and came to the conclusion that minke whales become sexually mature at 7–8 years old: judging by growth curve-when they are 7.9 years old, and by the rate of ovulation –about 8 years old.

In their subsequent work, S. Ohsumi and J. Masaki (Ohsumi and Masaki 1975) defined the age of sexual maturity by the ratio of mature and immature females, 3105 have been examined. They came to the conclusion that sexual maturity occurs at the age of 6.3 years. J. Masaki addressed the question in his work (Masaki 1976), determining that the ratio of mature and immature females appeared equal when the animals were 6.2 years of age. By rate of accumulation of ovulation traces the index fluctuated from 5.7 years in the second sector of the Southern Ocean to 7.8 years in the sixth sector.

P. Best (Best 1982) also defined the moment of approach of sexual maturity for female minke whales by correlating mature and immature animals, specifying the age of 7–9 years. However, according to the data used in this work, from the rate of accumulation of ovulation traces ($y = 0.6469x-2.8819$) it follows that the first trace on ovaries is formed at the age of 6.0 years.

How can this essential discrepancy of results be explained? In our opinion, there are several reasons. First, there was not enough data in the first years of catching minke whales. Secondly, the material was collected at different times and in various areas of the world's oceans. Thirdly, the procedure of discriminating between ovulation traces and pregnancy traces on the ovaries is not quite perfect. It is not clear whether the first ovulation always results in pregnancy, or whether traces of pregnancy and ovulation remain forever. Fourthly, the procedure for determining the age of baleen whales by ear plugs is imperfect, especially for minke whales. Therefore, to achieve a more exact assessment, it is necessary to compare and analyze the results recorded by using these different methods.

Judging by our minke whale postnatal growth curve (Fig. 3.21), retardation of growth rate is marked in the range of 5–7 years. The analysis of 206 primigravida females during seven voyages and correction of their age by means of microphotograms showed that the age of the females fluctuates in the range of 4–9 years, body length being from 7.9 m to 8.6 m.

Most of the primigravida females were 6–7 years old (48.6%). However, it is necessary to note that the selective nature of whaling, when larger animals were targeted, leads to an average overestimation of the age and length of females at the approach of sexual maturity. The average female length appears to be 8.5 m. The average length of primigravida females in various age classes remained constant and equalled 8.5 m. This fact testifies that the age of the animals has been incorrectly defined, and that the actual age of the females investigated fluctuates in a much narrower interval. But another assumption can be made: sexual maturity depends not on female age, but on reaching a certain length.

The percentage ratio analysis of mature and immature animals in various female age classes has shown that this correlation is equal (50% mature and 50% immature) when they achieve the age of 5.7 years.

By our calculations, the regression line of accumulation rate of ovulation traces (Fig. 4.1) is described by the formula:

$$Y = 0.852X - 3.588;$$

where

Y – the number of ovulation traces,
X – age.

Calculation by this formula shows that the first ovulation trace occurs when the average female age equals 5.385 years, or – in round figures – 5.4 years.

As not every ovulation leads to pregnancy, the most likely time of female sexual maturity approach should probably be within 5.5–6.0 years, the length of females being about 8.1 m, which almost coincides with S. Ohsumi's and J. Masaki calculations (Ohsumi and Masaki 1975).

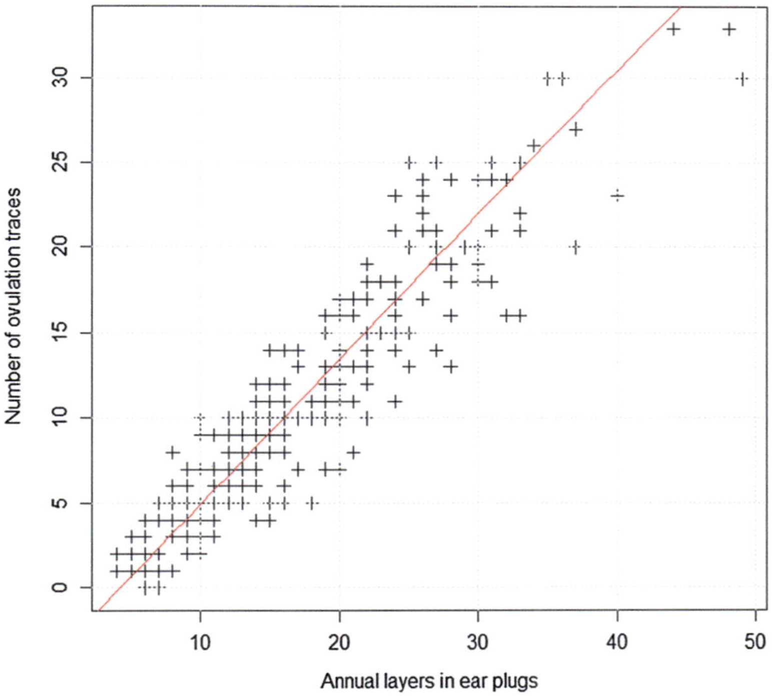

Fig. 4.1 Regression line of accumulation rate of ovulation traces

4.2 Definition of Sexual Maturity by Aural Plugs "Transphase"

In investigating aural plugs of baleen whales, T. Ichichara (Ichichara 1963, 1966) noticed that wide stratifications are replaced by narrower strips at the top of the plug. He connected this phenomenon with the moment of sexual maturity of the animals. Other researchers also noted the presence of a transition zone ("Transphase") in aural plugs (Best 1982; Masaki 1973, 1976; Lockyer 1974, 1977, 1979, 1981; Ohsumi 1986; Mikhalev 1990a, b, 1991). There were attempts to connect the change of stratification on registering frames with the growth periods for toothed whales as well (Berzin 1971; Mikhalev 1978; Cudgels, etc., 1982).

The analysis of the "transphase" position in whale aural plugs from 1960 to 1979 led K. Lockyer (Lockyer 1972, 1974) to the sensational conclusion that in 1958 fin whales reached sexual maturity at 6 years old, and in 1910 – much later –10 years old. According to similar retrospective calculations (data from 1971 to 1977) J. Masaki (Masaki 1976) defined that in 1944 females came to sexual maturity at 13.9 years old. Then the age of approach of sexual maturity dropped and in 1965 it was 6.4 years. The author tried to explain the cause of such phenomenon by the reduction of the whale population biomass owing to whaling.

P. Best (Best 1982) investigated Durban area minke whales using data from 1968 to 1975. His retrospective calculation has shown that in 1945 females reached sexual maturity at the age of 10.5 years and by 1973 it dropped to 7.7 years. S. Ohsumi (Ohsumi 1986) analyzed aural plugs from 360 female minkes taken in the Southern Ocean during the 1971–1982 seasons. His retrospective calculation shows that 35 years ago the sexual maturity of female minke whales came at the age of 13.5 years, and in 1972– at the age of 7 years.

This big discrepancy in the age of sexual maturity in minke whales many years ago and more recently gives rise to doubt. Really, modification variability – the influence of natural habitat conditions on physiological processes and the growth rate of organisms – is well-known. As an example, it is growth acceleration and earlier sexual maturing of humans has occurred in the recent years. It is also well known that artificial breeding of dwarf forms of plants and small aquarium fish can be achieved by changing the conditions of their natural habitat.

However in natural, as opposed to artificial, conditions in the same habitat this dependence is poorly expressed. Owing to the harshness of natural selection, animals with significant aberrations from the norm perish, as they do not meet infraspecific and interspecific competition. And if they do exist, it's only as an exception to the rule. Therefore the above mentioned conclusions about such a significant change in the terms of sexual maturity, based on the "transphase" position, seem to us rather doubtful, demanding more careful analysis.

In fact, comparing K. Lockyer's data (Lockyer 1974) and J. Masaki's (Masaki 1976) on fin whales and those by P. Best (Best 1982) and S. Ohsumi (Ohsumi 1986) on minke whales, the regularity breaks down at once and appears groundless. The time of sexual maturity approach for the same age appears different. Retrospective calculations are not grounded in research from the 1930–40s either. Baleen whales became sexually mature in those years not at 10–15 years old, as in the retrospective assessment, but at 3–6 years old (Mackintosh and Wheeler 1929; Nishiwaki et al. 1958; Ohsumi et al. 1958; Laws 1961; Ruud 1940; Mackintosh 1942a, b; Ruud 1945; Nishiwaki 1950, 1952, 1957; Symons and Weston 1958; etc.).

To understand the situation, we examined the aural plugs and ovaries of 8725 female minke whales from 1971 to 1982. What attracts attention first of all is the fact that in the aural plugs of old animals it is sometimes possible to see not one, but several, transition zones (Fig. 4.2).

The first "transphase" from the top of the tooth is most likely caused by growth retardation of the animal at the moment of approaching sexual maturity, but it is difficult to define its place. It is easier to define the transitive phase of larger animals older than 10 years of age, on whose ovaries there were several traces of ovulation.

The analysis has confirmed that, judging by the "transphase" position, the age of sexual maturity age increases as soon as we look into the past, when animals reached sexual maturity at an older age than they do now. This is true both for average data and for data from each season (Table 4.1).

However, for each separate age interval from year to year, as shown in the table's vertical columns, a regular increase of this index is not observed, which contradicts the hypothesis. Moreover, the tendency of the age of sexual maturity in female minkes to change is not grounded by the analysis of the rate of accumulation of ovulation traces, nor by the age of primigravida females and females with one ovulation body on the ovaries, nor by the percentage ratio of mature and immature animals. By none of these three methods does this index appear to be dependent on the whaling season, and in fact equaled 5–6 years (Table 4.2).

The same values have been recorded by other researchers, including supporters of the hypothesis of essential decrease in the age of sexual maturity when objective methods were used, rather than retrospective calculation on the transphase. It is important to remember, that according to K. Lockyer (Lockyer 1972), female minke

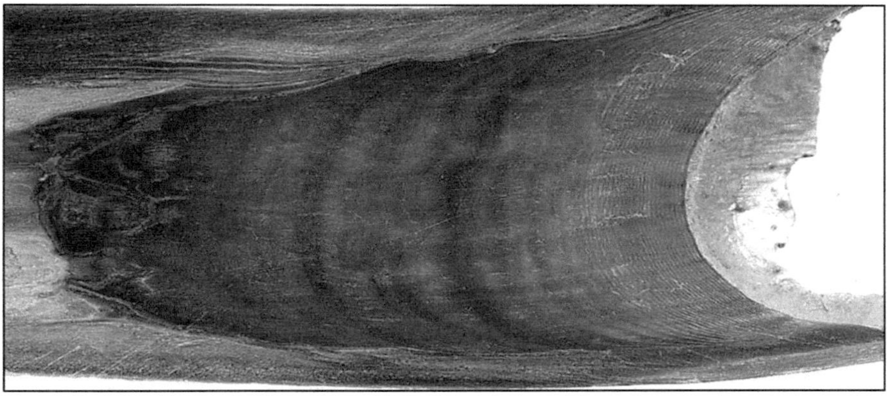

Fig. 4.2 Fin whale ear plug

Table 4.1 Retrospective calculation of age at sexual maturity in minke whales using the transition phase of earplugs

1.1 Whaling seasons	Year intervals of retrospective calculation			
	5 years	10 years	20 years	30 years
1971/72	4.9	5.8	7.6	9.4
1972/73	5.0	5.9	7.6	9.4
1973/74	5.8	6.8	8.8	10.9
1974/75	5.4	6.1	7.6	9.2
1975/76	5.5	6.4	8.4	10.3
1976/77	4.8	5.8	7.9	10.0
1977/78	5.8	6.4	7.7	8.9
1978/79	5.9	6.6	7.9	9.1
1979/80	4.9	5.9	8.1	10.2
1980/81	5.5	6.4	8.2	10.0
1981/82	4.7	5.5	7.1	8.7
Average	5.3	6.1	7.9	9.6

Table 4.2 Sexual maturity age of minke whales defined by different methods

1.1 Whaling seasons	Age (years) of sexual maturity		
	by rate of accumulation of ovulation marks (regression formula)	by primigravida females	by correlation of sexual mature and immature animals
1	2	3	4
1971/72	5.5 (y = 0.782x-3.301)	6.1	5.2
1972/73	5.4 (y = 0.901x-3.865)	6.2	5.3
1973/74	5.6 (y = 0.872x-3.887)	5.3	5.4
1974/75	5.7 (y = 0.779x-3.408)	6.3	5.5
1975/76	5.5 (y = 0.844x-3.637)	4.7	5.3
1976/77	5.5 (y = 0.911x-4.011)	5.4	4.9
1977/78	5.3 (y = 0.913x-3.866)	6.1	5.2
1978/79	5.7 (y = 0.844x-3.797)	6.4	5.4
1979/80	5.1 (y = 0.785x-3.000)	6.4	5.0
1980/81	5.6 (y = 0.852x-3.729)	4.2	5.5
1981/82	5.6 (y = 0.892x-3.982)	5.3	5.3
Average	5.4 (y = 0.852x-3.588)	5.7	5.3

Notes: y – number of ovulation marks; x – age in years

whales (using data from 1971) matured at the age of 5.5 years. S. Ohsumi and J. Masaki (Ohsumi and Masaki 1975) have analyzed data from 1967–74. By their calculations of the accumulation rate of ovulation traces on female ovaries, the age of sexual maturity of minke whales in those years was at 5.6 years of age.

According to the ratio of mature and immature animals, sexual maturity was at 6.3 years. When calculated by the rate of accumulation of ovulation traces (u = 0.883 h-3.430) according to H. Kato (Kato 1982), sexual maturity of female minke whales comes at the age of 5.0 years, and at 6.2 – by the ratio of mature and immature animals. P. Best (Best 1982) considers that Durban area female minke whales mature at 7–9 years, but according to the rate of accumulation of ovulation traces presented by him (y = 0.6469x-2.8819), the calculated age is 6.0 years.

Thus, both our data and data from the literature based on rather objective methods and the material which had been collected for many years, testify that the age of sexual maturity of minke whales and other species of baleen whales has not essentially changed in recent decades. In our opinion (Mikhalev 1990, 1990a, 1991, 2002a), the presence of transitive phases in aural plugs of baleen whales should become the subject of careful research, but their position cannot yet be used as a reliable basis for the retrospective definition of the age of sexual maturity of whales.

4.3 Definition of the Average Size of Newborns

International agreements forbid the taking of females with suckers. If suckers are taken, they are not always newborns by morphophysical features. As a rule, even the smallest suckers had cicatrized navels. Cicatrization is known to take at least 2 weeks (Sleptsov 1940). During this short period the size of newborns can increase significantly. Therefore, we have to judge the size of newborns by indirect reference points.

For sizing newborns we used several various indirect methods of calculation (Mikhalev 1972; Ivashin and Mikhalev 1978):

- Analogy to other animal species, the average size of whose newborns is known;
- The analysis of embryo lengths in variation lines;
- The analysis of the largest fetuses' length distribution (determination of modal value);
- The analysis of the degree of maturity of prenatal embryos and their mothers' physiological state;
- Comparison of size of the largest embryos and the smallest suckers.

When defining the average size of newborns by the size of the largest embryos, the latter were considered to be longer and heavier than average newborns by some percent. According to L.T. Kapralova (1967), the biggest of newborn lambs are heavier than the average by 30–32%. Newborn calves on the average weigh 40 kg, and the heaviest ones 60 kg, meaning they are heavier by 10.5%. The maximum weight of newborn fawns (Irzhak 1967) is 45% heavier than that of average ones. The weight of the heaviest newborn roe fawn (Wandler and Huber 1969) is 54% above the average. Among the pinnipeds, weight variations of nine neonatal Bearded seals (Potelov 1969) was within 120% of the average. K.K. Chapsky (1965) measured the length of 10 frozen newborns of the White Sea Harp seal, whose deflection from the average length was 7%. Fifty Harp seal newborns, measured by L.A. Popov and Y.K. Timoshenko (1965), had an average length of 90.2 cm, while the maximum length of newborns was 13% greater than the average. M.V. Ivashin (1965) specified that the deflection from the maximum lengths of average newborn northern seals and dolphins (porpoises) equals 10–20%. He considered that the same fluctuations of size are peculiar to newborn humpbacks.

From the examples it is clear that in various orders of mammals, both terrestrial, and marine, the dimensions of newborns vary widely, but, as a rule, only when the sample is big enough. Fluctuations of the maximum sizes from the average are lower than the latter from the minimum ones, being approximately 6–14% for length and 20–40% for mass. Sometimes extreme values strongly deviate from the average size. They are exceptions and are not taken into calculation.

On the basis of the aforementioned, it follows that when calculating the average size of newborn cetaceans by the maximum fetus size, it is necessary to consider the latter to be longer than newborns by approximately 10%.

Blue Whales According to the IWC (Tomilin 1957), the largest embryos of the real blue whales of the Southern Hemisphere often reached 700–800 cm. However, the records for taken suckers were much smaller, at 610–700 cm (Millais 1906). The average length of newborn blue whales R. Laws (Laws 1959) defines as 700 cm. T. Ichichara (Ichichara 1966) defines the maximum length of pygmy blue whale embryos as 630 cm. About the same maximum lengths (632 cm and 638 cm) are also noted by E.G. Sazhinov (1980), who defined the average neonatal size as 550 cm.

In our practice, among 608 embryos of pygmy blue whales, 98 have been attributed to prenatal on the basis of a series of features: 21 in the northwest part of the Indian ocean and 77 in other areas. The largest one was 680 cm long, but the basic group consisted of 500–600 cm fetuses, thus we define the average length of neonatal pygmy blue whales as 560–580 cm.

Fin Whales Out of our samples, which include 3500 embryos and the IWC data, the largest fin whale embryos were 690–719 cm in length, but they consisted of only 0.01%, and 0.03% for 630–660 cm embryos. If we take them as the statistical maximum, then, considering them to be 10% longer than average newborns, the average size of newborn fin whales will equal 580–600 cm, with a weight of about 1500 kg.

The analysis of the occurrence of fin whale embryos of the maximum size (larger than 520 cm) during various whaling seasons has shown the length of 577 cm to be most common (Mikhalev 1970b), which is very close to our calculated size of newborns.

The analysis of the size of fetuses taken from females with signs of prenatal state is interesting: swollen nipples and nipples coming out of skin pocket; thickened mammary glands, and colostrums secreting on being cut. A.G. Tomilin (1957) reports that on October, 18th, 1934 he investigated a female fin whale 19.8 m long, with enlarged mammary glands. Taken from it was a quite mature embryo 573 cm in length. N.E. Salnikov (1956) also examined a female with features of the prenatal state. The embryo found in it had a length of 608 cm.

Three of the females we investigated had features of the prenatal state in addition to the signs listed above, and of the adenohypophysis state (Mikhalev 1966, 1970b), which were enlarged and of an increased mass – 35 g on average. The embryos taken had a length of 540 cm (male), 563 cm (male) and 620 cm (female). Coloring of fetuses and ventral grooves were completely formed, baleens had a height of 4–5 cm with setas on the plate ends and were pigmented, the right baleen line being asymmetrically colored; blades of the tail fin were not elastic and were easily unwrapped from the "cam"; umbilical cords were narrowed at the umbilicus and parted easily. All this once again confirms the conclusion that newborn fin whales of the Southern hemisphere have an average length of 580–600 cm (Mikhalev 1972).

The average size of newborns in the Northern hemisphere is smaller. In the Northern Atlantic cases of extraction of suckers of 518 cm in length are known with baleens slightly showing from the gums, and the skeleton of a 525 cm long neonatal fin whale from the Mediterranean Sea is stored in the natural history museum of Paris. If, as is usually the case, the skeleton is shorter than the body by 1/10, the sucker must have been 580 cm long.

Sei Whales The only newborn taken was 427 cm long (Tomilin 1957). Among the 929 embryos of sei whales, the largest one was 530 cm long. Maximum lengths usually were within 480–500 cm (Budylenko and Mikhalev 1970). In our practice there were five sei whale female in the prenatal state. Quite mature fetuses of 407 cm, 412 cm, 432 cm, 458 cm and 480 cm were extracted from them. From the aforesaid it follows that the sizes of newborn sei whales are within the limits of 400–500 cm, the average being about 450 cm.

Minke Whales The analysis of the dotted graph of the sizes of 4500 minke whale embryos has shown (Ivashin and Mikhalev 1978; Mikhalev 1984) that during the season, apart from the main group of embryos, there is also another one, which is not as numerous and is within the 230–310 cm range (Fig. 3.10). According to the position of the group, these fetuses must be matured. The examination of such embryos has shown that the degree of their development corresponds exactly with prenatal.

Reduction of the maximum sizes of these embryos by 10% gives the average size of newborn minke whales of the Southern hemisphere 280 cm, which coincides with the size calculated by S. Ohsumi, Masaki and A. Kawamura (Ohsumi et al. 1970), and also by M.V. Ivashin (1976). The calculation by N.V. Doroshenko, etc. (1974) – 300–330 cm – seems to us overestimated.

The report on a 216 cm-long minke whale from the coast of Tasmania, (Devis and Guiler 1958), raises doubts about whether the species of the whale was correctly determined. If it really was a minke whale, it should have been a premature newborn, as its size is considerably below the smallest prenatal embryos. However it might have been aborted and cast ashore by the waves.

Humpbacks One case of extraction of a newborn humpback aged about 1 week is known, described by R. Chittleborough (Chittleborough 1958). Its length equaled 465 cm. The maximum size of humpback whale embryos, according to many years of IWC data, has been 518 cm. The length of the largest humpback embryos we measured fluctuated between 520 and 530 cm. But they were rare. More often prenatal fetuses were between 470 and 480 cm in length. This size reduced by 10%, makes the average size of newborns within 440–450 cm.

Right Whales The average size of a neonatal Greenland whale is considered to be 450 cm (Tomilin 1957), as the maximum registered fetus had that length. We will note, however, that the newborn whale taken in those waters was 407 cm long (Eschricht and Reinhardt 1866). As to southern right whales, the average size of their newborns is considered to range from 450 cm to 600 cm, as the largest fetus was 602 cm long. The skeleton of a sucker stored in the museum of natural history in Paris has a length of 450 cm, therefore, the zoological length of the whale was 500 cm. Newborn southern right whales should have about the same lengths.

Sperm Whales According to S. Ohsumi (Ohsumi 1965), who analyzed 2700 sperm whale embryos, the largest fetuses in the dimensional group were 490–520 cm, representing only 0.22% of the embryos. R. Gambell (Gambell 1966) reported that the largest sperm whale embryo from the Azores was 410 cm long. By R. Clarke's data (Clarke 1956), the greatest sperm whale embryo near Durban was 455 cm long. The data on measurements of the smallest sperm whale suckers are scarce. Their length fluctuates from 371 cm to 562 cm (Berzin 1971; Gambell 1966; Bennet 1840; Jackson 1840; Wheeler 1933; Matthews 1938; Matsuura 1940). The average size of ten measured suckers was 419 cm. Hence, the average size of neonatal sperm whales should be slightly smaller.

Out of more than 5000 sperm whale embryos measured by us the largest had a length of 512 cm. However, such large fetuses are rare and should be considered as exceptions. Much more often the maximum size was within the 440–480 cm range. Therefore, the probable size of the newborn should be about 400 cm. In our practice (Mikhalev and Shevchenko 1973, 1986) we happened to measure the newborn. This case is unique, so describing it in more detail makes sense.

In the 1971/72 voyage the whaling vessel *Bodry-25* took a feeding female. The sucker, which could hardly keep afloat, was also lassoed. It was a male 400 cm long weighing 800 kg. It had all the features of a newborn. Tuberosity of the skin, especially on the sides of the trunk, was expressed more strongly than that of adult whales. On the ventral side of the trunk, especially on the border of the head with the neck, there were a lot of skin folds, the greatest being 10 cm long and 3 cm deep. Two folds on each side of the head designated the border of the spermaceti pouch. On the head forefront there were three small fresh hollows located 4 cm, 5 cm and 5.5 cm from each other and slightly in front of them – three parallel scratches at the same distance from one another. Undoubtedly, they were traces from the teeth of the lower jaw either of the mother, or of "the female-midwife", who might have possibly supported it to stay afloat right after its birth. The navel was not cicatrized, and the spinal fin twisted to the side. There was a fold in the place where the tail part of the body bends and the cervical waist was accurately expressed. Dissections revealed milk mixed with ocean water in the stomach. By this set of features it is possible to assume that its age was no more than 3–5 days.

It follows from what has been said, that the size of neonatal male sperm whales is most likely about 400 cm, and females about 390 cm.

4.4 Correlation Between the Females and the Size of Their Newborn

The size of newborn cetaceans is 1/3 to 1/2 the lengths of their mothers. For example, the average length of a newborn blue whale is about 30% of the mother's body length; minke whales – 34%; sperm whales – 44% and porpoises – 51% (Scott 1948). The smaller the whale, the larger the size of its newborn.

V.E. Zaika (1970), comparing the maximum specific growth rate of various kinds of mammals (in particular, primates) and altricial birds, has come to a very interesting conclusion, deserving special attention from those who research the problems of growth: the size of newborns increases within every taxonomic group on a parabola "like different weight stages of a certain hypothetical animal". The author explains differences in the groups by the degree of development of newborns.

Let's consider this law taking marine mammals as an example. R. Laws (Laws 1959) for pinnipeds (real seals) and S. Ohsumi (Ohsumi 1960) for cetaceans pointed out the correlation between the sizes of newborn marine mammals and their mothers. S. Ohsumi (Ohsumi 1960) compared the average size of newborns and mature animals of seven species of baleen whales and eight of toothed whales. He also revealed correlation dependence of the size of newborns on the sizes of their mothers at the onset of sexual maturity, but within smaller taxonomic units – suborders and even families.

Unlike S. Ohsumi (Ohsumi 1960), we compared the size of newborns not with the female size at the age of sexual maturity, but with their definitive sizes, as they are practically at the final moment of linear animal growth and can be more precisely defined. For comparison we also correlated the size of newborns and their mothers for pinnipeds (Table 4.3). In addition to our own data, we used data from the literature (Tomilin 1957; Slijper 1962; Ichichara 1961, 1963, 1964; Zemsky and Boronin 1964; Sleptsov 1965a, b; Zimushko 1968; Sazhinov 1970; Omura and Sakiura 1956; Potelov 1969; Chapsky 1965; Naumov 1933; Ognev 1935; Essapian 1953; Belkin 1964; Krylov 1965, 1969; Fedoseev 1965; Tikhomirov 1894; Beloborodov and Potelov 1969; Pastukhov 1969; Sokolov, etc., 1969; Chugunkov 1969; 309; etc.).

The graphic analysis of the investigated correlation shows that in the logarithmic coordinate system, the points corresponding to the average size of newborns of various taxonomic groups in relation to the average size of their mothers, in general, lie on a straight line (Fig. 4.3). Deflection of some points from the straight line can most likely be explained by incorrectness of the definition of average size.

Table 4.3 Average size of newborns and definitive sizes of females for pinnipeds and cetaceans

		Length, m			
		Female		Newborn	
1 #	Species	*L*,	logL	l,	Logl
1	2	3	4	5	6
Pinnipeds					
1	Ringed seal	1.25	0.0960	0.60	1.7782
2	Fur seal	1.25	0.0960	0.64	1.8062
3	Baikal seal	1.30	0.1139	0.70	1.8451
4	Caspian seal	1.30	0.1139	0.83	1.9191
5	Harbor seal	1.56	0.1931	0.90	1.9542
6	Harbor seal (stejnegeri)	1.60	0.2041	0.83	1.9191
7	Ribbon seal	1.70	0.2303	0.85	1.9294
8	Harp seal	1.80	0.2553	0.83	1.9191
9	Beard seal	1.90	0.2788	1.02	0.0086
10	Gray seal	2.40	0.3802	1.10	0.0414
11	Bearded seal	2.50	0.3979	1.35	0.1303
12	Walrus	3.00	0.4771	1.30	0.1139
Tooth whales					
1	Finless porpoise	1.20	0.0791	0.52	1.7160
2	Harbour porpoise	1.50	0.1761	0.72	1.8573
3	Short-beaked common dolphin	1.70	0.2304	0.80	1.9037
4	Striped dolphin	2.20	0.3424	1.03	0.0128
5	Bottlenose dolphin	2.50	0.3979	1.20	0.0791
6	White-sided dolphin	2.30	0.3617	1.00	0.0414
7	Pygmy sperm whale	2.50	0.3979	1.10	0.0414
8	White-beaked dolphin	2.70	0.4314	1.10	0.0414
9	Gray dolphin	3.20	0.5051	1.50	0.1761
10	Beluga	3.70	0.5682	1.50	0.1761
11	Narwhal	4.00	0.6021	1.52	0.1818
12	Short-bodied killer whale	4.30	0.6335	1.60	0.2041
13	Pilot whale	4.40	0.6435	1.75	0.2430
14	Killer whale	5.50	0.7404	2.10	0.3222
15	Beaked whale	5.50	0.7404	2.70	0.4314
16	Bottlenose whale	8.70	0.9395	3.00	0.4771
17	Sperm whale	10.6	1.0253	4.00	0.6021
Baleen whales					
1	Minke	8.00	0.9031	2.70	0.4314
2	Gray whale	12.50	1.0960	4.50	0.6532
3	Humpback whale	12.80	1.1070	4.60	0.6628
4	South whale	14.30	1.1551	4.70	0.6721
5	Sei whale	15.50	1.1903	4.50	0.6532
6	Fin whale	22.50	1.3522	6.10	0.7853
7	Blue whale	26.00	1.4150	7.00	0.8451

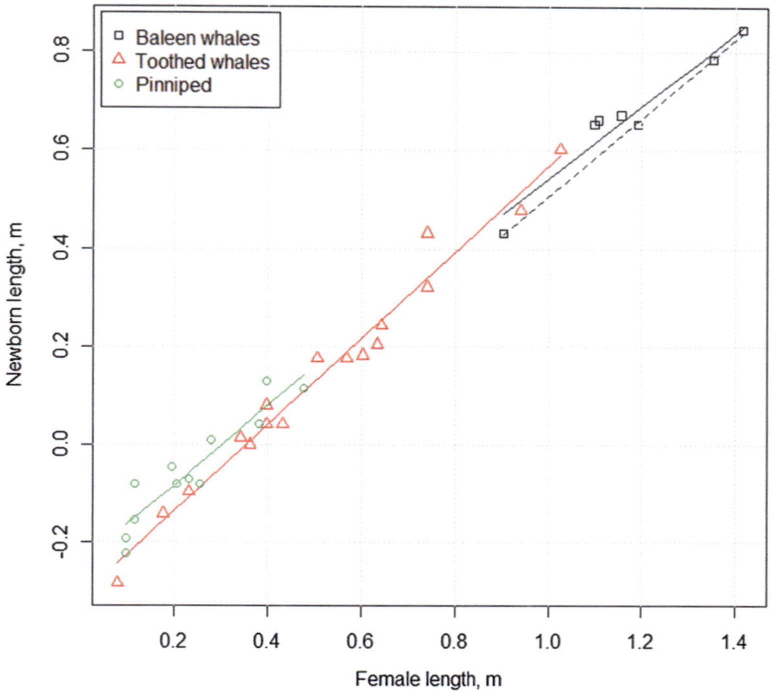

Fig. 4.3 Dependency between mother average size and newborn size

Nevertheless, despite some scattering of the points, regression lines both for pinnipeds and for baleen and toothed whales differ clearly, and correlation between the size of newborns (l) and their mothers (L) in various taxonomic groups of animals can be expressed by the following formulae with a satisfactory degree of reliability:

toothed whales: $l = 0.460\ L^{0.90}$;
baleen whales: $l = 0.636\ L^{0.74}$;
pinnipeds: $l = 0.569\ L^{0.81}$.

Thus, the correlations we considered using marine mammals as an example, also confirm V.E. Zaika's idea (1970) that the size of neonatals of various species among closely related groups increase with the increase of definitive sizes of their mothers, as the various dimensional growth stages of "a certain hypothetical animal", and that these correlations are specific for various taxonomic groups of animals. Zaika's second conclusion (1970) that the degree of maturity of newborns defines their relative size. If this thesis were true, there would be no distinctions between toothed and baleen whales, whose newborn have the same extent of maturity.

The correlation is clearly a characteristic feature of a certain taxon, which circumstancially allows differences within the suborder of baleen whales to be found.

As we see (Fig. 4.3, a dot-and-dash line) Minke whales form a straight line, different from the general line for baleen whales, which can be described by the following correlation:

$$1 = 0.506\ L^{0.80}$$

Values of the newborn of gray whales, humpback and southern whales, are also a little away from the straight line for real minke whales. This favors the conclusion that the aforementioned correlation is a regular sign. The correlation has both theoretical and practical value. It allows judgement of the dimensions of newborn of those species whose average length is not known yet, but of whose mothes definitive dimensions are known.

4.5 Period of Pregnancy, Peaks and Phases of Mating and Calving

Because of the improbably results derived from the indirect calculation methods, researchers do not have a common opinion about the duration of pregnancy of the various cetacean species, and further research is needed. The differeing opinions about this question are explained by the absence of a well-grounded theory of the prenatal growth of whales.

N. Hinton (Hinton 1925), who focused on embryo growth, believed that pregnancy of fin whales and blue whales lasts about 11 months. J. Fraser and A. Hugget (Fraser and Hugget 1959a, b), having examined the IWC data for a series of years, have come to the conclusion that the pregnancy of baleen whales, fin whales in particular, lasts 9–10 months. Some researchers (Risting 1928; Mackintosh and Wheeler 1929; Zenkovich 1935, 1952; Peters 1938; Sleptsov 1952, 1955; Kleinenberg et al. 1965; Tomilin 1957, 1962; Zemsky 1958, 1960; etc.) believe that the pregnancy of baleen whales lasts about 1 year. V.A. Zemsky (1961) supposed that pregnancy of baleen and toothed whales lasts more than a year. Summing up the previous research, for different baleen whales the duration of pregnancy is believed to last from 9 months to about 1 year or more; for toothed whales, sperm whales in particular, from 9 to 18 months. More exact data have been recorded for dolphins kept in dolphinariums. These data will be used for control in our analysis for other toothed whales.

The laws of prenatal growth (Mikhalev 1970b, 1984, 2007; Ivashin and Mikhalev 1978; Mikhalev and Shevchenko 1973) are based on calculations of duration of pregnancy of cetaceans. It is assumed that the latent period is absent for cetaceans (Kleinenberg et al. 1964), or it is short– less than 2 weeks (Sazhinov 1980); that the earliest growth period, from the beginning of zygote division to gastrula stage, proceeds more quickly – exponentially (MacDowell and Allen 1927), and that growth in the prenatal phase is a little retarded.

Table 4.4 Peaks of mating and birth, and gestation length for cetaceans

Whale species	Mating season peak	Whelping season peak	Pregnancy duration, months
Pygmy blue whale	Middle of May	End of April	10.5–11.0
Fin whale	Middle of July	Middle of June	10.5–11.0
Sei whale	End of June	Middle of May	10.0–10.5
Minke whale	Middle of September	Beginning of July	9.5–10.0
Humpback whale	End of September	Beginning of August	10.0–10.5
South whale	End of May	End of April	10.5–11.0
Sperm whale	October	October	Around 12.0

Defining the time for the peaks of mating and calving seasons was corrected by the data in cases of observation of mating and occurrence of females with little suckers; by the analysis of female testicular activity and male spermatogenesis; by the study of the process of ovaries and testicle weight change; by a rise in the number of prenatal females and by the dramatic decrease in the number of prenatal fetuses in samples. The speed of female migration to the areas of reproduction and the location of these areas were also taken into consideration. These calculations have given the following results (Table 4.4).

It can be seen that, on average, baleen whale pregnancy lasts from 9.5–11.5 months. Scientific opinion about sperm whales differs greatly, and, unfortunately, we did not succeed in recording more convincing arguments to specify the term of their pregnancy. From the analysis of the literature it follows that for the majority of toothed whales pregnancy lasts about 1 year.

It is known (Turner 1875) that the duration of pregnancy in animals essentially depends on the degree of perfection of ability to feed the fetus, from the placenta type. The uterus form, locating of vili on the chorion and allocation of placenta vessels in whales are very similar to those of the ungulates. These morphological features speak in favor of E. Slijper's hypothesis (Slijper 1966) about the distant relationship between cetaceans and even-toed and odd-toed ungulates, whose pregnancies also last about 11 months. It is impossible to ignore such a resemblance, as the duration of pregnancy is a rather conservative sign. Albeit indirectly, this makes the conclusion, that the pregnancy period of baleen whales and the majority of toothed whales not exceeding a year, more convincing.

4.5.1 *Duration of Reproductive Seasons, Areas of Mating and Calving*

Direct observation of humpbacks has enabled reliable recordings of the duration of mating and calving periods (Chittleborough 1954, 1958). More often researchers make judgments about the reproductive seasons of other whales by the dimensional variations in embryos using M. Hinton's method to determine their age (Hinton 1925).

A.G. Tomilin (1957) believed that the mating season of fin whales lasted for the whole year, because frequently in 1 month it is possible to come across embryos from the smallest to prenatal sizes. At the same time, he specified that the majority of whales couple in a shorter period of time.

Under the law of variation statistics it is known that when the distribution of dates is normal, 68.3% of mating will happen within "average ± o_", 99.5% – within "average ± 2o_". Proceeding from this, there is no need to define the age of all embryos, if distribution of their lengths is known. For many Southern hemisphere whales the distribution of embryo lengths is close to normal in December and January.

The calculations made according to that procedure show (Mikhalev 1970b, c; Budylenko and Mikhalev 1970; Mikhalev 1980) that about 70% of all mating and calving of baleen whales take place within 2.5 months, and 95% within 5.5 months. Exceptional cases can, certainly, be observed in a longer period of time. For example, according to N.E. Salnikov (1956), the captain of the whaling vessel *Slava*-1 L.A. Kalinin observed mating of fin whales in February, that is later than the main mating season, and in the Southern Ocean waters too. Observation in dolphinariums has shown that among the Black Sea bottlenose dolphins, which became pregnant both when they were free and in captivity, almost 90% of mating took place within 5 months, and half of all matings occurred during two summer months (June and July) (Ozharovskaya 1990).

It is necessary to note also that, depending on weather conditions and other seasonal factors, periods of mating and calving of whale populations can be shifted to this or that side. Therefore, the total data on the duration of reproductive seasons for a series of years appears more extended than in one particular year. That is one of those paradoxical cases when enlarging the sample leads to an enlargement in errors.

Remember that for some baleen whales –fin, minke and humpback–there exists, in addition to the main group, a less significant group of embryos with a 6-month delay in development (see Sect. 3.2.1). The presence of this group has, probably, given the impression of an excessively prolonged reproductive season for baleen whales (Tomilin 1957; Zemsky 1950, 1958; etc.).

Female humpback whale with its suckling

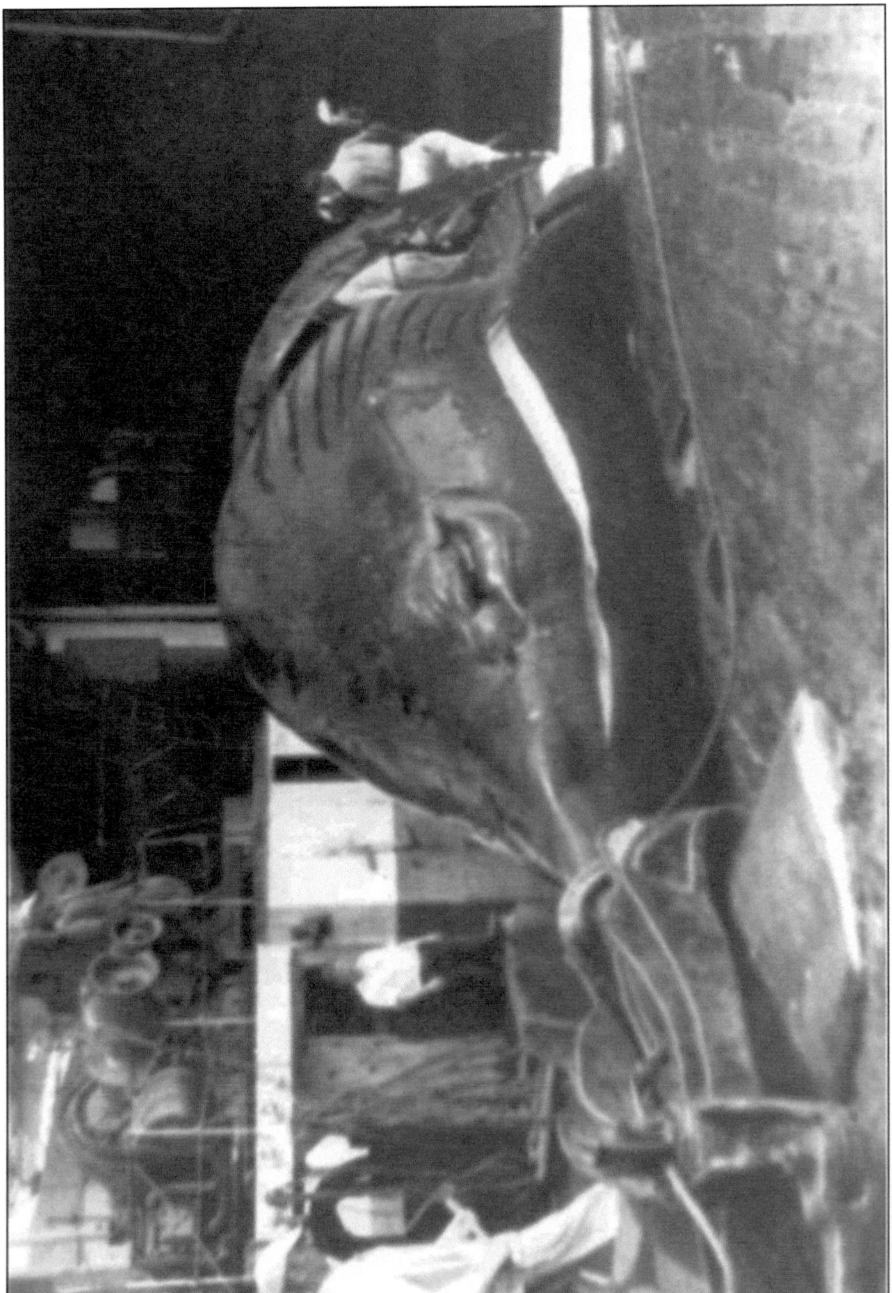

Lactating humpback whale female on desk

4.6 Lactation and Duration of Lactation Period

It is almost impossible to observe the process of whale calves feeding at sea, which is one of the reasons this matter has been poorly studied. Cetaceans kept in dolphinariums allow us to find out about some features of lactation and to specify the duration of the lactation period. But a lot of questions still remain.

4.6.1 Nutritious Value of Whale Milk

Relationship between the nutritious value of milk and the growth of suckers as a general biological law has been stated by N. Aron (Aron 1927). He asserted that the more nutritious the milk, the quicker weight of the newborn doubles. Other researchers have come to similar conclusions (Otha et al. 1955; Arshavsky 1968). For example, a newborn foal doubles its weight in 2 months when milk contains 2% protein. The neonatal lamb doubles its weight in half a month when milk contains 5.6% protein. The dependence on other components defining the nutritious value of milk, including fats is similar.

The nutritious value of whale milk is extremely high. According to the literature, (Berzin 1971; Yablokov et al. 1972; Bodrov and Grigoriev 1963; Kiesewetter 1953; Sleptsov 1952; Otha et al. 1955; Zenkovich 1938; White 1953; Tomilin and Plavsky 1962; Utreht-van 1968), the fat content of the milk of toothed whales–common dolphin, bottlenose dolphin, pilot whale, porpoise, beluga whale, bottlenose whale and sperm whale–is within 23.0% to 51.2%; protein – from 4.9% to 11.5%; lactose −1.3% to 3.9%. The fat content of the milk of baleen whales–gray, blue, sei, Bryde's, humpback and southern_is within 21.0% to 53.0%; of protein −3.6% to 13.6%; lactose - 0.2% to 5.6%.

Our analysis of milk fat content of the Southern Ocean whales (Mikhalev and Sventsitskaya 1970a; Mikhalev 1971a, b), using the butyrometric method, gave the following results: pygmy-blue whale – 35%; fin whale – 20%; 25.0%; 33.3%; sei whale – 16%; 21%; 22.8%, 28%; 29%; 31%; 32%; 33%; 34%; Minke whale – 43%; humpback – 32%; right whale – 31%; sperm whale – 19%; 22%; 25%; 26%; 27%; 28%; 30%; 30%; 30%; 36%; bottlenose whale – 23.0%. Protein in the milk of fin whales −13.0% and of sperm whales – 5.0%.

Taking data from the literature into account, the average fat content of toothed whales milk equaled 34.6%; protein – 8.1%; lactose – 1.5%. The average fat content of baleen whales' milk – 33.2%; protein – 10.3% and lactose – 1.6%. Thus, it is possible to say that the nutritious value of toothed and baleen whales' milk is almost identical. It is necessary to notice, however, that during the lactation period, the nutritious value of whale milk does not remain constant – by the end of the lactation period the milk fat content drops to 15–20%.

Such a high nutritious value of milk is not observed for terrestrial mammals, which, like whales, give birth to precocial calves. Naturally, the question arises about the sense of this milk characteristic of cetaceans, the only purely aquatic group of mammals.

M.M. Sleptsov (1940) explains the reason for the highly nutritious value of whale milk by the fact that feeding of the newborn happens rarely, and little milk accumulates in the mother's mammary glands. A.G. Tomilin (1957) believes that the highly nutritious value of milk promotes the rapid growth of whale calves, whose linear dimensions double during 4–6 months of lactation. According to B.A. Zenkovich (1938), the fact is that whale fetuses "being almost matured for birth, … have absolutely no fat" and for thermal insulation of a calf after delivery "fat should accumulate extremely intensively", which will promote the production of milk with a high fat content. Other researchers share the opinion that the hypodermic fat layer basically carries out a thermal insulation function, saving whales from hypothermia (Yablokov et al. 1972; Arsenyev and Zemsky 1973; etc.). However, J. Gray (Gray 1936) and R. Gawn (Gawn 1948), on the contrary, believe that overheating due to intensive energy release threatens whales more than hypothermia.

In unison with this discussion I will dare to cite one of problems from M. Jarman's book (1972) "Quantitative biology in problems and examples".

So, the problem: a harpooned blue whale is capable of towing a whaling ship at a 12 knots (about 6 m/s) for several hours. The power of the whale can be estimated at approximately 460 h.p.

Question: how much fat does the whale spend in 3 h 46 min 40 s (i.e. 10,000 s) if the caloric power of the fat equals 10 kcal/g, 3/4 of the chemical energy is wasted, and 1/4 reverts to mechanical work?

Answer: the general power consumption, taking into account the energy turning to heat, compounds 1840 h.p., or 328 kcal/s. As 1 h.p. = 0.178 kcal/s, therefore, a total of 328 kg of fat should be expended.

It is clear that the fat saved up by a whale is stored as energy, which is absolutely necessary in long migrations and times of starvation. Thermal insulation is another question.

But we will return to the problem of feeding neonatal whales. Our research has shown that in the second half of intrauterine development, fatty tissue appears under the embryo's skin and layers of fat at the side on the level of the spinal fin equals 3–5 cm in prenatal embryos with the fat content up to 5% (Mikhalev 1971a, b; Mikhalev and Sventsitskaya 1970b), which contradicts B.A. Zenkovich's statement (1938) about the absence of hypodermic fat in prenatal embryos. Fat content in the lard of cetacean embryos is also noted by N.E. Salnikov (1956). As to the frequency of calf feeding, observations in oceanariums have negated M.M. Sleptsov's opinion (1940) about calves being rarely fed. Newborn dolphins and bottlenoses appeared to be fed often enough, every 10–40 min (McBrde and Krizler 1951; Tomilin 1969). Feeding lasts a few seconds and goes on all day and night (McBrde and Krizler 1951; Arshavsky 1968).

As can be seen, the nutritious value of whale milk is extremely high, but meanwhile the relative growth rate of neonatal whales is considerably below what could be expected, if we were guided by the law illustrated by H. Aron (Aron 1927).

A.G. Tomilin's calculations (1946) have shown that newborn fin whales gain about 53 kg per day, beluga whales – 1.8 kg. That is, the mass of newborns approximately doubles in 1–3 months, instead of in some days, as would be the case if the above-mentioned law held true, as E. Slijper (Slijper 1962) assumed. Other prominent researchers share the opinion that whales do not grow faster than terrestrial mammals (Zemsky 1956; Yablokov et al. 1972).

The growth rate of cetaceans we calculated, along with the connection between the body weight and the animal's length allows the specification of the time necessary for newborns to double their mass. The results are the following: for minke whales – 25–30 days; for fin whales – 45–55 days; for bottlenose dolphins – 35–45 days; for beluga whales – 50–60 days. Thus, the connection between the nutritious value of milk and the growth rate of cetacean suckers does not correspond to law characteristic to terrestrial mammals. Obviously, it is necessary to search elsewhere for answers to the contradictions that have arisen.

Observation of cetaceans in oceanariums have shown that while being fed, the whale calf softly clips his mother's small nipple with the ends of his jaws. Then the female injects some milk into his mouth. However whale jaws are rigid and do not have soft lips, and it is not absolutely clear how suckers (especially in the random conditions that occur at sea) grope and keep the nipple. To understand this, we examined the "twin formations" on the jaw ends of cetaceans.

4.6.2 Constitution and Role of "Twin Formations" on the Jaw Ends of Cetaceans

Two small C-shaped hollows on whales have been known for a long time -("Stenson's dimples", nasopalatine dimples), located on the ventral side of the top jaw between the border of the baleen line and the tip of the snout. The function of these formations are not clear, although there have been assumptions (Yablokov 1961, 1966) that they are chemoreceptors. They are also referred to as rudiments of Jakobson's (vomeronasal) organ (Yablokov et al. 1972; Japha 1905; Mizue and Jimbo 1950; Quay and Mitchell 1971).

However the vomeronasal organ of cetaceans is reduced at the embryonic development stage (Herzfeld 1888). Besides, during the evolution of cetaceans the posterior nasal apertures were not displaced forward, but in a caudal direction, and consequently nasopalatine dimples should be displaced backwards. Hence, these formations cannot be rudiments of Jakobson's organ. From the description of the morphological constitution of palatonasal formations, made by V. Quay and E. Mitchell (Quay and Mitchell 1971), it follows that the hollows in the center of each dimple have 12 mm to 15 mm canals that terminate in the dermis. Histological research into these structures has revealed a considerable quantity of free and encapsulated nerve endings, Pacini corpuscles and thin formations with encapsulated plates. Neither olfactory epithelium nor taste bulbs have been revealed in this area.

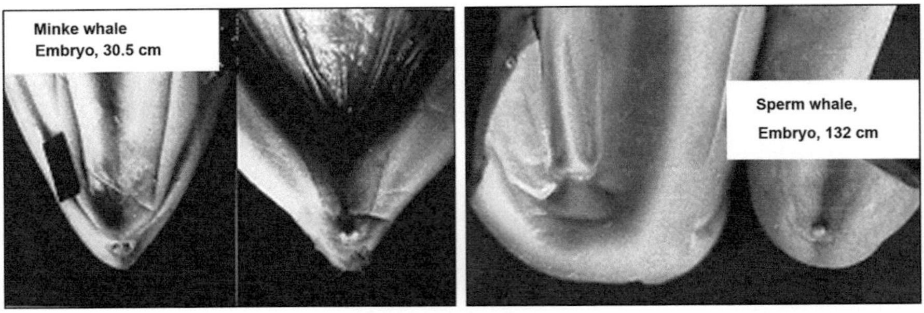

Fig. 4.4 "Twin formations"on the upper (left) and lower (right) jaws of Minke whale and sperm whale embryos

The data mentioned does not say anything certain about the functions of these structures. What is clear is that they are not chemoreceptors, as A.V. Yablokov considered them to be (1961, 1966). Further research is necessary to ascertain their role.

Let's note that earlier investigators knew about the presence of the "twin formations" only on the upper jaw of baleen whales and studied them only in adult animals. The research conducted in our laboratory has shown (Mikhalev and Zelenya 1986; Mikhalev 1991, 2000) that there are "twin formations" not only on the upper, but also on the end of the lower jaw (Fig. 4.4), and not only in adults, but also in embryos, and not only in baleen but also in toothed (at least, sperm) whales. The twin formations on the lower jaw look like knobs, rather than like the dimples on the upper jaw, and are located precisely under the "twin formations" of the upper jaw in such a manner that when the jaws are joined, the knobs coincide with the dimples on the upper jaw.

"Twin formations" become noteceable rather early. The larger the embryo, the more visible are they, being completely formed by the neonatal stage. Individual variability in their form can be observed. On the lower jaw more often they look like two rather large white knobs pulled together or pressed to each other, their cranial side being rounded, and caudal side pointed. Or they can be in the form of two small dots located at some distance from each other and connected by the rounded bridge of a horseshoe form. These two forms occurred in our data with almost the same frequency. Out of the 47 minke whale embryos examined, 53.2% were of the first group and 46.8% of the second one. The "twin formations" of prenatal embryos with considerable skin pigmentation skin, darken on the lower jaw and are less clearly expressed. Because of strong pigmentation these knobs are not appreciable on adult whales, and for this reason might have not been noted by previous researchers.

It is interesting that artists have appeared more attentive than scientists. In T. Ogawa and T. Shida's work (Ogawa and Shida 1950, fig. 2), investigating sei whales and fin whales, the upper and lower jaws of a baleen minke whale are represented. The "twin formations" are presented on the lower jaw in the form of two dots located in relation to the nasopalatine dimples of the upper jaw. In H. Kato's work (Kato 1979), describing the ugly shortened upper jaw of a minke whale, the artist also noted the "twin formations" on the lower jaw. But the authors of these works didn't say a word about them!

"Twin formations" in the form of two white dots on minke whales become discernible when the embryo length is about 17–18 cm, which corresponds to the age of 3–4 months. The larger the embryo in the course of growth, the more clearly discernible they are, being completely formed by the neonatal stage. Nasopalatine "twin formations" of minke whales on the upper jaw become visible earlier – when the embryo length is about 7 cm and aged about 2.5 months. It might happen because they are of a different form – not dimples, but hollows, which makes them easier to see.

It is very important that similar "twin formations" are found not only on baleen whales, but also on toothed whale embryos – sperm whales (Fig. 4.4). However, the form of these formations differs a little, undoubtedly because of the distinctions in the jaw formation of baleen and toothed whales (Mikhalev and Zelenaya 1986; Mikhalev 1991, 2000). On the upper jaw of sperm whales they look like two symmetrically located dimples, but, differing from minke whales, there is a pestle-form outgrowth between them. These formations become discernible in the pre-fetus period of development and clearly visible in the fetal and neonatal periods.

In the postnatal period, the dimples turn into slit-like grooves, located on the palatal side of the cranial end of the top jaw, to which the lower jaw adjoins when the mouth is closed. At some distance from the two main groove-deepenings there are similar but smaller grooves. On the lower jaw of sperm whales, the "twin formations" look similar to those of baleen whales. They become visible when the fetus is 20 cm and look like two small light knobs. A section of jaw has shown that, as with baleen whales, a canal runs from the "twin formations" deep into the jaw, blindly terminating in the dermis (Fig. 4.5).

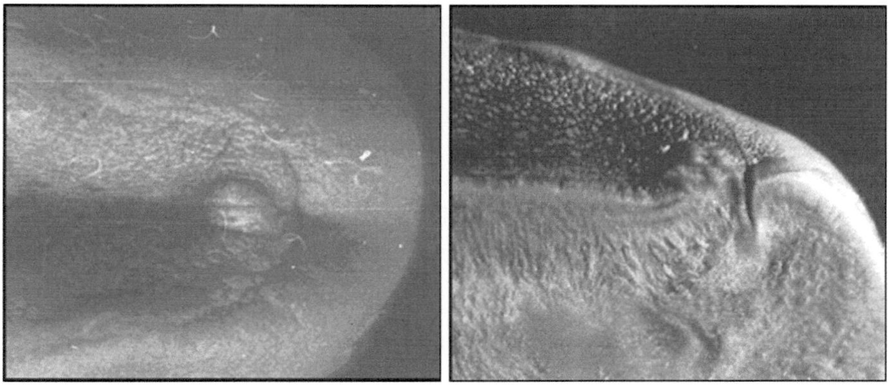

Fig. 4.5 "Twin formation" on lower jaw of sperm whale embryo (left) and its vertical section (right)

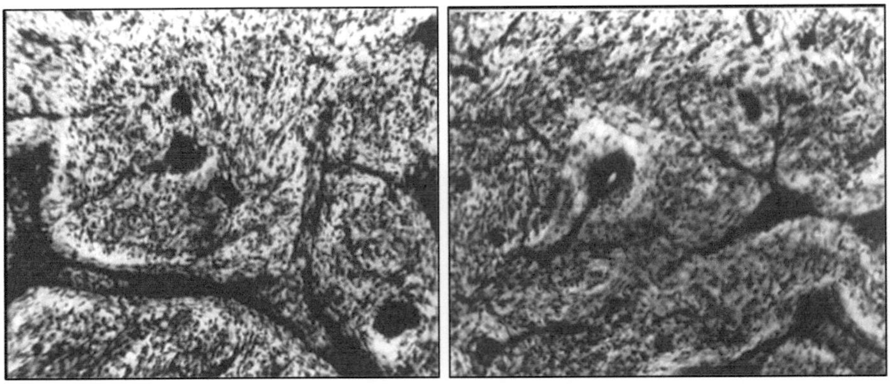

Fig. 4.6 Histological section of "twin formations" of lower jaw of sperm whale (left) and Minke whale (right)

The histological analysis (coloring of sections with hematoxylin-eosin) did not reveal any essential differences in the structure of the "twin formations" of minke whales and sperm whales (Fig. 4.6).

What attracts attention is the strong vascularization, the penetration of capillaries into the epithelium, which proves the active embryogenesis of these formations in relation to nearby tissues. Their epithelium is multilayered, flat and horny. Its difference from the epithelium of the neighboring area is its greater height, and also in its havingmore veins. Olfactory epithelium, bipolar neurons and mucous glands have not been revealed. Nor do not appear in pair structural formations in the postnatal period of development. Only more Vater-Pacini corpuscles are marked.

Electronic microscopy of the "twin formations" of the upper jaw of minke whales has been made at Lomonosov Moscow State University (by the operator V. Kuznetsov). The 200 cm male embryo and the 9.2 m adult female were investigated. It is necessary to stipulate that the material was not fresh, therefore in the pictures chains of bacteria and slime clots are sometimes seen. Nevertheless, the microstructure of the formations is seen well enough (Fig. 4.7).

In the top left picture (50 times magnified) the usual integumentary epithelium is visible on the surface of the "dimple" bottom, lateral view. At the bottom of the "formation", below, some round cells can be seen. In the picture on the right, 1000 times magnified borders of integumentary cells are visible. On the surface of the cells there are numerous microvili (thin outgrowths, approximately one micron thick and 5–7 microns long). In the lower pictures (2500 times magnified), microvili are clearly visible on the surface of one cell. Their length is 3–5 microns, width 0.5–1.0 microns (on the left picture the chain of bacteria is visible).

Summing up the research of the macrostructure, histological and electronic microscopy of "twin formations" on the upper and lower jaws of cetaceans, one can ascertain a resemblance of these frames of representatives of baleen and toothed whales (sperm whales, at least). This resemblance once again testifies to a relation-

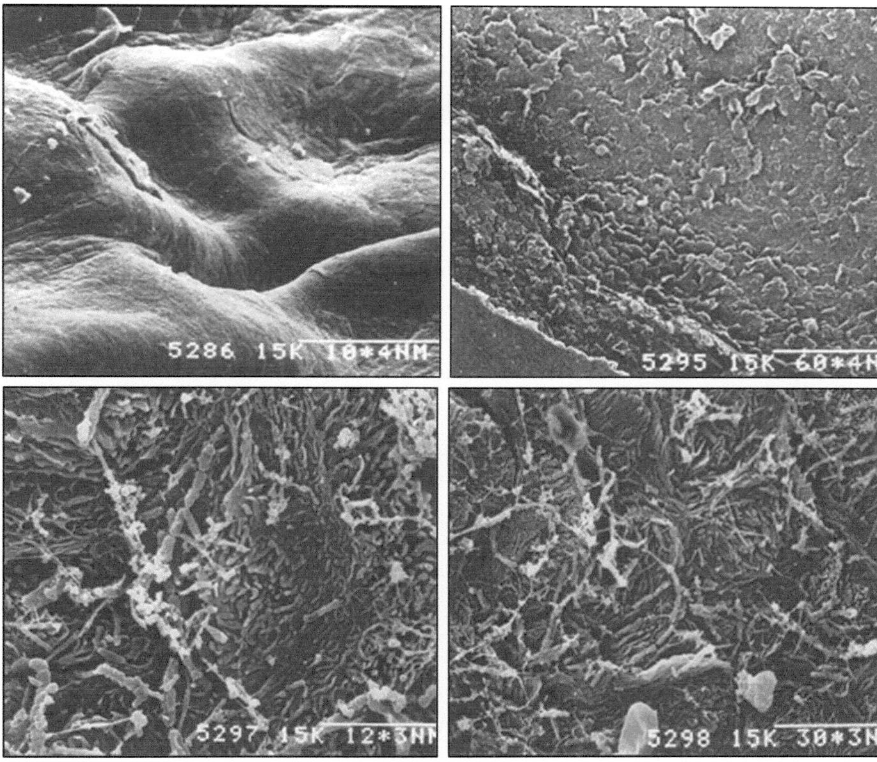

Fig. 4.7 Electronic microscopy of the "twin formations" of the upper jaw of minke whales

ship between the two cetacean suborders, and cannot be, as A.V. Yablokov believed (1964), an argument in favor of their convergent origin.

The role of the described "twin formations" is not absolutely clear. Nevertheless, considering the form, location and histological constitution – presence of microvili on the cell, large number of Pacini corpuscles, which are quickly adapting receptors that can be provoked by touch, i.e., defining tactile sensitivity – and also the fact that "twin formations" are completely developed by birth, it is possible to assume that the "twin formations" play their part when a sucker searches for and seizes its mother's nipple on its mother's belly when feeding.

4.6.3 Process of Feeding Cetacean Calves

It is known that whale milk jets out under the impact of special ring muscles when the area round nipples is pressed (Tomilin 1957; Berzin 1971; Sleptsov 1940; Mikhalev 1971a, b; Uralskaya 1957; etc.). It is possible to assume that in rough seas and when the female moves, the calf clasps the nipple. Indeed sperm whale suckers were observed trying to trap the mother's nipple by the corner of the oral fissure. Pressing the udder provokes a jet of milk. Considering feeding conditions in the wild, the morphology of the nipple and lips, mixing of milk jet with water seems likely. Sometimes milk may jet into the water and the calf engulf the mixture, as we have heard from skilled whalers. Incidentally, in dolphinariums adult dolphins were sometimes observed sucking up food, rather than biting it (Tomilin and Morozov 1969).

In December 1969, the pilot of the helicopter from *Sovetskaya Ukraina*, V.G. Denisenko, observed a group of sperm whale females with suckers. He twice noted a white "cloud" behind the females, into which their suckers swam. The pilot assumed that it was milk, on which the calves were feeding. A similar event was observed in the Odessa dolphinarium, "Nemo", (reported by the head of the dolphinarium, S.V. Kelushock).

When pushing at mammary glands of lactating female whalethe milk is thrown out

In connection with the assumption that a milk-and-water mixture is sucked, it was interesting to carry out an analysis of the stomach contents of recently fed calves, which we had the opportunity to do three times (Mikhalev 1971a, b). They were suckers of a sei whale and a fin whale, which were still feeding on milk alone, and a neonatal sperm whale. Milk from the suckers' stomachs did not have appreciable traces of curdling. The volume was insignificant −3–5 l. Fat content was 7.0% for the sei whale, 7.2% for the fin whale, and 6.9% for the sperm whale. In the course of 5 min, while the analysis was carried out, the milk in the test tube settled, and 2/3 of its volume was oceanic water, which was easily defined by tasting. After shaking, the milk quickly settled again.

It is well known that, as a rule, there is no water in the stomachs of dissected whales. They are only occasionally filled with water, in animals having endured the death throes. There was little water mixture in the stomachs of any of the suckers, which means that they had not experienced the death throes, and mixing of milk with water had happened at the time of feeding, especially in the case of the neonatal sperm whale lassoed alongside the whaling vessel.

If it is true, the exclusively high nutritious value of cetacean milk can be interpreted not only as the means of providing fast growth and fat accumulation in hypodermic cellulose (Zenkovich 1938; Zemsky 1958), but also as an ecological device for calf feeding. In that case, the nutritious value of milk mixed with ocean water on which calves are fed, appears close to that of the milk of other terrestrial mammals, including odd-toed ungulates, and the growth rate of whale suckers will not be an exception from H. Aron's rule (Aron 1927).

Besides, the assumption that milk has been diluted by ocean water when female whales suckle their young, could answer the question posed by A.V. Yablokov (Kleinenberg, etc. 1964), why cetacean embryos in the final development stage and young whales have kidneys of relatively large dimensions. It is clear that if sea water gets into the organism of the newborn, the role of the kidney is to regulate aqueous and salt metabolism. To check this hypothesis, it would be beneficial to analyse milk from the stomachs of cetacean suckers kept in dolphinariums. Unfortunately, we have no such data.

To be fair, it is necessary to note that our hypothesis contradicts the following fact: pinnipeds do not feed their calves in water, but on firm substrate, such as onshore or on ice floes, where mixing milk with water is excluded. However, the fat content in the milk of pinnipeds reaches 40 and even 60%. But, unlike cetaceans, pinnipeds give birth to immature pups. Subsequent research will have to solve this contradiction.

4.6.4 Duration of Lactation and Reproduction Areas

On the basis of indirect calculations, different terms of lactation in cetaceans have been determined: from 2 to 3 months for beluga whales (Kleinenberg, etc., 1964), to 2 years for pilot whales and sperm whales (Berzin 1963). More often lactation has been supposed to last 4–5 months for minke whales and gray whales; 5–8 months for blue whales, fin whales and sei whales (Yablokov et al. 1972); and about 1 year for sperm whales (Berzin 1971). Direct observations of dolphins in oceanariums have shown that in simulated conditions lactation lasts 8–12 months (Tavolga and Essapian 1957).

It is important to note that lactation does not interfere with mating or new pregnancy in baleen whales (Chittleborough 1958; Ivashin 1959) or toothed whales (Tomilin 1957; Berzin 1961; Sleptsov 1941, 1952; Yablokov 1959). This conclusion can be substantiated by frequent cases of pregnant females that are still feeding, the presence of dead-ripe follicles on ovaries and the functioning yellow body of pregnancy. According to A. Jonsgard (Jonsgard 1951), among Minke whales of the Northern Atlantic, and according to H. Omura and H. Sakiura (Omura and Sakiura 1956) Minke whales of the Northern Pacific, almost all pregnant females were simultaneously lactating. The number of such female humpback whales was 39% (Chittleborough 1958), which also corresponds to our data. We have come across lactating, pregnant female humpbacks in the Southern Ocean waters (Fig. 4.8).

Lactating female fin whales reached 29% (Laws 1961). According to S.E. Kleinenberg and A.V. Yablokov (Kleinenberg and Yablokov 1960), almost half of the pregnant beluga whales were lactating. However, this index varies greatly according to the whaling area.

Mating and calving have rarely been observed, feeding has been seen a bit more often (Berzin 1971; Budylenko 1970a, b; Mikhalev 1971a, b; Dudley 1725; Scoresby 1820; Bennet 1840; Brown 1868; Collett 1911–1912; Ruspoli 1955; Pervushin 1966; Tormosov 1970). The reproduction areas of some whale species can be judged by detection charts of embryogenesis stage embryo, prenatal fetuses and lactating females.

Blue Whales Not much data was collected for allow definitions of reproductive regions to be made: 34 lactating females and 4 females with prenatal embryos – four animals. They occurred in isolated areas: in the Bellingshausen sea, around South Georgia, to the south of Kerguelen Island, and also to the north of the Balleny Islands and the Ross Sea.

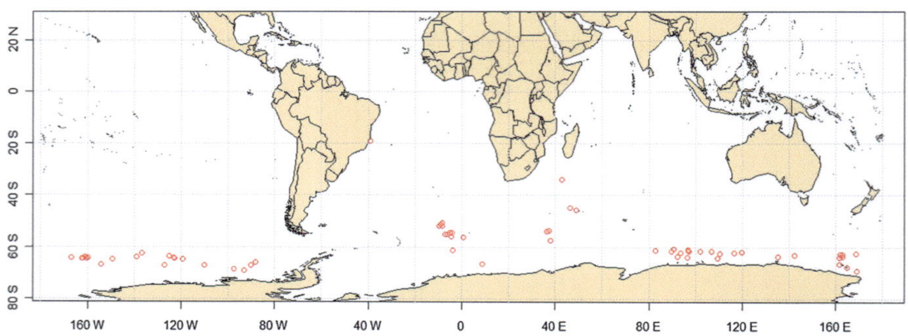

Fig. 4.8 Distribution of lactating pregnant humpack whales

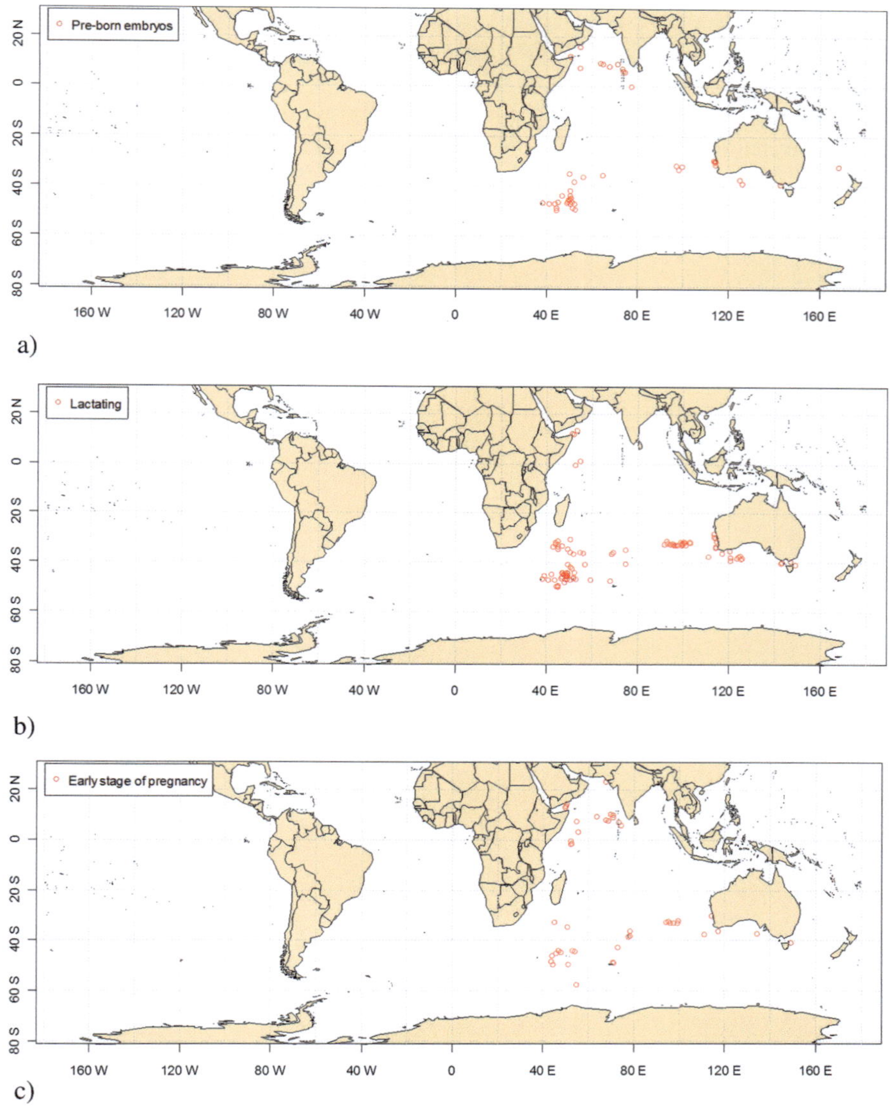

Fig. 4.9 Pygmy blue whale females: (**a**) with pre-born embryos, (**b**) lactating, (**c**) with EDS embryos

Pygmy Blue Whales In comparison with blue whales, they prefer warmer waters. The charts were compiled on the basis of data from about 57 early development state (EDS) embryos, 98 prenatal fetuses and 142 lactating females (Fig. 4.9). The following regions of reproduction can be designated on the charts: the area south-west of Australia (probably three local herds: to the north-east of Amsterdam and

Sen-Pol island, southwest coast of Australia and the area around Tasmania), and to the north of the Prince Edward and Crozet islands.

Fin Whales Coordinates of 102 females in early pregnancy state (EPS) have been plotted on charts, 33 females with prenatal fetuses, and the largest group of all – 426

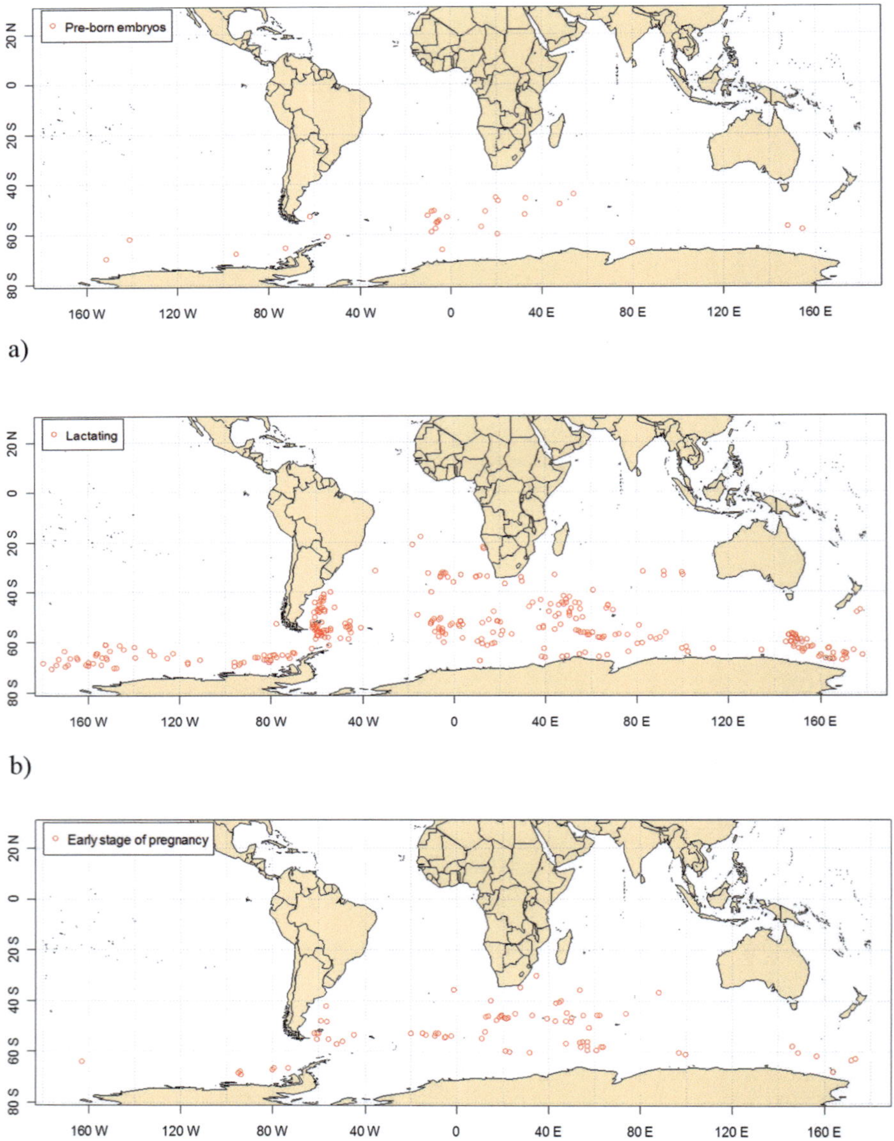

Fig. 4.10 Fin whale females: (**a**) with pre-born embryos, (**b**) lactating, (**c**) with EDS embryos

lactating females. Three regions of reproduction for fin whales have been distinguished in a latitudinal direction (Fig. 4.10). In the thirtieth latitudes it is the area of Tristan-da-Cunha, also the region near the southern extremity of Africa and the area to the north of Sen-Pol and Amsterdam islands. In the fortieth-fiftieth latitudes, the Falkland Islands – Argentina area and an area near the Balleny Islands are distinctly visible, and less distinctly – the "island" area of the Indian Ocean sector of the Sothern Ocean –Bouvet, Crozet and Kerguelen islands.

Bryde's Whales They also prefer warmer waters. As there is not enough data on these whales, charts are not shown. There were only 12 EPS females, 10 with prenatal fetuses and 7 lactating females, at our disposal.

Sei Whales The charts were compiled from the data on 94 EPS females, 139 females with prenatal fetuses and 657 lactating females (Fig. 4.11). The distribution of sei whale reproductive regions is very similar to that of fin whales: the west and east coasts of South America, Tristan-da-Cunha, and the thirtieth latitudes of the Indian Ocean, Balleny Island.

Minke Whales The Soviet whalers actively took Minke whales (about 15,000 whales) in the Southern Ocean in high latitudes and mainly in the summer season – December, January, February. Possibly, owing to this, there were no feeding females, 27 with prenatal fetuses, and 412 EDS embryos. Reliably enough, two areas of reproduction can be singled out by these features: Commonwealth Bay and the Bellingshausen Sea. Another region of minke whale reproduction is the coastal waters of Brazil (Fig. 4.12).

Humpbacks Many lactating females (278), a few EDS embryos (63) and prenatal fetuses (11) were detected. On the basis of this data the following maps (Fig. 4.13) were composed. The reason is the same: whaling does not coincide with humpback mating or calving seasons. In warm waters, regions of reproduction are marked near the coasts of Brazil and southwest to the coast of Africa, to the south of Madagascar, off the southwest coast of Australia, and near the coasts of Tasmania and New Zealand. In the Southern Ocean waters reproduction areas occur around Bouvet Islands and Crozet and Kerguelen islands. In the Southern Ocean waters it is possible to delineate humpback reproduction areas in the Bellingshausen Sea, to the east of Commonwealth Bay and to the east of Balleny Island.

Southern Right Whales In spite of the fact that there is only data from about 13 females with EPS embryos, 16 with prenatal fetuses and a small number of lactating females at our disposal, the reproductive areas of the southern right whale are accurately defined (Fig. 4.14). They are: the coast of Argentina, waters adjacent to Tristan-da-Cunha, the southern extremity of Africa, the Crozet Island (females with prenatal fetuses were fixed to the south -west of the island, they probably had not had time to move to the calving area), the east part of Commonwealth Bay, Tasmanian waters and the Aucklands, Campbell and Macquarie islands area.

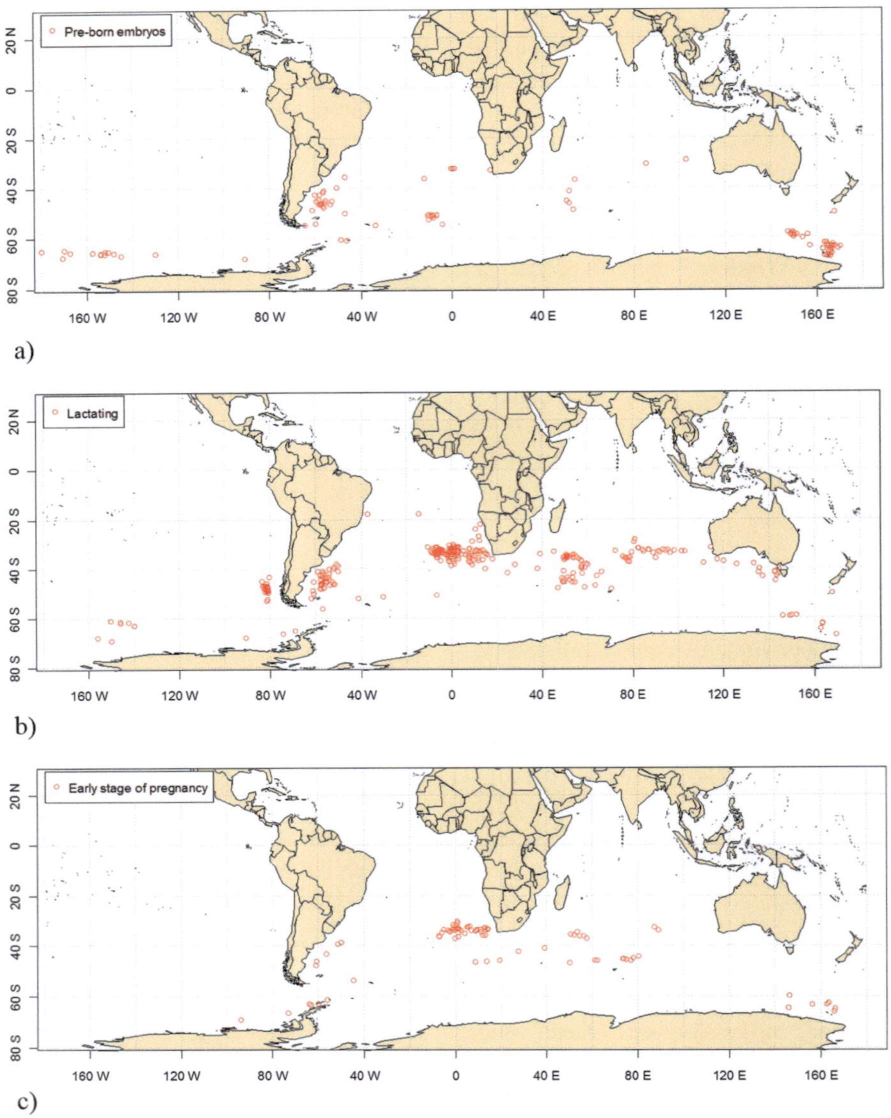

Fig. 4.11 Sei whale females: (**a**) with pre-born embryos, (**b**) lactating, (**c**) with EDS embryos

Sperm Whales While talking about spermwhale breeding it is important to take into account that, contrary to males, females rarely move to the high latitudes and cross convergation zone (actual orthern border of the Southern Ocean). They prefer warm waters and mostly stay in moderate, sub-tropical and tropical latitudes. 450 females with embryogenesis stage embryos, 575 with prenatal fetuses and 2302 lactating females were collected, which enabled us to compile charts of their areas of reproduction (Fig. 4.15). It is first of all the rather isolated waters in the northwest

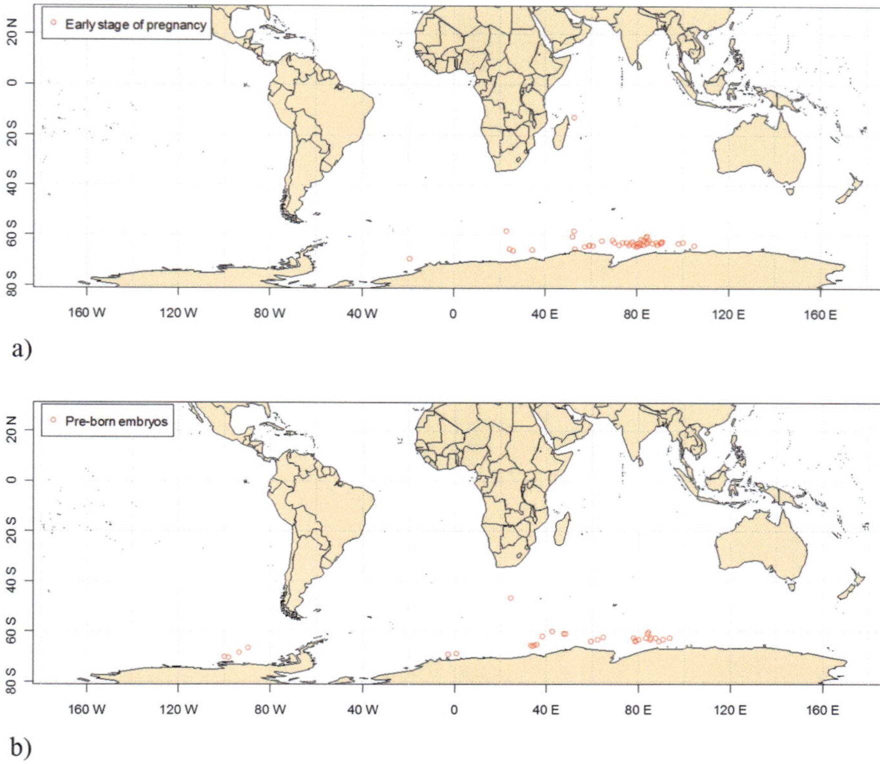

Fig. 4.12 Minke whale females: (**a**) with EDS embryos, (**b**) with pre-born embryos

part of the Indian ocean. Reproduction regions accurately come to light near the coast of Peru and the east coast of South America, the southwest coast of Australia, Tasmanian waters and the Tasman sea; less isolated regions lie between Tristan-da-Cunha Island and the southern extremity of Africa, and the region southeast of Madagascar. As a whole, the region of reproduction of sperm whales to the south is limited by 45°S lat. The main region of reproduction is the area between the thirtieth and fortieth parallels.

***

As can be seen, the mating, calving and feeding areas of baleen and toothed whales, generally coincide, forming reproductive regions mostly situated in the 30–40 latitudes, but for some whales these regions appear to be have moved closer to the Southern Ocean waters. It is important to note that if their reproductive regions coincide, then the timing of the reproduction seasons differs (Fig. 4.16), which reduces inter-specific competition.

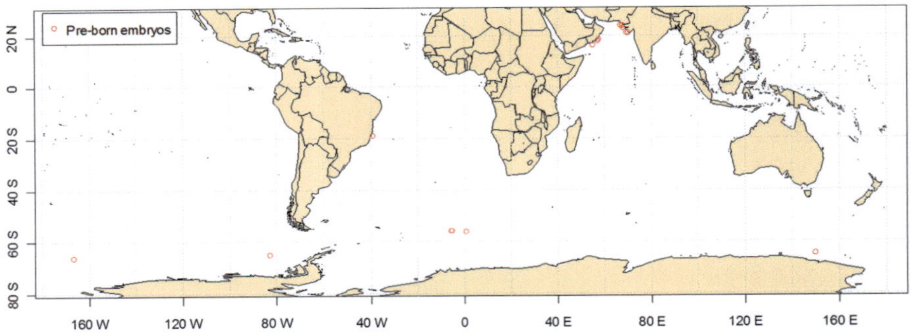

a)

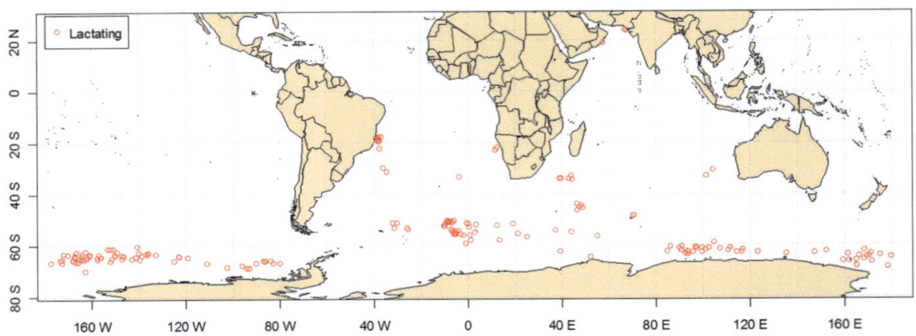

b)

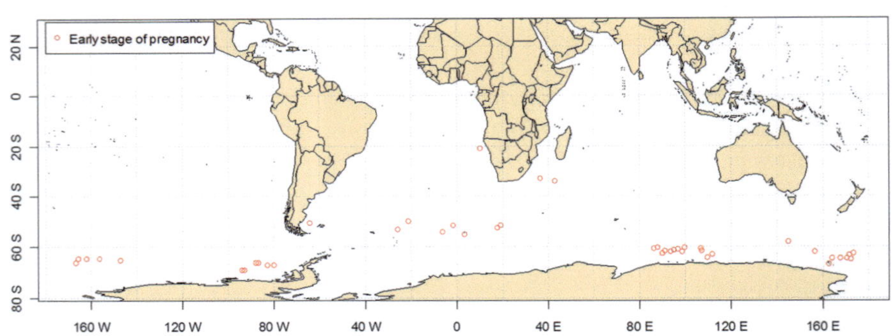

c)

Fig. 4.13 Humpback whale females: (**a**) with pre-born embryos, (**b**) lactating, (**c**) with EDS embryos

4.7 Reproductive Capacity of Whales

The reproductive capacity of whales can be estimated by the percentage of pregnancies among mature females. If females give birth annually, there will be 100% pregnancies; if once in 2 years −50%; etc. In our research the percentage of pregnant

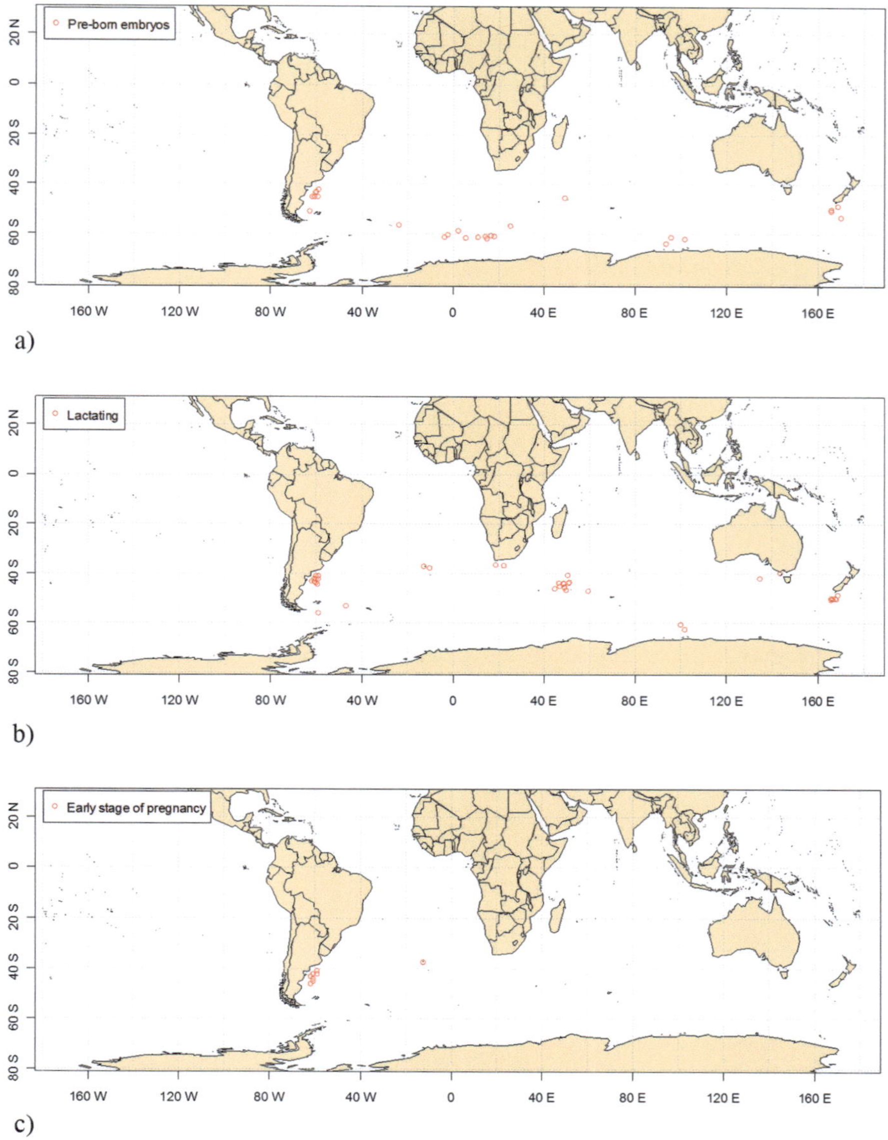

Fig. 4.14 Southren right whale females: (**a**) with pre-born embryos, (**b**) lactating, (**c**) with EDS embryos

females was defined by the interval between the end of the mating season and the beginning of the calving period. Migration of pregnant females to and from the feeding areas was differentiated. As for whales, unlike terrestrial mammals, lactation does not interfere with pregnancy, so some pregnant-feeding females were found with pregnant females.

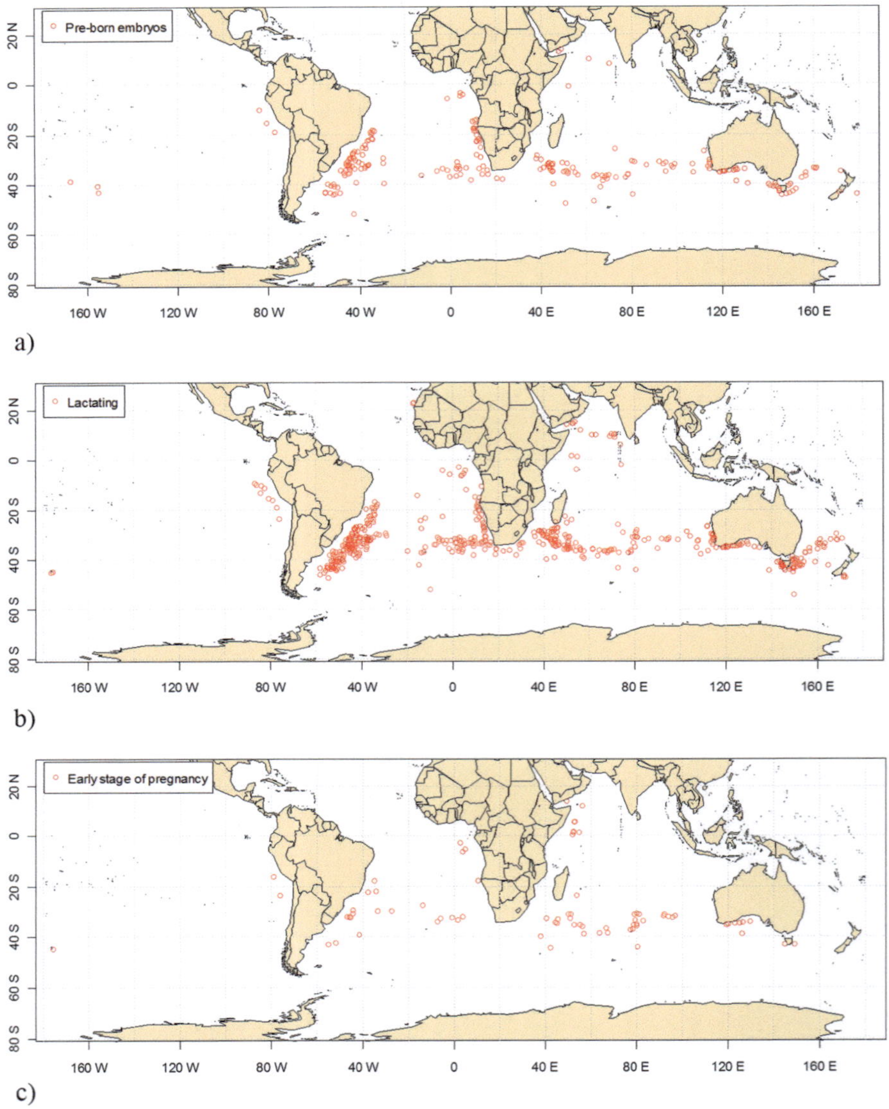

Fig. 4.15 Sperm whale females: (**a**) with pre-born embryos, (**b**) lactating, (**c**) with EDS embryos

According to A. Laurie (Laurie 1937), pregnant females account for 54–61% of blue whale females. According to T. Ichichara's data, 42–57% (Ichichara 1961, 1966). Therefore, female blue whales give birth once in 2 years. According to our data, 40% to 55% of pregnant mature, female pygmy blue whales in the north-western part of the Indian Ocean were pregnant in November–December (Mikhalev 2000, 2002a). There were 30–35% pregnant females in the "island" (Crozet – Kerguelen) and "Australian" areas, according to T. Ichichara (Ichichara 1961, 1966),

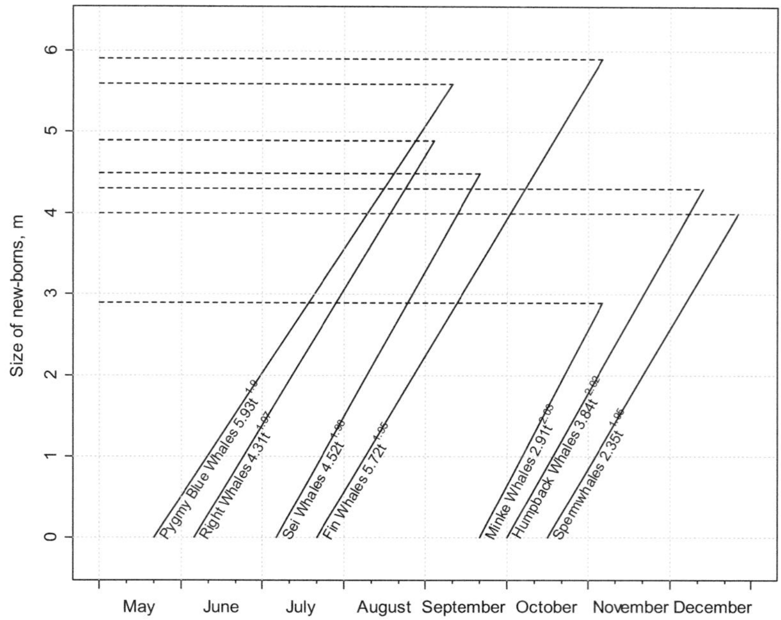

Fig. 4.16 Period of pregnancy of different whale species

E.G. Sazhinova (1980), N.V. Doroshenko (1996). Undoubtedly, these figures are underestimated. So, near Australia in November 64% of the females were pregnant and to the south of Madagascar in area of Walters Shoal – 78%. In February and in March in the Crozet - Kerguelen area the number of pregnant females reached 60–70% in some years.

The number of pregnant female fin whales, according to both our data, and that of V.A. Arsenyev and Zemsky (1973), is close to 50%. Among the largest female sei whales the numebr of pregnant females reaches 65% (Budylenko 1970a; Budylenko and Mikhalev 1970). In the middle of the whaling season in high latitudes of the Antarctic, more than 80% of female minke whales appear to be pregnant, however by the end of the season the number drops considerably, because of migration to the feeding areas (Zinchenko 1986). The interval between deliveries for young animals can be longer, and more mature females can sometimes give birth once in 18 months, or even annually. This is primarily true for sei and humpback whales and (Ivashin 1959; Chittleborough 1958).

The aforesaid percentage of pregnant females is evidence that baleen whales have a biennial reproduction cycle. Moreover, A.V. Yablokov, V. M. Belkovich, and V.I. Borisov (1972) believe it to be true for all cetaceans.

The reproduction capacity of cetaceans estimated by the rate of pregnant and dry females is confirmed in other ways: the analysis of the rate of accumulation of ovulation and pregnancy traces on the ovaries.

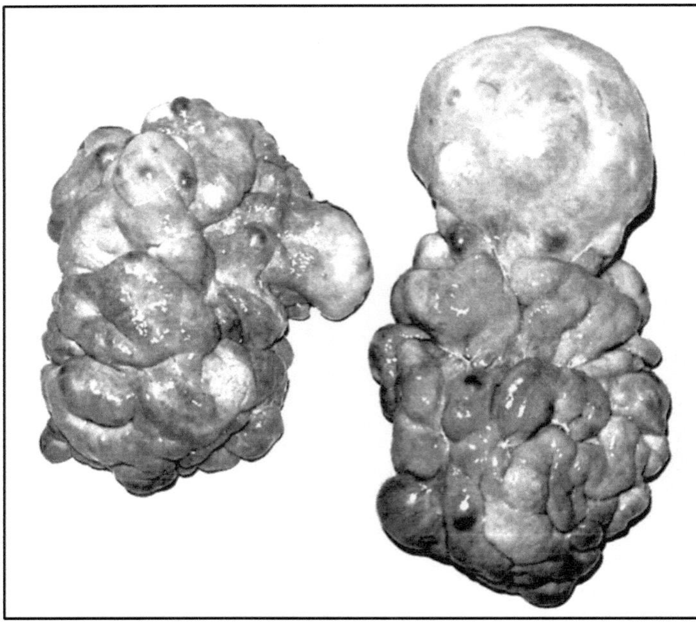

Fig. 4.17 Ovaries of sei whale female with traces of ovulation bodies, traces of pregnancy and functioning yellow body of pregnancy on the right ovary

Remember that as a result of each ovulation one, or several, yellow bodies of ovulation is formed. Two or 3 weeks after ovulation these bodies are reabsorbed, and a scar is formed in their place – the trace of an ovulation body. If the ovulation results in pregnancy, the functioning yellow body of pregnancy is formed in place of the mature follicle (Fig. 4.17), which is reabsorbed within 8 months after pregnancy (Laws 1958).

Regression analysis has shown that there is a correlation between the number of traces of ovulation on female whale ovaries and the age of the animals. For example, the correlation coefficient for minke whales is 0.998. The correlation can be expressed by:

$$N_{OV} = 0.856t \quad 3.588;$$

where,

N – the number of traces of ovulation,
t – the age of the females.

A very similar conclusion was reached by J. Masaki (Masaki 1973) and S. Ohsumi, J. Masaki (Ohsumi and Masaki 1975). It follows from the correlation that on average 0.852 traces of ovulation are formed in the course of a year on the ovaries of mature minke whales. As not every ovulation always results in pregnancy, the rate of accumulation of traces of pregnancy is a little below – 0.6–0.7. That is, one trace is formed in approximately 18 months. The intensity of reproduction of other baleen whales is slightly lower than that of minke whales. So, from our calculations, this index for fin whales is 0.52 of a trace per year; for sei whales and Bryde's whales – 0.55 trace of a trace per year.

Thus the average rate of accumulation of traces of pregnancy for minke whales is approximately one trace in 1.5 years. The age of sexual maturity is 5–6 years. Females participate in reproduction for 10–15 years and for this time are capable of giving birth to 6–7 calves.

In spite of the fact that the rate of accumulation of traces of ovulation for other minke whales is slightly lower, their reproductive ability appears to be the same, so they also give birth to 6–7 calves. This happens because larger Minke whales have a longer lifespan and so, consequently, is the period of participation in reproduction.

The question of sperm whale reproduction capacity is debatable. They participate in reproduction having reached age 8–10 years. According to A.A. Berzin (1963), one trace of ovulation on female ovaries is formed in 2–3 years. S. Ohsumi (Ohsumi 1965) believes it to be 3.4–3.8 years, and D.D. Tormosov (2002) –4 years. By our calculations considering their shorter term of pregnancy, which is based on the analysis of embryo growth (Mikhalev 1971a, b), one trace of pregnancy is formed during no more than 3 years. This conclusion is also confirmed by the ratio of pregnant (33%) and dry (67%) females in one sexual cycle. During their lives, female sperm whales participate in reproduction also about 6–7 times.

From all the data it follows that during participation in reproduction, female baleen and toothed whales are capable of giving birth to 5–7 calves. Of course, not all of them live to the age of sexual maturity and participation in reproduction. Suckers and teenagers are victims of killer whales to a greater degree.

In the wild there are many and varied causes of the increased death rate of the young. Unfortunately, the degree of survival (death rate coefficient) of whales has not been studied enough at this stage, hence, we can only speak about the reproductive capacity of whales as a rough approximation.

Therefore, when defining the terms of moratoriums on whaling with the purpose of restoration of the whale population, it is necessary to use a lower estimate of reproduction index – not more than 3–4 animals.

4.8 Summary

Comparing the results of several methods the age of sexual maturity of whales was defined:

- by the moment of decrease of growth rate when analyzing the growth curve;
- by the regression line of the rate of accumulation of ovulation traces on female ovaries;

- on the basis of percentage ratio between mature and immature females;
- on the basis of the analysis of primigravida female occurrence.

Defining the age of sexual maturity by the position of the "transphase" in aural plugs of baleen whales, which a number of researchers have tried to do, does not present reliable results. Comparison of long-term data has shown that the age of onset of sexual maturity in whales is a conservative enough biological index, which has not essentially varied during the last five decades of stopping whaling.

The correlation between the size of a female and her newborns is species-specific and is expressed by the following dependence: toothed whales – $L = 0.480\ l^{0.90}$; baleen whales – $L = 0.636\ l^{0.74}$, rorqual whales – $L = 0.569\ l^{0.81}$.

This regularity can be used for practical purposes. It allows judging the average size of newborns of the species, when the average size of mature animals is known and the length of newborns is not known.

The opinion about the excessive duration of reproduction seasons is not correct. About 70% of all mating and calving takes place during 2.5 months, 95% – within 5.5 months. The duration of pregnancy of different baleen whales is 9.5–11.0 months. Pregnancy of sperm whales lasts not less than a year. There is an additional shift of 6 months in the reproduction season for fin, minke and humpback whales.

A hypothesis has been put forward about the mixing of milk with ocean water, which can happen when suckers are feeding. In this case the fat content of the water and milk mixture corresponds to the law of growth rate of suckers, characteristic to terrestrial mammals ("Aron's rule"). This assumption was in accordance with peculiarities of constitution and the growth of cetacean kidneys.

Specific "twin formations" have been determined to be not only on the ends of the upper jaw, as was considered earlier, but also on the ends of the lower jaw, not just of baleen whales, but of toothed whales as well. They become appreciable in the pre-fetus period of development and are completely formed by the neonatal period. Their role in searching for and seizure of a mother's nipple during feeding is assumed.

Resemblance in morphology and position of these formations of baleen and toothed whales proves the relationship between the two suborders and is contrary to the argument in favor of their convergent parentage, as A.V. Yablokov (1964) thought.

Mating, calving and milk feeding areas of Southern Ocean whales essentially coincide. Reproduction seasons of different whales are differentiated, which reduces inter-specific competition. Reproductive regions are situated in the 30–40 latitudes, but for some whales they move closer to Antarctic waters.

Chapter 5
Whale Distribution

Mating games of right whales

Y. Mikhalev, *Whales of the Southern Ocean*, Advances in Polar Ecology 5,
https://doi.org/10.1007/978-3-030-29252-2_5

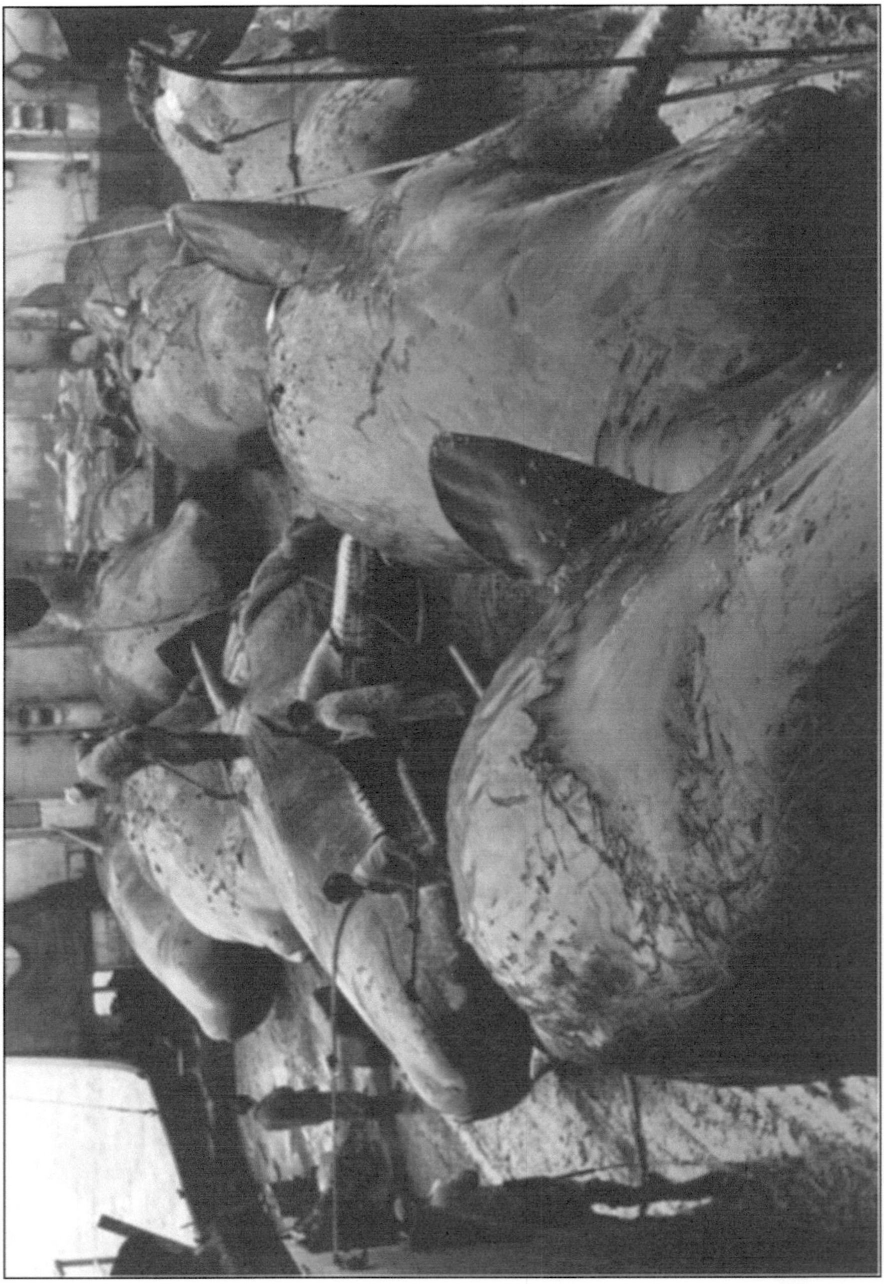

Sperm whales on desk

There have been attempts to summarise the habitats of different whale species in the world's oceans by K. Townsend, 1931, 1935, M. Sleptsov, 1955b, A. Tomilin, 1962, S. Kleinenberg and others, 1964, A. Berzin, 1971, F. Hershkovitz, 1966, M. Nishiwaki, 1966, D. Rice and V. Scheffer, 1968. However, as A. Yablokov reasonably notes in his book "Whales and Dolphins" (Yablokov et al. 1972), there are many disputable, grey areas in this question, which require the ongoing accumulation of factual data.

The distribution maps are based on actual Soviet whaling data, which had been concealed from the public (Zemsky et al. 1994, 1995, 1996; Mikhalev 1995, 1997, 1997a, 2000, 2002a, b; Tormosov et al. 1998; Zemsky and Mikhalev 2000b, c; Yablokov et al. 1998), and also on the results of whale observations from scientific and hunting whaler vessels. Data collected from tagging whales by all countries that belong to the International Whaling Commission (Mikhalev and Tormosov 1997) have also been included. The data from tagging were not connected with whaling and not fabricated; they are authentic, which cannot be said about data from returned tags. The latter were included after being thoroughly checked and corrected. Very important is the fact that information collected from tags can define whale distribution in months and in regions in which whaling did not take place, which is especially noteworthy.

5.1 Distribution and Migrations of Baleen Whales

The following map (Fig. 5.1) gives the general picture of baleen whale distribution in the Southern Ocean and adjacent waters. It shows 56,360 captured whales and 17,399 tagged ones.

As we see, the greatest concentration of baleen whales is observed in highly productive zones of the world's oceans: in zones of water convergence and divergence; where currents join; where temperatures jump and in areas of atmospheric cyclonic activity (Mackintosh 1942a, b, 1965; Wheeler 1934; Tomilin 1940, 1965; Salnikov 1953; Clarke 1954, 1957; Arsenyev 1958; Beklemeshev 1959, 1961;

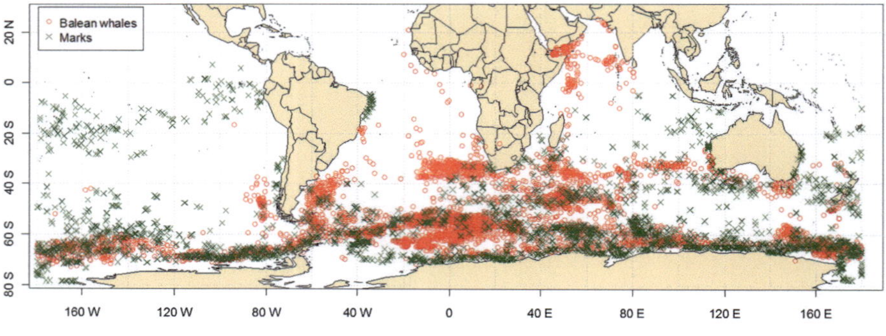

Fig. 5.1 Blue whale distribution in the Southern hemisphere

Nemoto 1959; Zemsky 1965, 1974; Mikhalev 1978; Mikhalev et al. 1981). The unproductive or "dead" zones where whales are almost never seen, even during migrations, which have not been marked by previous researchers, are also visible on the map.

Here are distribution maps of the main commercial whale species and the direction of their migrations compiled from our tags and from the literature (Chittlbotough 1965; Ivashin 1959; Mikhalev and Tormosov 1997; Mikhalev et al. 1981; Brown 1968, 1976, 1995; Dawbin 1964; Dowbin 1966; Gambell 1967; Ivashin 1971, 1973, 1986, 1990; Tomilin 1980).

Blue Whales The real blue whale distribution map is compiled on the basis of occurrence of this whale species from November till April. It includes information about the locations of catches of 1891 blue whales, and also the data from 699 tagged whales. They were seen only in Antarctic waters. During that period real blue whales were never registered north of 40° latitude South. However, there could have been pygmy blue whales among those taken between the 50th and 40th parallels (Fig. 5.2a). On the basis of tag results (44 returned tags) we can observe the migration of blue whales from the South Georgia region and the South Sandwich

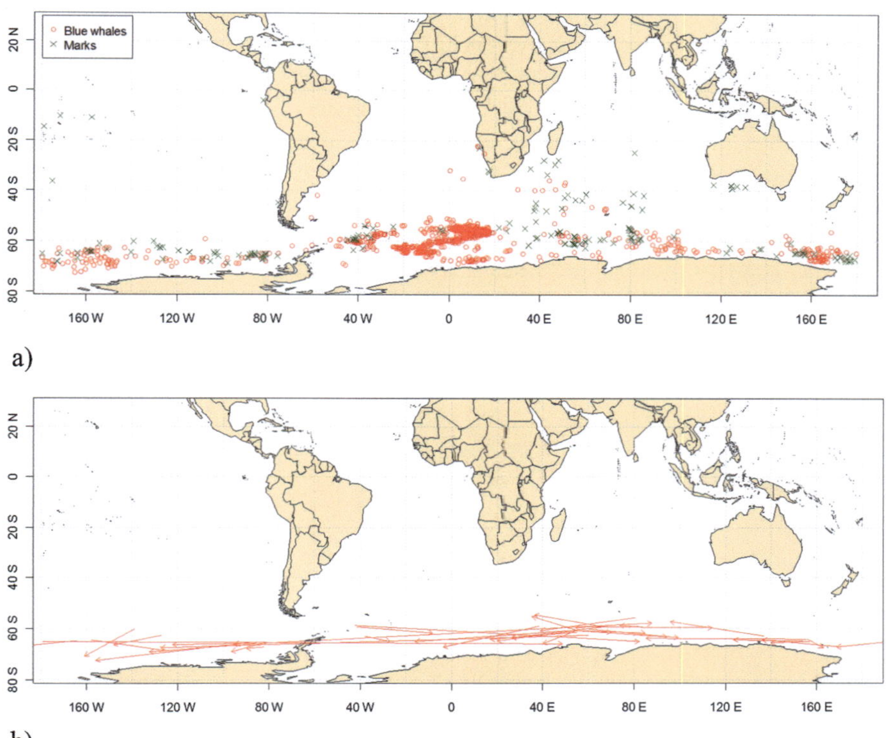

Fig. 5.2 Maps of distribution (**a**) and migration routes (**b**) of Antarctic blue whales

Islands to the Weddell Sea and the Concord Sea, and from the Bellingshausen Sea to the Ross Sea (Fig. 5.2b).

Pygmy Blue Whales The map includes the coordinates of catches of 7613 whales of this species taken between October and May (Fig. 5.3).

As can be seen, the area of pygmy blue whales is practically limited by the Indian Ocean and the Tasman Sea where they can be met from the coasts of Africa, Eurasia and Australia almost up to 60° latitude south. These are waters beyond convergation zone, so they are in the Southern Ocean. But till 1960-x of the nineteenth century these tow species of bluewhales were not recongnized, so it seems the whales to the south of 50 latitude were not pygmy, but real bluewhales. Aggregations of these whales, which are rather isolated from the main part in the northwest part of the Indian Ocean present special interest. The gatherings of pygmy blue whales are observed to the south of Madagascar, to the north of Amsterdam and Saint-Paul Islands, and in the coastal areas of western and south-eastern Australia.

Outside the main areas, pygmy blue whales (there are doubts about the definition of this species!) have been observed along the coast of Brazil (Wheeler E. 1945; Della Rosa and Secchi 1997; Pinedo and Barreto 1996), the Chilean (Aguayo 1974; Rice 1977), Peru (Donovan 1991), and even near Galápagos Islands and the Mexican Coast (Berzin et al. 1976). Soviet whaling data do not prove their occurrence in these regions. They weren't observed near Chile or the Peruvian Coast in the summer seasons of 1973–1975 when these areas were inspected by expedition members of the scientific and search whaler *Bodry-25* (Mikhalev 1975, 1978).

Nowadays genetic research to determine the distinctions between real blue whales and pygmy blue whales (Brownell Jr. et al. 1997) is being actively conducted. From the report of the IWC Scientific Committee (Leduc et al. 1998) it follows that beyond the Indian Ocean it is necessary to relate the blue whales of the Brazilian coast to pygmy blue whales. In the Atlantic, pygmy blue whales were registered at 70°S, which is much further south than Bouvet Island where they had been observed earlier. Identification of pygmy blue whales in such high latitudes

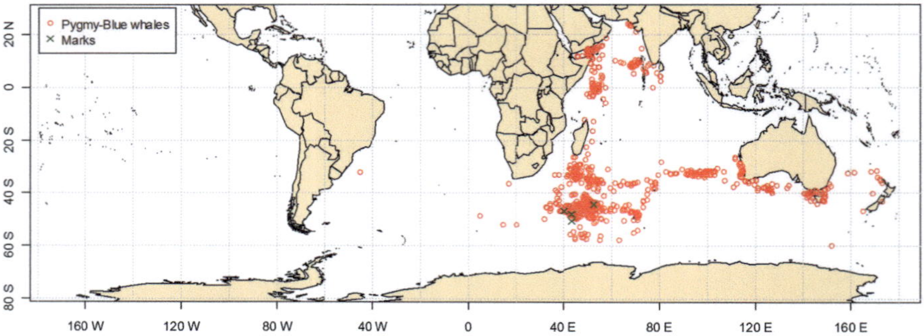

Fig. 5.3 Distribution map of pygmy blue whales

raises doubts. However we do not exclude the possibility that it could be one more subspecies of blue whales which has not yet been described or investigated!

The analysis of monthly distribution shows that in the first half of the southern summer, Madagascar and Amsterdam gatherings of pygmy blue whales relocate to the area of Crozet and Kerguelen Islands, sometimes reaching latitude 60° South. As to tagging, there are data from only about one tag that came back: the whale had been tagged at 1.12.1962 in latitude 56° South, longitude 48° East, and taken on 4.04.1963 in latitude 44° South in longitude 49° East. West-Australian congestion is, apparently, displacing to the Big Australian Gulf, while East – Australian – to the area of the Bass and Cook straits. Return migration of pygmy blue whales begins in the second part of summer.

Fin Whales In the Southern hemisphere the distribution of fin whales is wide – from the 20th parallel up to the Antarctic ice edge (Fig. 5.4a).

The large number of both caught (82,171) and tagged (3755) fin whales from October till May were found in two longitudinal zones, the distance between which was about 40 degrees. In the middle of summer all of them migrated a little to the south. In March and April a return migration was observed.

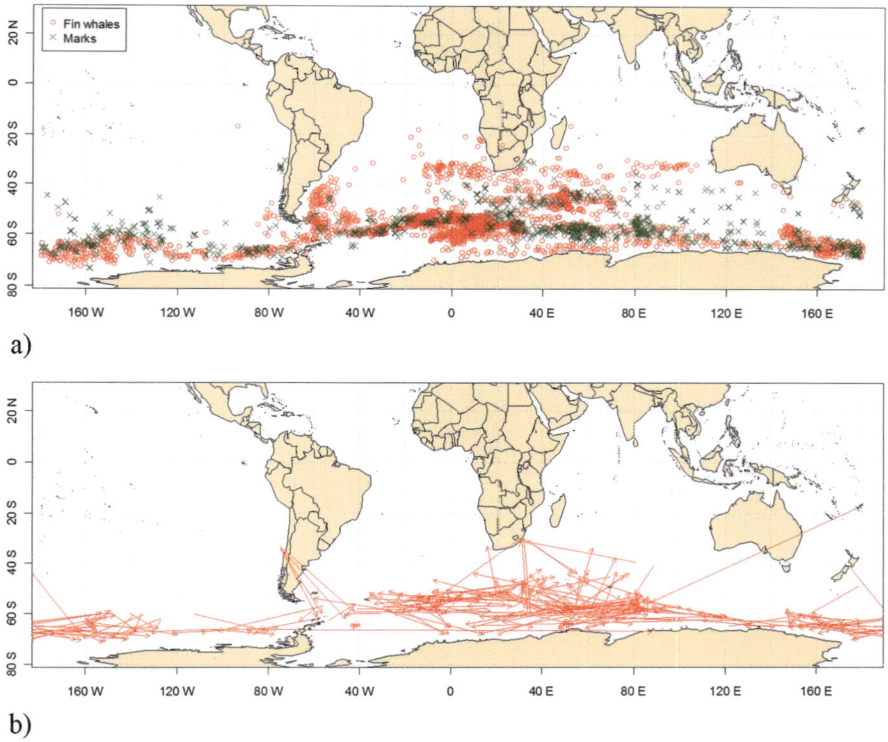

Fig. 5.4 Distribution maps (**a**) and migration routes (**b**) of fin whales

Tagging has shown (340 marks recovered, Fig. 5.4b) that fin whales from the Chilean aggregation migrate through the Drake Passage to the Weddell Sea. Fin whales from coastal waters of southern Africa migrate to Bouvet, Crozet, Kerguelen and Amsterdam Islands. The whales from the coastal areas of New Zealand feed near the Balleny Islands and in the Ross Sea. Fin whales are also known to migrate from Fijian waters to Commonwealth Bay.

Sei Whales The distribution of sei whales is presented according to the catching of 38,769 and tagging of 1395 whales (Fig. 5.5).

The distribution picture is very similar to that of fin whales. Their migration routes do not essentially differ. What is different is that sei whale gatherings in the area of reproduction are more precisely separated. However, this distinction can be caused by the fact that sei whales were actively taken during a rather short period of time – no more than two decades – while fin whales were taken for more than five decades. Longer period also give also a greater dispersion of hydrobiological conditions, and, hence, a greater disorder in the distribution of whales.

The majority of sei whales migrate to subantarctic waters a little bit later than fin whales, but the former migrate to the subtropics and enter the reproductive season

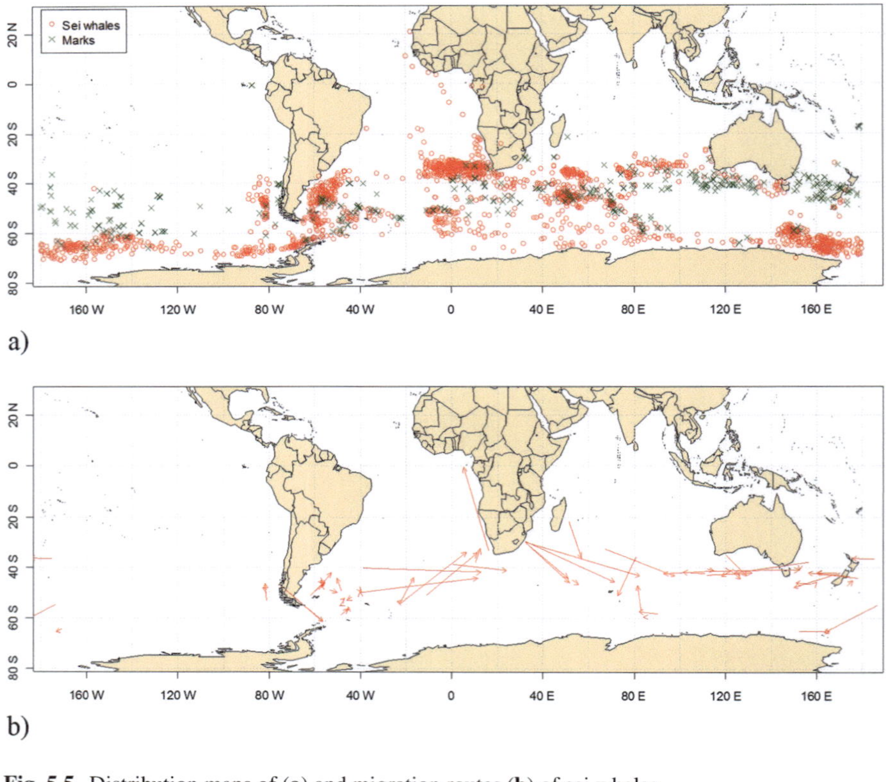

Fig. 5.5 Distribution maps of (**a**) and migration routes (**b**) of sei whales

earlier. This is another reason (apart from differences, say, in nutrition) for the reduction in competition between fin whales and sei whales, and for the two species' virtual coexistence in the same area. As to sei whales to the north of 20° South, they are, in fact, mostly Bryde's whales, the two being difficult to distinguish. The same is true for the data from tagging in the area of Africa's Liberian coast.

Bryde's Whales These are a rather small thermophilic species of rorqual whale. In the Southern hemisphere Bryde's whales keep within the limits of isotherm 20 °C, and are almost never seen to the south of the fortieth latitude. Their habitats are known near the coast of Brazil and the western coast of South America; near the west, south and east coasts of Africa (two races) and around the islands of Indonesia. The area is limited to tropical and subtropical zones where they were basically taken as bycatch. For this reason Bryde's whale has been less investigated than other rorqual whale species. The existence of some subspecies or even similar species is possible. S. Wada (Wada 1998), on the basis of osteological and genetic analysis, considers that Bryde's whales are represented by two subspecies (*B. edeni* and *B. brade's*). He distinguishes Bryde's whales of the western part of the Indian ocean as an independent species, *Balaenoptera olseni*. The question is being discussed and remains open.

The coordinates of 897 caught and 577 tagged Bryde's whales are shown on the maps (Figs. 5.5 and 5.6).

Aggregations are marked on the map: off the coasts of Liberia and Argentina; in the northwest part of the Indian ocean; near the southern extremity of Madagascar; to the north- east of Australia; off the coasts of Tasmania and New Zealand; in the island areas of Fiji, Samoa and Tonga; in the waters of the Galápagos and off the coasts of Ecuador and Peru. Bryde's whales of the north-west part of the Indian Ocean need a separate study. The whale tagged not far from New Zealand was also caught there. Data from the retrieval of other tags are not present, apart from two tagged whales found near the coasts of Guinea and Liberia, which were registered as sei whales.

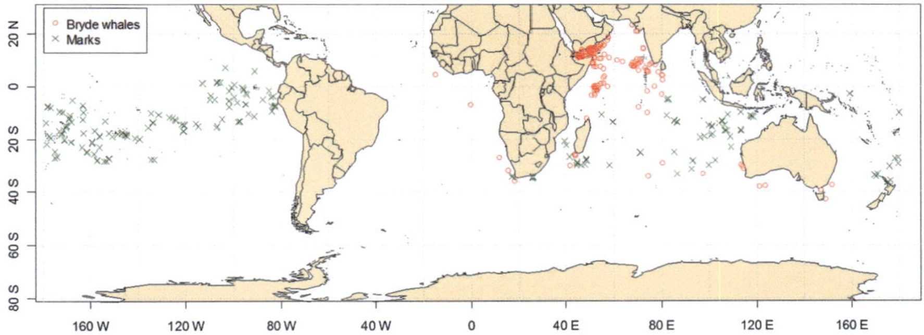

Fig. 5.6 Map of Bryde's whale distribution

Minke Whales The coordinates of 4777 caught and 6761 tagged Bryde's whales are included in the maps (Figs. 5.5 and 5.7a). Out of 112 returned tags the majority testifies to minke whales' moving along the coast of Antarctica (Fig. 5.7b), as basically whales were marked in the Southern Ocean waters during the summer period, and only a few tags were returned from minke whales marked in the tropics and subtropics.

The analysis of monthly distribution shows that not all minke whales migrate annually to the food-rich waters of the Southern Ocean. During the summer period minke whales can be found both in warm and cold waters. It is possible, though, that there is not one but several forms of minke whales, morphologically and genetically distinguished from one another (Arnold et al. 1987; Arnold 1997). Whales of this species migrate from the Balleny Islands area, and also from the Bellingshausen Sea and from the Cook Island area to the Ross Sea. Whales of the Brazilian aggregation migrate to the Weddell Sea.

Humpbacks Maps of humpback whale distribution are made on the basis of coordinates of places where they were caught (12,382 animals) and places where they were tagged (3996 animals). For the areas of Australia and adjacent waters there are data on their distribution year-round as humpbacks were taken there by coastal

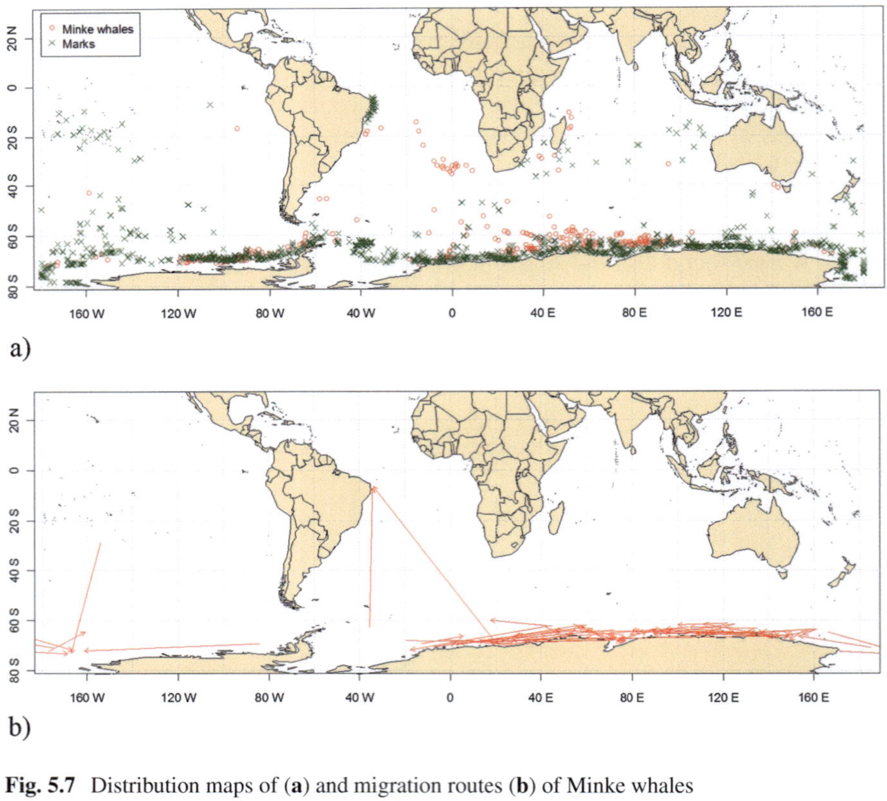

Fig. 5.7 Distribution maps of (**a**) and migration routes (**b**) of Minke whales

stations all year. From June till October they were tagged off the west and east coasts of Australia, the coast of New Zealand and in the waters to the south of New Caledonia. In the Atlantic, gatherings of humpbacks were recorded off the coast of Brazil, near Tristan da Cunha and off the southwest coast of Africa; in the Indian Ocean – in шеы north-west part and to the south of Madagascar. In the waters of the Southern Ocean humpback herds occur to the east of Gough Island, to the east of the Commonwealth Bay, near the Balleny Islands; to the north of the Ross Sea and in the Bellingshausen Sea (Fig. 5.8a).

The analysis of monthly distribution has shown that humpbacks migrate from the waters of Brazil to the Bellingshausen Sea and to the west of it to feed. West African whales feed in the area near Gough Island and to the south. West Australian herds migrate to Commonwealth Bay. East Australian herds of humpbacks basically migrate to the Balleny Islands and also to Commonwealth Bay and to the Ross Sea. New Caledonian herds fatten in the waters near the Balleny Islands and up to the Bellingshausen Sea. Active migration of humpbacks to their reproduction areas is observed in April, but some whales linger in the high latitudes (Balleny Islands and the Bellingshausen Sea). In May whales were observed in the Cook Strait and near the southern extremity of Africa (180 marks recovered, Fig. 5.8b).

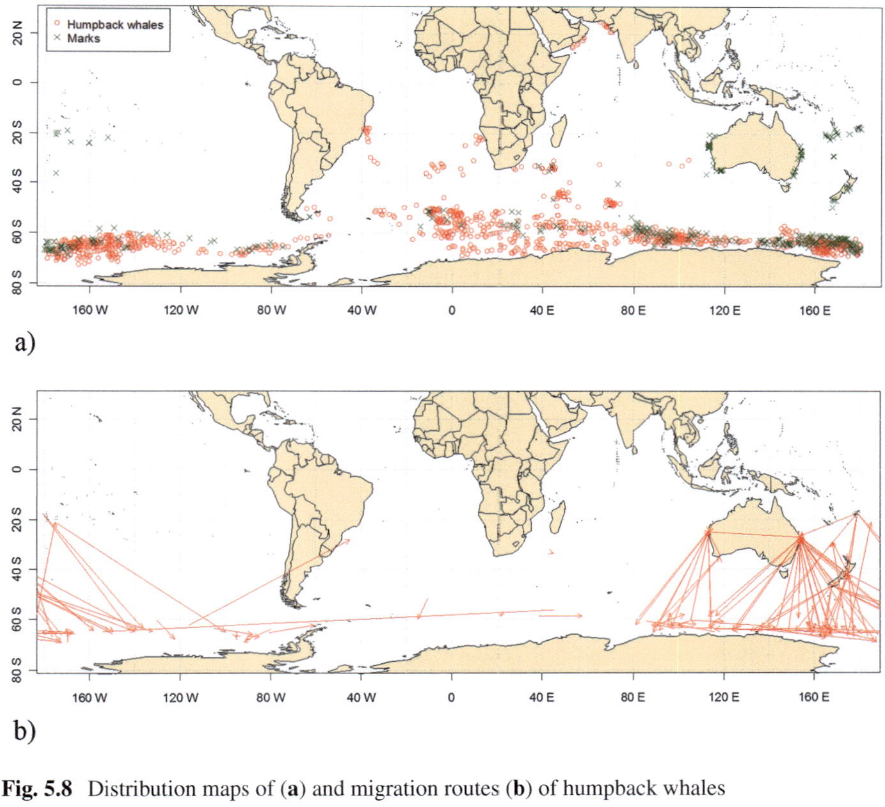

Fig. 5.8 Distribution maps of (**a**) and migration routes (**b**) of humpback whales

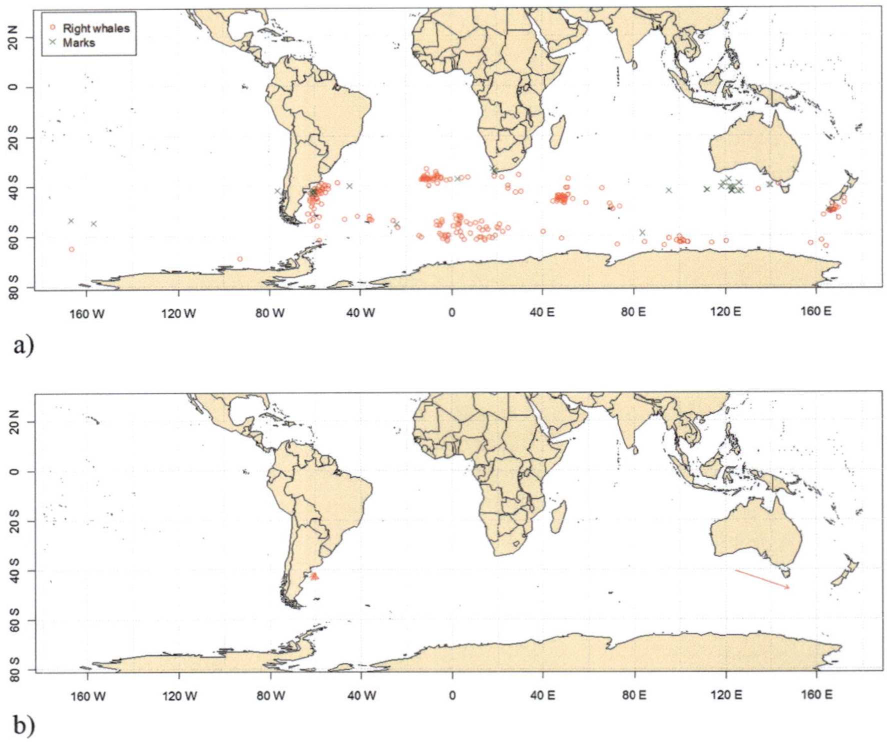

Fig. 5.9 Maps of distribution (**a**) and migration routes (**b**) of right whales

Right Whales Distribution of southern right whales of this species is shown on the basis of coordinates of 2774 caught animals (Fig. 5.9a) and the tagging locations of 179 animals (7 marks recovered, Fig. 5.9b).

The following areas of aggregation come to light: the Argentine coast; Tristan da Cunha; southward of Madagascar; the Great Australian Bite and Tasmania; the waters of Auckland, Campbell and Macquarie islands. In October and November an aggregation appears near the Chilean coast. In August whales were observed and registered off the southern extremity of Africa.

From seven tags, which were retrieved, six were found in the same area where they had been tagged – off the Argentine coast. One tag was found in the whale taken to the south of Tasmania on March, 27, 1970 on 48 °S.lat and 146 °E.long. The whale was marked in the Great Australian Bite on November, 26, 1969 on 41 °S and 122 °E. The analysis of monthly distribution shows that southern right whales from the Tristan da Cunha area migrate to the waters of Gough Island and further to the southeast. Some of the southern right whales from the Argentine aggregation stay in this area during the summer period too. The rest migrate to the Falkland Islands and further on to the southeast to South Georgia.

Two sperm whale males in waters of the Southern Ocean

Sperm whale male on the desk

5.2 Distribution and Migrations of Toothed Whales

Among the toothed whales only the sperm whales had trade value in the Southern hemisphere. In the recent years some attempts to catch killer whales were also made, but only several hundred animals were taken. More often they were "bycatch". We shall consider some features of the distribution of these two toothed whale species.

Sperm Whales Maps of sperm whale distribution have been compiled on the basis of an enormous amount of material – 34,166 caught and 4590 tagged animals (recovered 62 marks) (Fig. 5.10a, b). For some areas the material covers the whole year.

Sperm whales are found in all zones of the Southern Hemisphere: tropical, subtropical, subantarctic and the Soutern Ocean. Females were mostly extracted in the thirtieth and fortieth latitudes, only occasionally in the fiftieth latitudes; large males – in the sixtieth latitudes. The distribution of males in the sixtieth latitudes is closely connected to the character of distribution of their basic source of food – primarily various species of squid and fish (Berzin 1971; Yukhov 1971; Mikhalev et al. 1981).

During the southern winter and early spring (from June till September) sperm whales were found off the coast of Ecuador and Peru, the southeast extremity of Africa, and off the west and southwest coasts of Australia. In October sperm whales

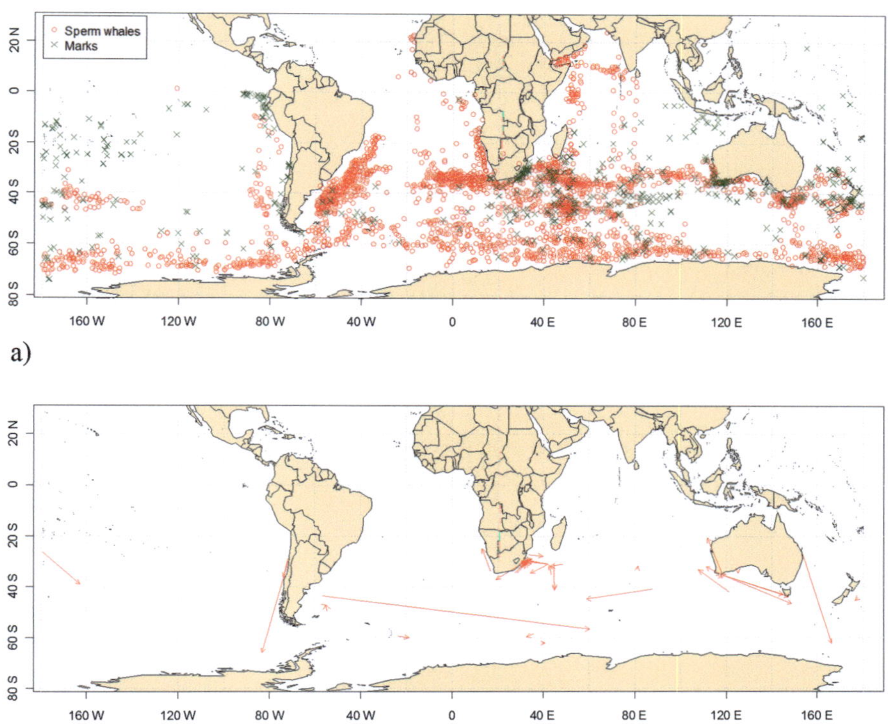

Fig. 5.10 Maps of distribution (**a**) and migration routes (**b**) of sperm whales

were taken near the coast of Africa to the east of Cape Verde; in the Gulf of Guinea; to the north of Tristan da Cunha; in the north-west part of the Indian Ocean; along all the western coast of Australia up to Java In November and December sperm whales migrate to the thirtieth latitudes, and some males even penetrate into waters of the Southern Ocean. In January and February their migration to the south comes to an end and the northward migration begins.

As to the migration routes (Fig. 5.10b), judging from the return of 62 tags, sperm whales of the Chilean pod migrate to the Bellingshausen Sea. One sperm whale from the Argentine pod was caught to the south of Crozet and Kerguelen islands. Sperm whales (mostly females) from the waters of the western coast of Australia displace to Tasmania and males of the east coast – to the Balleny Islands.

Killer Whales As it was already mentioned above, the active whaling of killer whales was not conducted, and data on their whaling is small. They were caught only occasionally, simultaneously with the main whaling. Therefore, to date, killer whales are relatively poorly studied animals. Some of the information on their distribution, size and features of biology are given by Tomilin (1957), Sleptsov (1965a), Shevchenko (1975), Budylenko (1978), Doroshenko (1978). We have processed the data on 323 killer whales caught by Odessa whaling fleet "Slava" and "Soviet Ukraine." As well as the results of observations by the scientific search whalers of these fleets (Fig. 5.11).

Analysis of the distribution of killer whales showed (Mikhalev et al. 1981) that in November, they were more common along the coast of Brazil and Argentina, in the area of Islands Three da Cunha and Crozet (see Fig. 5.12).

In December killer whales were more common by the islands of Saint-Paul and Amsterdam, as well as observed off the coast of New Zealand and Tasmania. In December part of killer whales penetrates into higher latitudes (see Fig. 5.13).

In January, the main part of killer whales migrates to the Southern Ocean. The greatest concentrations are noted in the Commonwealth Sea, Ross and Bellingshausen (see Fig. 5.14).

In February, killer whales are constantly found in the waters of the Balleny Islands to the Antarctic Peninsula (see Fig. 5.15).

In March (see Fig. 5.16) killer whales migrate north in the 30th and 40th latitudes, where they are also found in April and May (see Fig. 5.17).

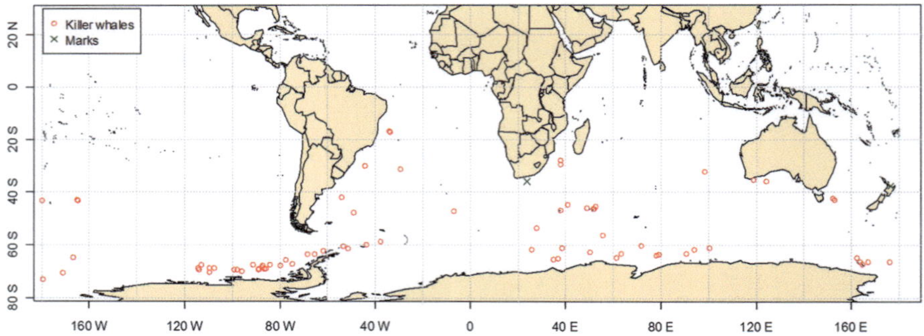

Fig. 5.11 Map of the distribution of killer whales in the Southern Hemisphere (237 caught, 6 tagged)

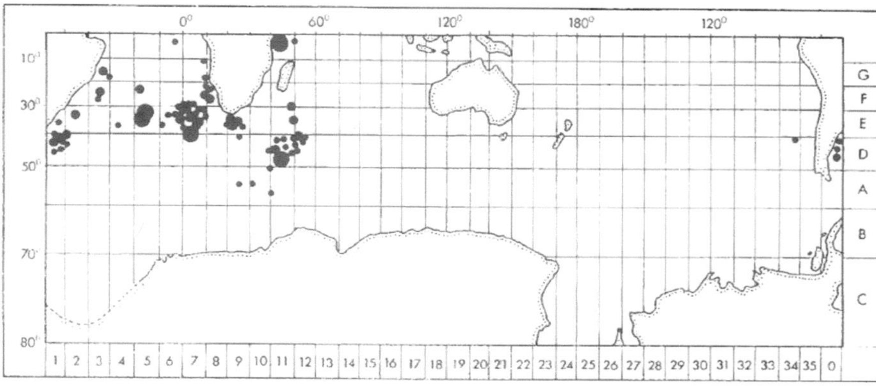

Fig. 5.12 Distribution of *killer* whales in November

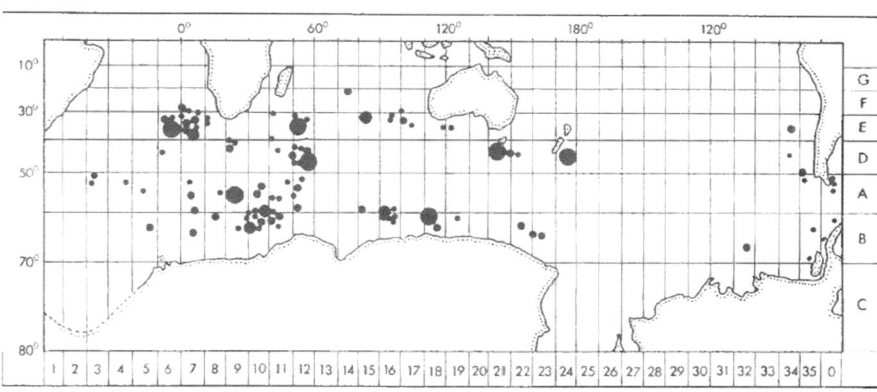

Fig. 5.13 Distribution of killer whales in December

Of 480 registered groups of killer whales (about 3000 individuals), the most numerous clusters marked in the region of the Rio Grande, on Ob' tablemount, in Prydz Bay and the Ross Sea (Mikhalev 1978). Analysis of the stomach of 231 caught killer whales showed that in waters of the Southenr Ocean the basis of their food (54%) is Minke whales. North of 50S Minke whales serve them as food only in 30% of cases. The rest of the food consists of small toothed whales, pinnipeds and fish.

5.3 Characteristics of the Migration of Different Groups of Whales

The research has been carried out on the basis of the analysis of the amount of coating by diatoms, which in the waters of the Southern Ocean is species *Cocconeis ceticola*. Whales that were coated in diatoms to a large extent were seen in the

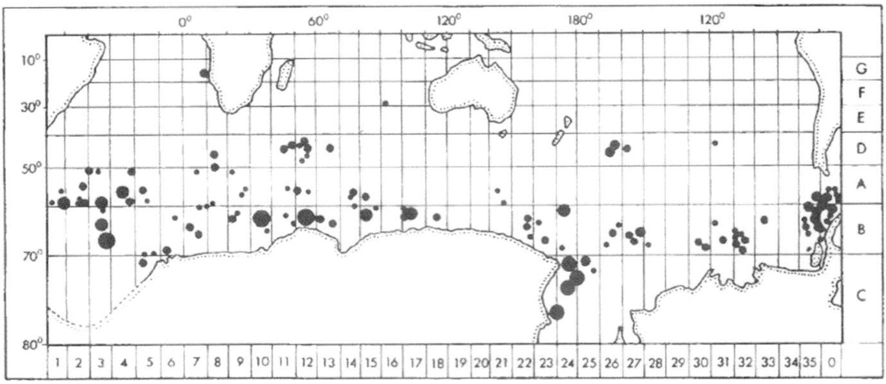

Fig. 5.14 Distribution of killer whales in January

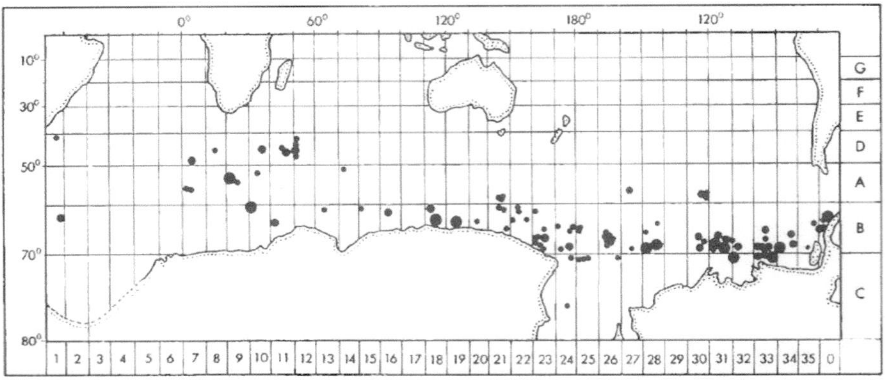

Fig. 5.15 Distribution of killer whales in February

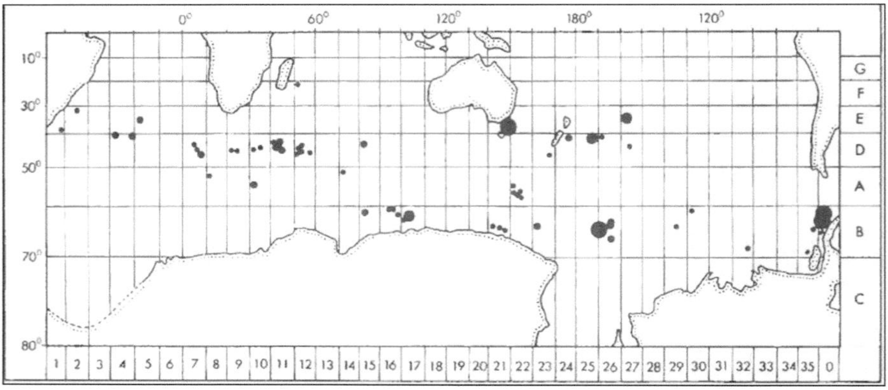

Fig. 5.16 Distribution of killer whales in March

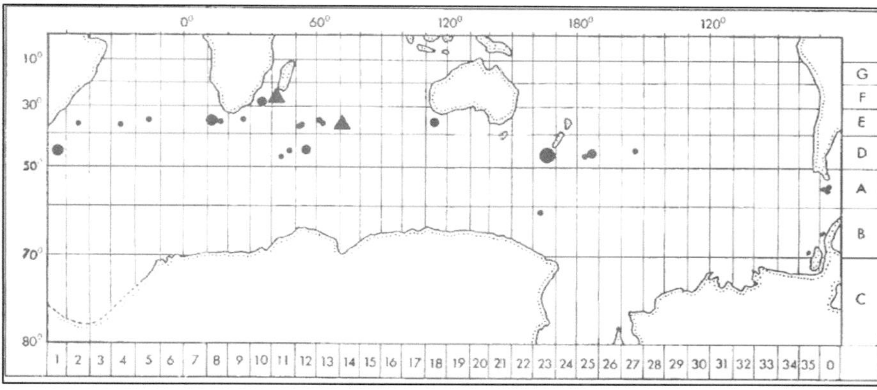

Fig. 5.17 Distribution of killer whales in April and May

Southern Ocean in February, March and April (Mackintosh and Wheeler 1929; Hinton 1925; Ivashin 1965; Bennett 1932; Hart 1935; Matthews 1937). There is evidence of such whales in the waters of low latitudes (Mackintosh and Wheeler 1929; Best 1969). But diatoms usually die off when the whales leave the Southern Ocean and reach warm waters.

T. Hart (Hart 1935) assumed without any argument or proof, that whales are coated quickly and that the layer of diatoms is appreciable a month after the whales arrive in high latitudes. Subsequent researchers haven't set criteria for the establishment of the speed at which coating occurs.

When we established the precise correlation between the degree of female whale coating and the average sizes of the fetuses found in them, the answer to that question became clear. That is, embryo growth rate has been used as an original measure of time, or as a biological clock (Mikhalev 1984; Zinchenko 1986; Zinchenko and Mikhalev 1986). We shall use minke whales to illustrate this method.

Only pregnant females have been included in the analysis for the development of the method. The following groups can be selected by the degree of coating with unicellular algae: not coated; minimally coated; medium coating; thickly coated. During the whole hunting season, the average embryo length of uncoated females equaled 57 cm, of minimally coated ones – 90 cm, medium coating – 120 cm. That is, the more coated with diatoms, on average the larger the embryos were, and they were larger by approximately 30 cm.

As the monthly average gain of minke whale embryos during the spring-and-summer period is approximately 22 cm, the first degree of whale body coating comes in a month with a half, and they reach "medium coating" in 3 months. Thickly coated whales by our calculations have spent 4.5–5.0 months in Antarctic waters. Thus, the whale coating rate appears to be considerably slower than that assumed by T. Hart (Hart 1935).

The analysis of the degree of diatomic coating of various biological animal groups (males, dry and pregnant females, immature animals, senior age groups) has shown that the migration of minke whales to Antarctic waters and back has a complex character. It appeared that after a mass migration of whales into the Antarctic region during the entire spring-and-summer period (October–January) new small groups, distinguished by the absence of a diatom layer on their body, proceed to approach the high latitudes. The first males arrive in the feeding areas as early as October. Then dry females, pregnant and immature ones migrate to the Antarctic region. In February–April some of the already fattened whales migrate to lower latitudes. As to the mass northward migration, the first to leave the feeding areas are large pregnant females, then males and dry females. Some minke whales stay in the Antarctic region for the winter period.

Coating on lower jaw of sperm whale

In generalizing about the distribution and migration of different whale species of the Southern Ocean and adjacent waters it is necessary to say that there are zones or centers of greatest concentration. They are the east and west coasts of South America; a zone between the southern extremity of Africa and Tristan da Cunha; the bank of Walfish Bay (south of Madagascar) and the subantarctic island zone of the Indian Ocean; the southeast and southwest coasts of mainland Australia; Tasmanian and New Zealand waters; and the northwest part of the Indian ocean. During migrations whales concentrate in the zones of the Southern Ocean, due to their high productivity.

On the other hand, there are "dead zones", in which whales are practically never seen. Areas of mating, calving and the feeding of suckers coincide more often, forming reproductive zones. Basically they settle down in the thirtieth – fortieth latitudes, and some whales breed in Antarctic waters as well. It is necessary to take these characteristics of distribution into account when monitoring and making recommendations for periods of interdiction and recommencement of catch for various whale populations.

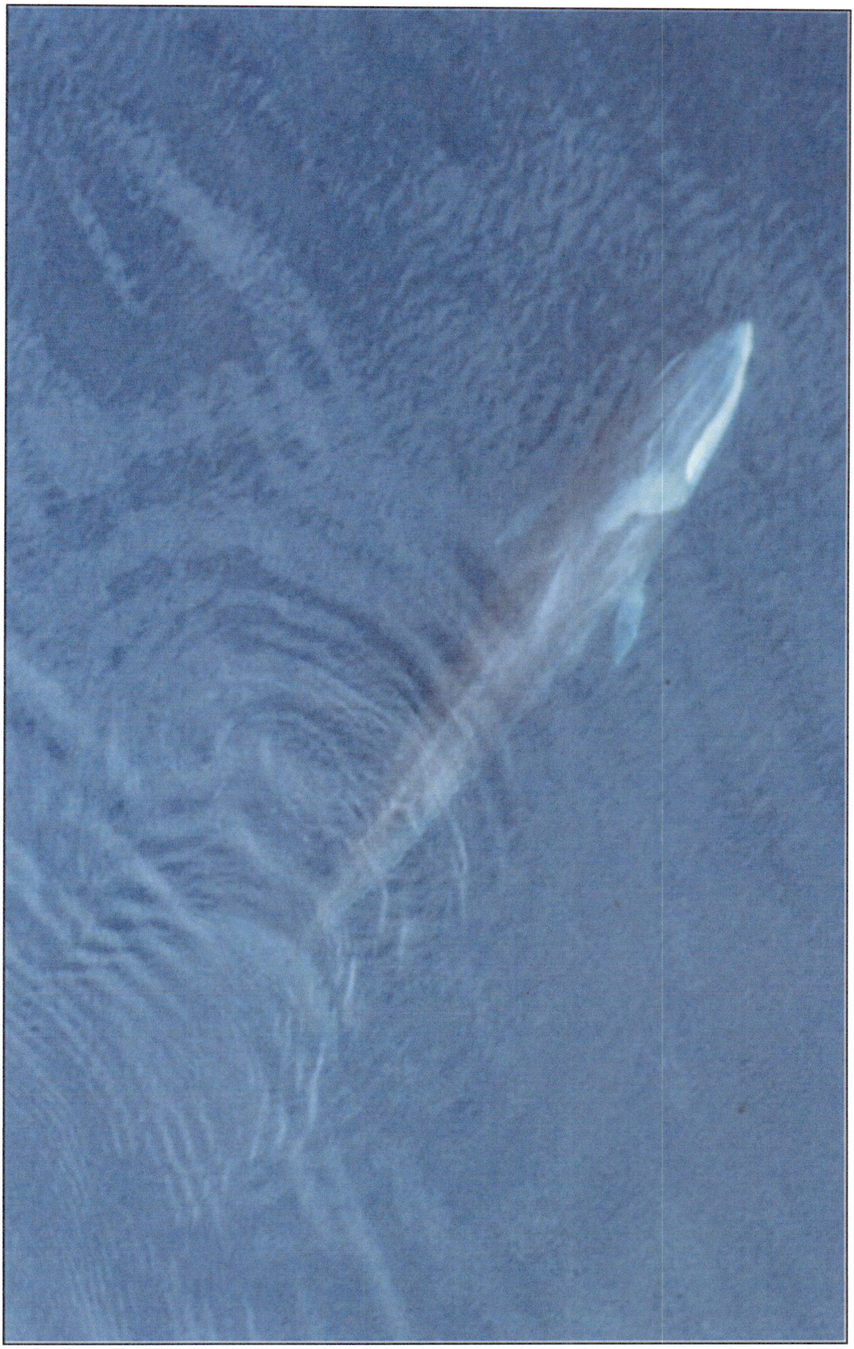

Fin whale under water

Fin whale on the desk

5.4 Determining Differences Between Whale Populations

For a long time all whale types have been considered as cosmopolitans distributed in both hemispheres. Determination of distinctions in average size and in proportions of body parts of whales of the Northern and Southern hemisphere, discrepancy between biological cycles, as well as the results of tagging gave ground to consider them as independent subspecies (Tomilin 1953, 1957; Kleinenberg, etc. 1964;, etc.) or even species (Zemsky 1972; Zemsky and Budylenko 1995). Within the subspecies (species) "local herds" adhering to the same areas, with their own migration routes and areas of reproduction (Klumov 1955) have been distinguished. These herds were not identified as populations, because at that time morphological distinctions between these whale herds had not been revealed. Moreover, it was stated that there could not have been any distinctions "as in the ocean there are no obstacles to isolate herds" and, consequently, whale species represent homogeneous uniform herds (Zenkovich 1966).

Further research has contradicted this opinion. Thus, on the basis of more than 20 various distinctive metric and non-metric attributes of sperm whales (coloring characteristic of the body, the form of the fork of tail blades, number of lumbar and tail vertebrae, fastening of sternal ribs, the shape of the spleen, liver, the "marks" and the grain of the tooth osteodentin, etc.), population distinctions have been revealed for the whales of both Northern and Southern hemispheres (Klevezal and Tormosov 1969; Veynger 1974; Berzin et al. 1975). Distinctions amongst killer whales have also been revealed (Evans et al. 1982).

Similar methods were applied to distinguish between baleen whale herds. Body coloring, fin form, the form of "pair formations", average animal sizes, a series of standard body measurements, body weight, amount of vibrissa on the jaws, the number of baleen plates, number of belly strips, "white" scars on the body, etc. (Yablokov, 1966 were used; Shevchenko 1971, 1975, 1977; Mikhalev et al. 1975; Bushuev 1986). Some positive results have been recorded on the basis of acoustic and genetic research (Leatherwood, Donovan/Ed./, 1991).

In our search for distinctions the following attributes – or phenes – have been used: the form and position of "pair formations" on the ends of baleen whale top jaws, the degree of integument injury, the structure of polished sections of sperm whale teeth, average embryo size, features of whale body coloring. Such biological characteristics as the age structure of whale gatherings, rate of accumulation of ovulation traces on female ovaries, the degree of coating with diatoms and other parameters have been also used.

5.4.1 Distinction of Whale Herds by the Form and Position of "Pair Formations"

Studying of "pair formations" on the top jaw of baleen whales has shown that different species of whale have various pair formations These distinctions are so essential that they can even be used as a species characteristic. Minke whales are easily distinguished by phene characteristics as early as the pre-fetus stage of development when it is still difficult to distinguish them by other features.

For example, fin whale crescent-shaped recesses are inverted to each other by the convex side, and that of sei whales, on the contrary, the concave side (Fig. 5.18). In minke whales, unlike fin whales, the cranial ends of crescent recesses are at a greater distance from each other, than caudal ones.

Within the same species the form and position of "pair formations" (especially on the top jaw) very much vary both in size and appearance. They are more or less flat; either with or without a small deeper dimple inside, the arrangement of this dimple also varying; pigmented to a greater or lesser degree; and are either closer to baleen line or to the rostrum end. We tried to use these distinctions as attributes – phenes – for defining populations of baleen whales of the Southern Ocean and adjacent waters.

We measured the distance from the rostrum tip to the baleen line (L) and also the position of "pair formations" in ratio to the beginning of baleen line (l_1) and to the rostrum end (l_2) in more than 5000 whales.

The average l_1 of fin whales appeared to equal 5.5 cm, and l_2–12.0 cm; $l_2/l_1 = 2.2$. l_1 of sei whales = 5.0 cm, and l_2–9.5 cm; $l_2/l_1 = 1.9$. In females of different whale species of identical length, these measurements and their sum, i.e., the distance from the end of the snout to baleen line L, were greater than those of males. Therefore, further analysis was made for each sex separately. During January and February dense gatherings of fin whales, sei whales and minke whales were investigated in the areas of the Chilean coast, Bellingshausen Sea, Bouvet and Gough islands, and the islands of Prince Edward, Crozet and Kerguelen.

Fin Whales Herds of fin whales within the limits of each ocean showed no distinctions from one group to another. At the same time, fin whales from the Bellingshausen

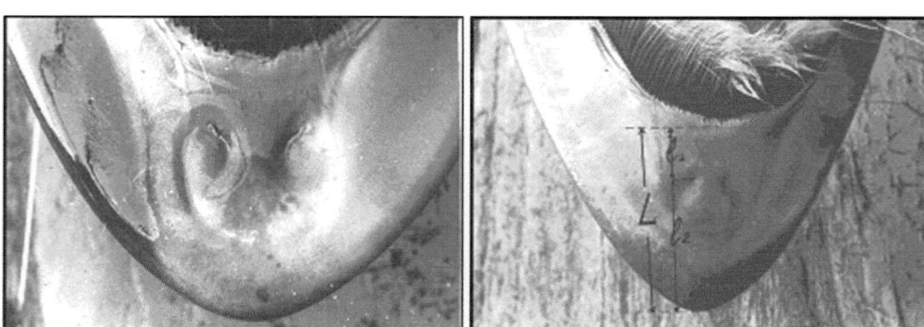

Fig. 5.18 "Pair formations" on upper jaws of fin whale (left) and sei whale (right)

Sea differ in these attributes from whales of Bouvet and Gough islands and area of the islands of Prince Edward, Crozet and Kerguelen. For both males and females, distinctions between the Atlantic and Indian Ocean fin whale herds (Fig. 5.19) are also appreciable.

Sei Whales As with fin whales, herds of sei whales from the Bouvet and Gough island area (Atlantic sector of the Southern Ocean) showed precise differences according to this attribute from both the geographically closely located herds of the Indian sector of the Southern Ocean, and the Chilean aggregation of sei whales (Fig. 5.20). The established distinctions are also confirmed by the research of "white scars" on whale bodies from small pelagic shark bites (Shevchenko 1977). No distinctions between the sei whale gatherings of either of the oceans have been established.

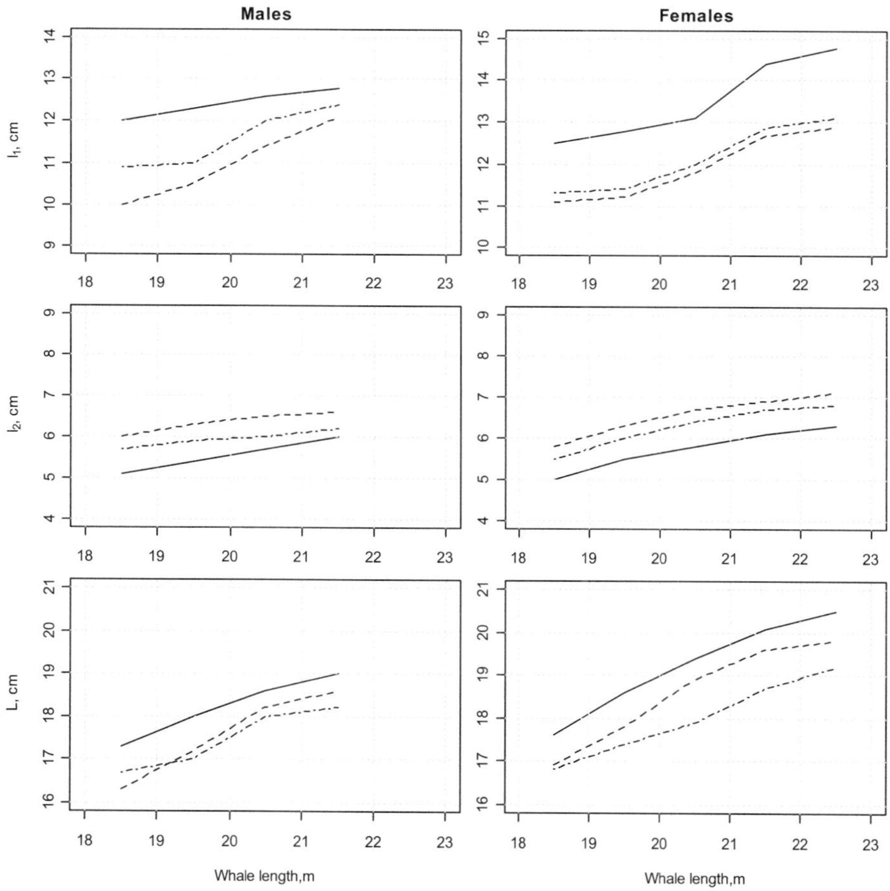

Fig. 5.19 Differences between finwhale herds based on position of "pair formations": Solid line – Peter the First Island region. Dot-dashed line – islands of Prince Edward, Crozet and Kerguelen. Dashed line – Bouvet and Gough islands

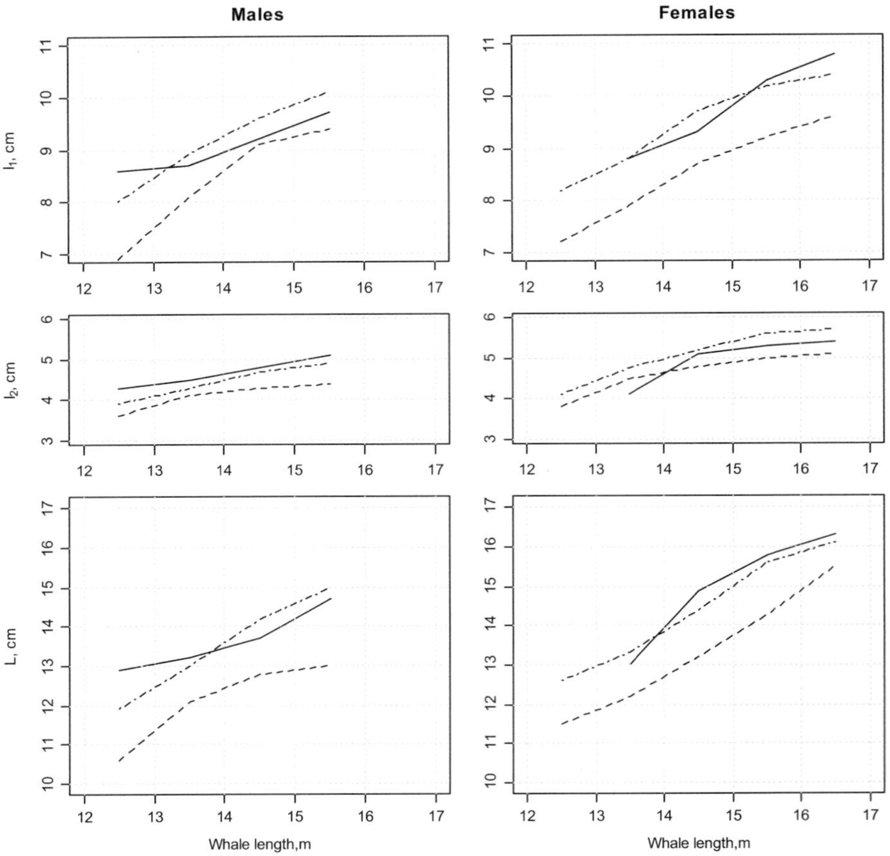

Fig. 5.20 Differences between seiwhale herds based on position of "pair formations": Solid line – region off Chili and Antarctic Peninsula. dot-dashed line – islands of Prince Edward, Crozet and Kerguelen. dashed line – Gough islands

Minke Whales The material from two areas has been analyzed: the Bellingshausen and Weddell seas. Precise distinctions using the measurement l_2 (Fig. 5.21) were observed between minke whale gatherings of these seas for both males and females.

5.4.2 *Distinctions of Herds of Whales by the Form and Degree of Whale Cover Injuries*

The possibility of using traumatic variability of cetacean integument in population research was specified by A.V. Yablokov (1966). The most positive results were obtained when fresh oval ulcers on the whale body – so-called "white scars" – were analyzed.

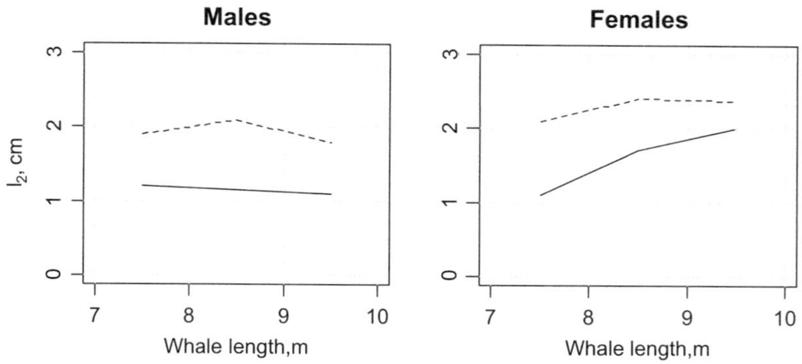

Fig. 5.21 Differences between Minke whale herds based on position of "pair formations"
solid line – Weddell Sea
dashed line – Bellingshausen Sea

D. Lillie (Lillie 1915) observed that "white scars" are formed in warm waters and consequently their presence on whales in waters of the Southern Ocean proves their migration from warm to cold waters. Mackintosh and Wheeler, 1929, noted the fact that the number of white scars increases. On this basis they concluded that whales regularly migrate from cold to warm waters and back. H. Omura, 1950, attempted to connect the quantity of white scars with the existence of various sei whale herds, migrating by various routes. Probably due to the fact that in those days there was no distinction between sei whales and Bryde's whales, no authentic results were recorded. M. Ohno and K. Fujino (Ohno and Fujino 1952) also tried to use the presence of ulcers on whales' bodies when studying the laws of migration of whales to waters of the Southern Ocean.

An employee of our laboratory, V.I. Shevchenko (1970, 1971, 1975), was the first to show that "white scars" from small thermophilic shark bites remain on a whale's body. Jones, 1971, studying spherical pieces of stomach tissue of cookiecutter shark *Isistius brasiliensis* came to the same conclusion in a different way.

In one of his subsequent works, V.I. Shevchenko (Shevchenko 1977) analyzed the frequency of occurrence of fresh ulcers and old scars on the bodies of 890 sei whale bodies and established that the sei whales from the waters of Crozet, Kerguelen and Prince Edward islands share this attribute. Whales have few "white scars", fresh ulcers are noticed on few animals. On the contrary, there are a lot of both old "white scars" and fresh ulcers, especially on the caudal peduncle, on sei whales of the southeast Atlantic in the area of Gough Island,. On the basis of these distinctions, southeast Atlantic sei whales and sei whales of the "island" zone of the Indian Ocean were divided by Shevchenko into separate populations.

We continued this research and analysis of the occurrence of scars and fresh ulcers from the bites of small pelagic sharks on the bodies of 570 pygmy blue whales, 324 fin whales, 50 Bryde's whales, 30 humpbacks and about 1000 sperm whales (Mikhalev 1995, 1997). The quantity of "white scars" was estimated by a five-point scale: “very few”, “few”, “average “, “many “, “very many”. The following results were recorded.

Pygmy Blue Whales Ichichara, 1963, noted that there are more “white scars” on pygmy blue whale bodies than on real blue whales of the same length. It is clear that, as pygmy blue whales of the same body length are older than real blue whales, hence, pygmy blue whales were exposed to the attack of small sharks for a longer period. Additionally, pygmy blue whales are more thermophilic, and the sharks that attack them are found in the subtropical waters of the southern hemisphere, and in this zone fresh injuries on the bodies of whales were found especially frequently. For example, in the Amsterdam-Australian congestion at the end of December fresh ulcers were registered on almost every pygmy blue whale that was caught.

As opposed to this area, fresh injuries and cicatrized ulcers on the body of pygmy blue whales of the northwest part of the Indian Ocean were less frequently found. There was the cicatrized trace from a shark bite on the tail fin of a 20.8 m-long female pygmy blue whale of the Aden-Oman aggregation. Some ulcers were marked on the tail stalk of a 19.7 m-long male from the Pakistani-Indian aggregation.

Ulcers on pygmy blue whales of the equatorial-Seychelles and Laccadive – Maldives gatherings were seen a little bit more often, not only on large animals, but also on yearlings 16.5 m long, but still much more rarely than on pygmy blue whales of the Amsterdam-Australian aggregation, with whom authentic distinction has been established by this attribute.

Fin Whales The presence of scars and fresh ulcers was observed on fin whales from the areas of: Crozet, Kerguelen, Prince Edward, Gough and Bouvet islands; Enderby Land; Scotia sea; coast of Chile; Bellingshausen Sea; the Balleny Islands; southwestern Australia and South Georgia. Authentic distinctions have been established for five groups: the Chile-Peruvian coast (on average 7.8% fresh wounds; “few” scars – 78%); South Georgia area (on average 9.1% fresh wounds; “few” scars – 48.3%, “moderate” – 41.4%); Gough Island (30.1% fresh wounds; “moderate” – 9.1% (please check this entry); “many” – 30.2%); islands of Prince Edward, Crozet and Kerguelen (5–7% fresh wounds; very few old scars); Balleny island area (33% fresh wounds; “very many” scars 51.6%).

Sei Whales Traumatism of integument of sei whales was noted from groups of the Indian ocean “island zone”; waters to the east of Gough island; Scotia sea; coast of Chile; Bellingshausen Sea; southwestern Australia; Tasmanian waters; and the Balleny islands. It appeared that Atlantic sei whales (Gough Island) have many more “white scars” (“many” – 75.6%), than sei whales of Crozet, Kerguelen and Prince Edward islands (“many” – 3.2%). In comparison with sei whales from the waters of Tasmania (“many” – 60.0%) southwestern Australian sei whales are less injured (“many” – 4.2%), and the same applies to sei whales of theBalleny Islands.

It is possible to assume that sei whales from Tasmanian waters migrate to the Balleny Islands, as they have similar parameters for the traumatism of integuments.

Similarly, there are grounds to believe that Chilean coast sei whales migrate to the area of Graham Land by the middle of summer. In favor of this assumption, we note the fact that *entozoon Lecithodesmis* was not found in the livers of Drake Passage sei whales, by which about 30% Falkland Islands sei whales have been struck. The hypodermic fat of both Chilean and Graham Land sei whales has bright, citric coloring, which was not the case for sei whales from other areas.

Bryde's Whales In our practice, Bryde's whales were extracted from large groups in only two regions: the southern extremity of Madagascar and in the Gulf of Aden. V.I. Shevchenko (1971) specifies that in the Arabian Sea in 1965 there were scars on the bodies of three Bryde's whales (ostensibly taken for scientific purposes), but not fresh ulcers. Actually in the 1964/65 and 1965/66 voyages not three, but 868 Bryde's whales were taken from the Arabian Sea. We examined more than half of them. There were few scars on their bodies and no fresh ulcers. The whales near the Madagascar coast are usually considerably injured, and traces of fresh stings are frequently found.

Minke Whales Earlier it was considered (Tomilin 1957) that "white scars" are absent from the bodies of Southern hemisphere minke whales. The Antarctic catch has shown that this opinion is erroneous. We examined 395 Minke whales from the area of Prydz Bay (Commonwealth Bay) and Peter I Island (the Bellingshausen sea) and observed both old scars forming a "marble" pattern on the skin and fresh ulcers. Scars are typically oval in shape. The size of most fresh ulcers is approximately 2.5 × 4.5 cm. There is a cleft in the center of the ulcer.

In spite of the rather small sample, it is nevertheless certain that minke whales from the Bellingshausen Sea, in comparison with those from Commonwealth Bay, have many more oval-shaped fresh wounds, and old scars. Research by S.G. Bushuev (1989) confirms this conclusion about minke whales from these areas.

Sperm Whales Fresh ulcers on sperm whale bodies are quickly pigmented while healing and, consequently, are less appreciable than those on baleen whales. Additionally, sperm whale dermis is thicker and stronger. Small sharks leave a bite mark, rather than biting the integument off. Besides, it is more difficult to distinguish the traumas inflicted by sharks from those made by the suction cups and claws of squid, or those caused by other means.

Nevertheless, a distinctive difference in integument damage has been established between the east Atlantic sperm whale (82% of injuries) and those at Enderby Land (11% injuries) The border between these populations lies at approximately 20°–22°E. The sperm whales of the north-west part of the Indian Ocean differ greatly from other populations. Neither fresh injuries nor cicatrized ulcers were revealed on their bodies.

Thus, our research has confirmed the conclusion made by V.I. Shevchenko (1971, 1975, 1977): that Southern hemisphere whales are generally injured by small sharks in the subtropical zone. In November–December the southern border of their area in the southwest part of the Atlantic ocean is 23°–25°S, and in the southeast, 34°–35°S. In the area of the Tasman sea, the border of occurrence of these sharks is 38°–40°S. The southern border of their area in the Indian ocean is more displaced to the north, at 19°–20°S.

5.4.3 Distinguishing Between Whale Pods by the Average Embryo Length

Distinctions between the average embryo length of different fin and minke whale groups of the Southern Ocean have been observed by a number of researchers (Mikhalev 1984; Mikhalev and Ivashin 1978; Ivashin and Mikhalev 1978; Zemsky 1960). However, K. Ash (1953) and M.V. Ivashin (1976) stipulated that the reliability of differences revealed by this attribute is low.

We processed data, including 1480 minke whale embryos from various groups, in the feeding grounds in high latitudes of the Antarctic Region. In reproductive zones of this kind, almost no minke whales were taken, and reliable data was not able to be collected. For greater comparability of the average embryo sizes a small interval of time – not more than half a month – is taken.

The analysis shows that average embryo lengths in Mawson Sea and nearby Commonwealth Bay essentially differed. For whale groups more remotely located from each other (the Lazarev Sea and the Amundsen Sea), the difference in average embryo lengths was insignificant. Further, more remote gatherings of minke whales from Commonwealth Bay and the Amundsen Sea have shown a sharp difference in this attribute. It is necessary to note, however, that average embryo sizes for various trade seasons were compared, which reduces the correctness of research, and the results can be accepted only as approximate ones.

5.4.4 Distinctions of Sperm Whale Herds by the Structure of Teeth

As we have already specified, lamination of sperm whale teeth is complex. Some teeth include grains of osteodentin. We attempted to use features of teeth structure for revealing distinctions between sperm whale herds. One thousand and seventy-nine sperm whale teeth from 11 dense groups, rather isolated from one another, were selected. The nature of stratifications in dentin was estimated by profilograms of decalcified polished sections of teeth. The number of osteodentine grains was also compared.

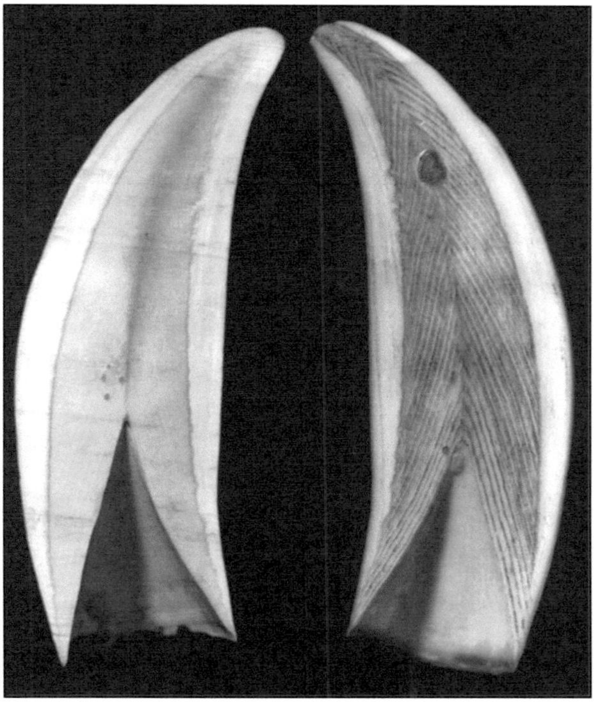

Two sections of sperm whale teeth: left – original, right – decalcified

For better comparability, the comparison was made between animals of uniform size. The analysis shows 41.1% of teeth with osteodentine grains in sperm whales of the Chilean coast; 26.6% – in sperm whales from the aggregation to the south of Tristan-da-Cunha; 38.1% in the sperm whale aggregation to the north of Crozet and Kerguelen islands. That is, some distinctions between sperm whales of the Atlantic, Indian and Pacific oceans are appreciable using this attribute. D.D. Tormosov (1982) compared not only quantity but the sizes and character of osteodentine grains distribution in sperm whale teeth. As a result, distinctions of sperm whale herds of these three zones have been revealed with a greater degree of reliability.

As opposed to G.A. Klevezal and D.D. Tormosov's (1971) technique, the nature of stratifications in sperm whale teeth was defined by us not directly by decalcified teeth polished sections, but by profilograms of them, and the degree of clearness of lamination was estimated not by five but by only two criteria – "precise" and "indistinct". We investigated 279 teeth from sperm whales taken in the same areas, and for which comparison of the number of osteodentine grains has been carried out. Indistinct clearness of stratifications in teeth dentine was observed in sperm whales from the Indian ocean in 61.9% of cases. In the Atlantic sperm whales – in 35.7% of cases; in the Pacific ocean ones – in 28.2% of cases.

It can be seen that sperm whales of all three oceans differ according to osteodentin grain analysis and the nature of lamination. The biggest differences are also appreciable in the sperm whales of the Chilean coast. G.A. Klevezal and D.D. Tormosov's (1971) research confirms that distinctions between the Atlantic and Indian ocean sperm whales are less defined, and so does G.A. Klevezal (1988), who took into account not only clearness, but also uniformity of stratifications. Analysis of the nature of stratification of each sperm whale group within each ocean has not yet revealed any authentic distinctions.

Cutting teeth from the lower jaw of sperm whale

Whaling boat model made of sperm whale teeth

5.4.5 Distinctions Between Killer Whales in the Southern Hemisphere

Killer whales as predators, undoubtedly, play an essential role in marine ecosystems the Southern Ocean and adjacent waters. They were not actively caught, but were occasionally taken, in passing with other whales. Therefore killer whales remain rather insufficiently studied amongst captured whales. Some data on their distribution, sizes and biological features are presented in a number of works (Tomilin 1957; Sleptsov 1965a; Shevchenko 1975; Doroshenko 1978; Budylenko 1978; etc.).

We (Mikhalev et al. 1981) collected data from 323 killer whales. The size of 205 males varied from 4.5 m up to 9.0 m, average length being 7.26 m; and the size of 118 females – from 3.7 m up to 7.7 m, with an average length of 6.42 m.

In this above-mentioned article I wrote a section titled "Reproduction", which included a map of the distribution of killer whales. Analysis of the weight of seminal glands shows that a dramatic acceleration of their growth occurs when the animal's body measures 7.1–7.5 m. However, an essential increase in the weight of seminal glands of smaller killer whales 6.1–6.5 m long has also been observed. A similar, but better defined law has been revealed while analyzing the weight of seminal glands. These observations allow an assumption of the existence of shorter body species among killer whales.

Our assumption has been also confirmed in the course of analyzing the numbers of pregnancy traces and ovulation on ovaries. Females 5.0–6.0 m long appeared to have a few more such traces than those whose sizes were longer by 1.5 m (Fig. 5.22).

The same pattern was observed in the relation graph of percent of pregnant females in different size classes for female killer whales (Fig. 5.22a).

As we can see, the number of pregnant females with length 5.5–6.5 m significantly higher than the number of pregnant animals with length 8.7 m (the larger and,

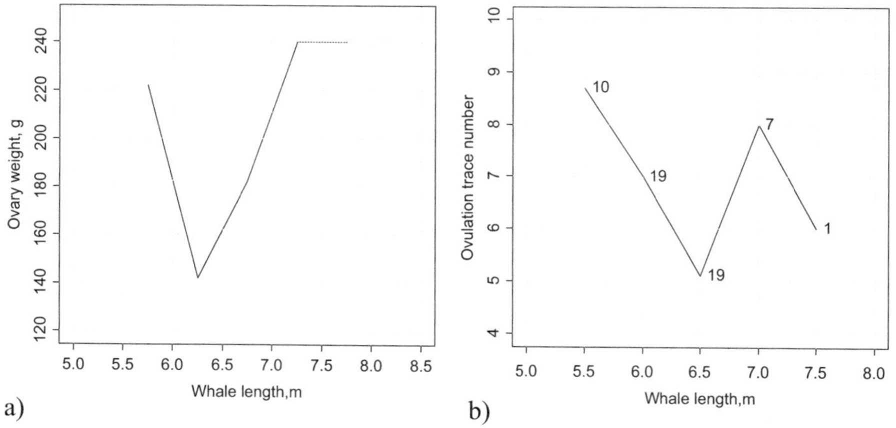

Fig. 5.22 Weight of ovaries (**a**) and number of pregnancy traces (**b**) depending on killer whale female length

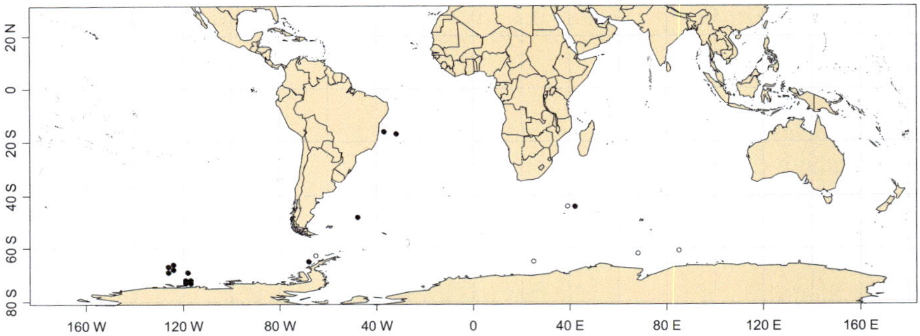

Fig. 5.23 Distribution of shorter-bodied killer whales

consequently, the older groups of females). This can only be explained if we assume that the killer whales of the Antarctic have not one but two forms (perhaps two kinds) of animals – regular and short-body ones.

Also of importance is the fact that in January–February shorter-bodied but mature killer whales were taken in the area of the Amundsen Sea during a number of different seasons. In spring and at the beginning of summer they were seen off the Argentine coast and in the Drake Passage near the coast of Graham Land (Fig. 5.23).

Measurements of 19 ordinary and five short-bodied killer whales have been analyzed to reveal distinctions in the proportion of body parts. The small sample does not yield sufficiently reliable results. Nevertheless, it is appreciable that the proportions of the head of short-bodied killer whales are bigger than those of ordinary killer whales, and the chest fins are larger The dorsal fin of short-bodied killer whales is also bigger and is positioned slightly closer to the head. The tail part is shorter, and tail blades are relatively larger (Table 5.1).

Distinctions in the average size of animals of about the same age, in body proportions, sizes at the onset of sexual maturity, dramatic distinctions in relative weight of sexual glands and in the numbers of traces of pregnancy and ovulation on ovaries (in females of the same length), as well as isolated habitats have allowed us to draw the conclusion that in the Southern Ocean and adjacent waters not one but two killer whale species exist: the ordinary – Orcinus orca, and a new kind – the short-bodied

Table 5.1 Standard sizes of killer whales (percentage of length)

Measurements	Ordinary killer whale	Short-bodied killer whale
From end of snout till blowhole	11.70 ± 0.23	12.40 ± 0.47
From end of snout till month corner	9.40 ± 0.26	9.67 ± 0.30
From end of snout till eye center	10.40 ± 0.26	10.94 ± 0.42
Back fin height	10.40 ± 0.34	10.84 ± 0.68
Back fin length	10.20 ± 0.35	11.34 ± 0.58
From end of snout till back fin	41.50 ± 0.90	37.99 ± 3.04
From tail blades split till back fin	49.50 ± 1.24	51.57 ± 3.89
Small length of chest fin (till antrum)	10.64 ± 0.31	11.64 ± 0.39
Large length of chest fin	7.90 ± 0.21	8.29 ± 0.26
Length of tail blade	12.60 ± 0.25	13.11 ± 0.60
Width of tail blade	8.00 ± 0.79	8.33 ± 0.43

killer whale (Orcinus nana, Mikhalev 1981). With a tinge of regret I should say here that the article, prepared by me with employees of the Odessa laboratory, was rpresented to the IWC Scientific committee by a VNIRO employee M.V. Ivashin, who without any correspondence with us, entered his name into the list of authors of the article (something he had done repeatedly). Even worse, he entered his name next to mine in the naming of a new species of killer whale, both in the section titled "Reproduction", written by me, and in the conclusions of the article. Such were the morals and customs in VNIRO and the Ministry of Fisheries of the USSR!

Further research (Berzin and Vladimirov 1982, 1983) confirmed the existence in the Antarctic region of killer whale variations, which differed in the form and size of teeth. Moreover, subsequently, R. Pitman and P. Ensor (Pitman and Ensor 2003), while investigating the form of a white oval stain on the killer whale body, that as many as three variations exist in the Antarctic Region. It is absolutely clear that the Southern Ocean and adjacent waters killer whales are still rather poorly investigated.

Killer whales in their glades

Killer whales on the desk

5.4.6 Pygmy Right Whales

Systematics A very small 5–6 meter baleen whale – *Caperea marginata* Gray, 1846 – has been studied less than other whales. Therefore, we should pay more attention to it. According to the structure of its skull and baleens, the diamond shape of its ear bones *Bulla tympani*, and a number of other features, this type of whale belongs to the family of right whales (*Balaenidae*), subfamily *Neobalaeninae*, monotypic genus *Caperea* Gray, 1846 – pygmy right whales (Beddard 1901; Tomilin 1957).

In the course of studying *Caperea marginata*, it was found that, unlike regular right whales, *Caperea* has a distinct dorsal fin, which among baleen whales characterizes rorquals, and the presence of two throat grooves, so they share some characteristics with gray whales. Similarities with the latter, indicated by molecular analysis, gives grounds to move *Caperea* into a separate family – *Neobalaena* (Sasaki et al. 2005; Deméré et al. 2008). However, by the sum of its main features, *Caperea* clearly belong to the right whale genus the overall shape of the body, arcuate curvature of the maxillary bones of the skull and the shape and structure of the baleen plates, are similar to those of regular right whales. In our opinion, the selection of pygmy right whales into a separate family, *Neobalaena*, does not seem sufficiently convincing. If it were the case, with no less justification, the sum of the distinguishing features of humpbacks could justify allocating them to a separate family, and undersized minke whales to a separate genus. Therefore, according to the present state of knowledge, in our view, it is enough to allocate *Caperea* into the family of right whales (*Balaenidae*), subfamily *Neobalaeninae*, or even just a separate genus *Caperea*, as defined in the taxonomy of cetaceans by Yablokov A.V. (Yablokov et al. 1972).

Distribution In the oceans, organisms (plants or animals) usually have bipolar distribution (Berg 1920), that is, they can be found in both the Northern and Southern hemispheres. Many cetacean species have bipolar distribution. But *Caperea* is only found in the warm waters of the Southern hemisphere. There was no commercial whaling for *Caperea*. Objects of study were usually only individuals accidentally caught by nets in coastal waters, or "desiccated" beached whales (Fig. 5.24).

Caperea were most often observed near the coasts of Africa, Australia, Tasmania, New Zealand and South America (Fig. 5.25). However, there were observations of *Caperea* around the Falkland Islands, Crozet Islands and Stewart Islands (Gray 1846; Kellog 1928; Davies and Guiler 1957; Guiler 1961; Hale 1964; Ross et al. 1975; Kemper 2002a, b, 2009; Matsuoka et al. 2005, etc.)

Another area of *Caperea* habitat was identified by Soviet whalers – waters east of Tristan da Cunha. This is notable for the fact that *Caperea* were rarely recorded in open waters. Probably due to their small size and the presence of a dorsal fin, these whales were confused with minkes (Davies and Guiler 1957). It is possible that they could also be confused with the medium-sized representatives of toothed whales (*Odontoceti*). In this regard, Hector's recording Hector (Hector 1875) of an

Fig. 5.24 *Caperea* "desiccated" on a sandbank

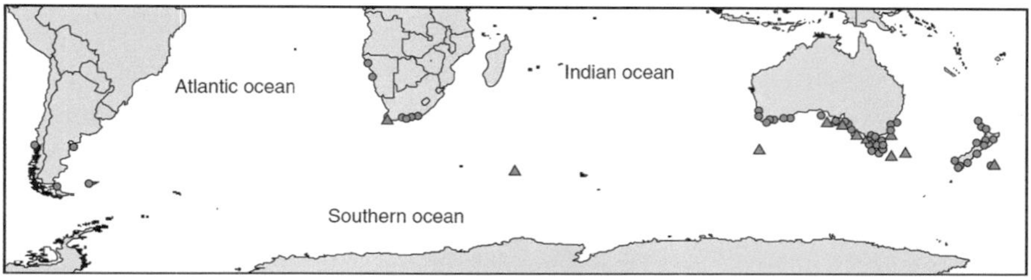

Fig. 5.25 Observation of *Caperea* (by C. Kemper 2009)

accidentally caught *Caperea* female (length 465 cm) in a large group of pilot whales is symptomatic. Further, we need to note that in the 60s –1963/1964 in particular – Soviet whalers caught southern right whales in the area of Tristan da Cunha. These whales are three times larger than *Caperea*. Therefore, it is possible that *Caperea* could be mistaken for right whales' calves.

So, during November and early December 1970, Soviet whaling covered its traditional area off the coast of Africa to the islands of Tristan da Cunha. In November the whaling flotilla *Soviet Ukraine* caught here 924 sei whales, 196 sperm whales, 20 fin whales and one minke whale. In early December, caught 137 sei whales, 12 fin whales, 7 sperm whales were caught. At the same time and in the same area, but somewhat to the west the flotilla *Yury Dolgoruky* was working. It caught 707 sei whales, 199 sperm whales, 29 fin whales and six minke whales. As can be seen, the main target species were sei whales, to a lesser extent – sperm whales, fin whales, and then rarely minke whales. In this area killer whales, bottlenose whales, pilot whales and smaller dolphins were also observed.

Among this variety of cetaceans *Caperea* were discovered thanks to observations by the *Soviet Ukraine* helicopter pilot, V.M. Ivanov. During a reconnaissance mission on 29th November 1970, he noticed small baleen whales with a dorsal fin, but that did not look like minke whales. He found two groups of these whales – the first had seven, and the second eight individuals. Not far from these whale aggregations he spotted a female sei whale and calf. The whaler *Bezuprechny* (Captain A.A. Stepanenko, harpooner G.S. Mogilny) was sent to the location of the unusual whales, and one whale was caught at 32°56′S and 12°42′E. It appeared to be a *Caperea* male, 5.47 m in length.

The presence of *Caperea* in the area was reported to the *Yury Dolgoruky* flotilla, which caught a *Caperea* whale at 33°50′N, 7°50′W on the same day. It was a pregnant female, 6.35 m in length. The whale was caught from a group of three individuals. A helicopter search yielded another group of five individuals. All whales were calm; they were feeding on zooplankton in the surface layer of water. In the area investigated, the water temperature ranged from 18.4 °C to 21.1 °C, with an average of 18.8 °C. A few days later, on fourth December, the *Buyny* whaler from the *Soviet Ukraine* flotilla (Captain A.F. Gostilov, harpooner I.E. Obukhov) at 33°49′S 0°56′E caught another *Caperea* – a pregnant female with a length of 6.21 m (Fig. 5.26).

Hot on the heels of these discoveries, brief reports about these whales were published (Shevchenko et al. 1971; Mikhalev and Shevchenko 1972; Ivashin et al. 1972a, b; Budylenko et al. 1973). Such unique material deserves special attention and a more detailed description.

The captured whales, plus those that have been reported by shipboard observers, totaled at least 25–30 individuals. Without doubt, there were many more in the area. The cluster was about 2500 km across in its extremity (Fig. 5.27).

All whales were calm and remained in one place. They dived in a shallow depth of two to three meters. Similar *Caperea* behavior has been observed in other regions (Gill et al. 2008). The interval between dives was 3–4 min; the whales were on the surface for 4–5 s. As from whaling ships and from helicopters, observers testified that *Caperea* were grazing in fields of plankton. According to some reports, the whale blows were small. According to other reports, no blows were observed. Most likely, the blows were not noticed, since *Caperea* blows are weak and fuzzy (Fig. 5.28), especially because in warm waters, even large whale species have weak blows.

Fig. 5.26 Caperea female, length 6.21 m, on the deck of Soviet Ukraine. (Photo by A.I. Flerov)

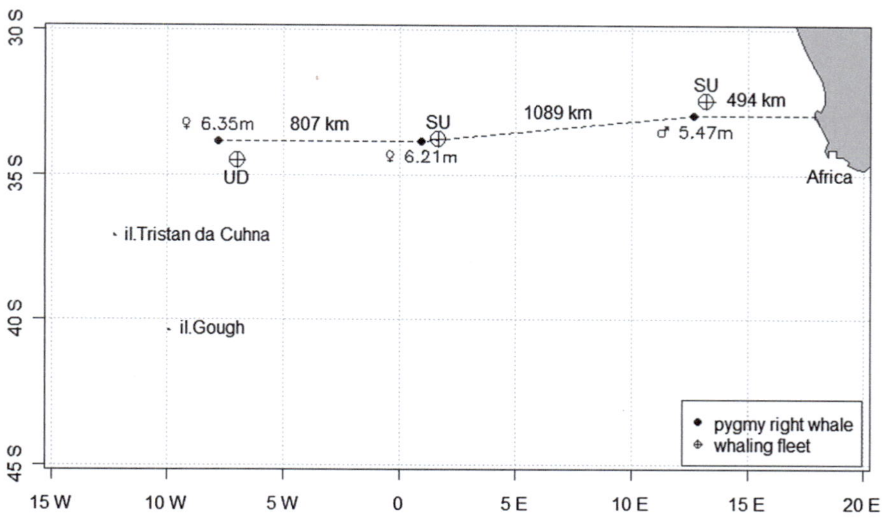

Fig. 5.27 *Caperea* whaling area by flotillas *Soviet Ukraine* and *Yuri Dolgoruky*

Fig. 5.28 Characteristic *Caperea* blow

External Morphological Features The captured *Caperea* whales were inspected, measured and weighed on the flotillas' decks (Fig. 5.29). One male had was 5.47 m in length and weighed 2850 kg; a female was 6.21 m long and weighed 3200 kg, and the second female had a length of 6.35 m and weighed 3430 kg.

Fig. 5.29 Weighing of 6.35 m Caperea on the deck of Yury Dolgoruky. (Photo by B.A. Leonov)

External inspection of the captured whales did not reveale any special features other than those of *Caperea* from other areas. However, the bodies of all three whales were marked by "white scars", 35–60 mm in size, which is usual for whales in warm waters. Fresh wounds were the traces of teeth bites from small pelagic sharks (Shevchenko 1970, 1971, 1977). Hale (1964) shows photos of *Caperea* with barnacle fouling on the lip of the mandible. In our case, no barnacles or whale lice were observed. All the typical signs of the form were noted: a curved line of the mouth (but less curved than that of regular right whales); the skin on the back was almost black in colour; a white or white-gray colored abdomen; a characteristic white stripe on the bottom edge of the upper jaw along the edge of baleens; a white mouth and white tongue; two characteristic throat grooves; a distinct dorsal fin somewhat shifted to the rear of the body.

The blowhole has the form of two slightly curved slots with a length of about 10 cm each, lying at a slight angle to each other. It was positioned in a small cavity in the uppermost part of the head and is biased forward (about 40 cm) from the section line through both eyes. From the blowhole to the end of the rostrum in the middle of the upper jaw was a slight but well-defined ridge. A prominent ear opening was located in a small notch some 30–40 cm behind and just a little below the eyes. Eyeball diameter was 9 cm, the eye slit 3 cm and pupil diameter 1.3 cm. At the end and sides of the snout were from 2–3 to 5–6 vibrissae and there were 2–3 more tactile hairs in front of the blowhole slots. The 42 centimeter embryo we inspected had 18tactile hairs. The length of the vibrissae ranged from 0.5 cm to 1.3 cm. Coverslip fat in the conventional place (the sides of the body at the level of the dorsal fin) had a thickness of 3 cm for the 5.47 m male, 3.5 cm for the 6.21 m female 2.5 cm for the 6.35 m female and 0.2 cm for the 42 cm embryo. The thickness of the epidermis in the area close to the dorsal fin was 2 mm for adults.

Pectoral Fins The shape of the pectoral fins in *Caperea* resembles truncated pectoral fins of regular rorquals (Fig. 5.30). For the measured whales the length of the pectoral fins was $1/8$ of the body length and the width was $1/4$ of the total length of the fin. The outside of the fins is black in colour and stands out against the lighter background of the sides of the whale. The underside of the fins is light gray, and the tips are black. The caudal fin is relatively large, the median notch being expressed. The blades are black on top and light gray with a dark rim on the rear edge on the underside.

Straining Apparatus The form of the baleen plates had nothing in common with rorquals or gray whales. They were typical of right whales: narrow and long. However, the color of the plates was not dark, but yellowish-white, with a dark brown rim on the outer edge of the plates. The periodical waviness reflecting the growth of animals was visible on the plates (Fig. 5.31).

Caperea baleens usually have 210–230 plates on each side (Kemper 2009). In our case, the number of plates varied in a bigger range. Thus, the 5.47 m male had 213 plates on the right side and 221 on the left. The 6.21 m female had the same

Fig. 5.30 *Caperea* pectoral fin (female, 6.35 m)

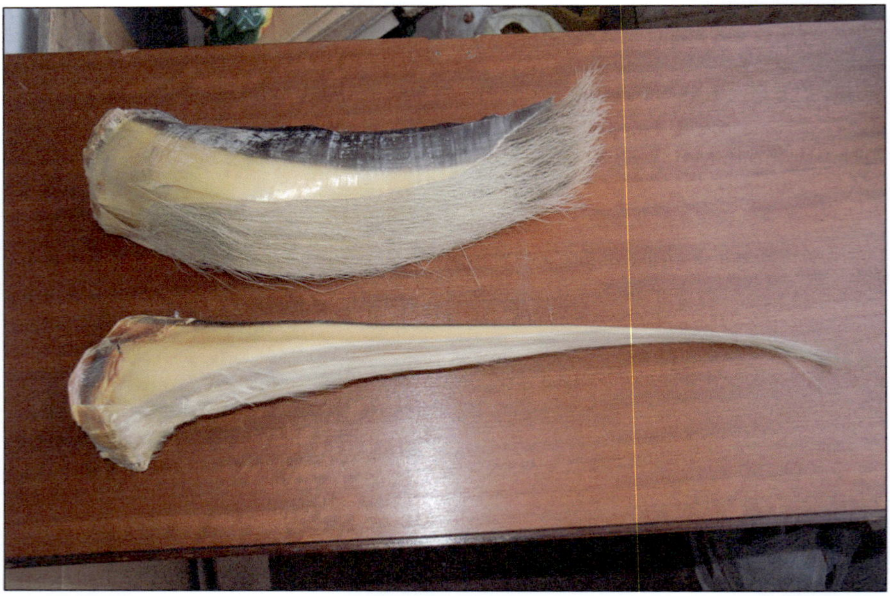

Fig. 5.31 Baleen plates of fin whale (top) and of *Caperea* whale (bottom)

number of plates on the right and left sides, being 230. The larger female with a length of 6.35 m had 262 plates on the right side and 261 plates on the left. That is, larger animals had more plates. If this is not accidental, it is possible that small baleen plates grow with age and become equal in size to the main ones. The largest of the plates for the male was 55 cm, with a width of 13 cm, for the 6.21 m female it was 62 cm with a width of 12 cm. The largest female had plates with a width of 10.5 cm and a height of 65 cm. The maximum width of the plates according to

Fig. 5.32 Fringe at the top of the *Caperea* baleen plates

Kemper (Kemper 2009) was 10.5 cm, the height was 65 cm. As can be seen, the height and width of the plates in different individuals also vary widely. The average depth of the flesh of the plates in our case was 2.7 cm.

The inner side of *Caperea* baleen plates splits into a thin silky fringe (Fig. 5.32). In our case, the length of the hairs at the top of the fringe reached 15 cm, with a thickness of 0.2–0.3 mm. Such a thin fringe shows that these whales feed on small planktonic organisms. And indeed, the autopsy of the whales revealed small *Calanus* species in their stomachs. According to the biologists from *Yury Dolgoruk*, *Calanus tonsus* were also present, and according to the biologists from *Soviet Ukraine* – *Calanus simillimus*. Literary sources also point to food items such as *Caperea* copepods, small euphausiids and larvae of larger euphausiids (Arnold 1987; Sekiguchi 1992; Kemper 2009). On the fringe of the baleen plates of the 6.36 m female, an iridescent anchovy of the *Myctophidae* family was found, probably accidentally trapped with the plankton. In our case, the degree of filling of the *Caperea* stomachs was from "a little" to "a lot", which indicates a good food base in this area.

Body Proportions All three whales and the embryo were measured by a series of standard measurements (Table 5.2). Because of the small number of whales, no information has been analyzed and has been presented as cumulative material for future generalizations.

Skeleton Generally speaking, desiccated skeletons of *Caperea* served as objects of study, rarely *Caperea* caught by nets in coastal waters. As a result, several dozen skeletons have been collected by museums in Australia, New Zealand and South Africa,. But any new data are always interesting, especially from a new area.

Table 5.2 *Caperea* body measurements

#	Measurement	Male		Female		Female		Embryo-♂	
		cm	%%	cm	%%	cm	%%	cm	%%
1	Zoological length (from the end of the upper jaw to the notch of the caudal fin)	547	100	621	100	635	100	42	100
2	Distance from the end of the upper jaw to the front edge of blowhole slits	73	13.34	77	12.40	–	–	–	–
3	Distance from the end of the upper jaw to the middle of blowhole slits	79	14.44	86	13.85	90	14.2	5	11.9
4	Distance from the end of the upper jaw to the corner of the mouth	105	19.19	116	18.68	125	19.7	7.2	17.1
5	Distance from the end of the upper jaw to the center of the eye	115	21.02	133	21.42	130	20.5	8.5	20.2
6	Distance from the center of the eye to the ear holes	37	6.76	33.5	5.39	–	–	–	–
7	Height of dorsal fin	26	4.75	31	4.99	28	4.40	1.3	3.1
8	Length of dorsal fin	47	8.59	37	5.96	40	6.30	2.7	6.4
9	Length of pectoral fin from the top to the axilla	43	7.86	65	10.47	55	8.70	3.2	7.6
10	Length of pectoral fin from the front edge to the top	72	13.16	85	13.69	80	12.6	5.2	12.4
11	Maximum width of the pectoral fin	19	3.47	20	3.22	23	3.6	1.5	3.6
12	Length of caudal peduncle (from recess of tail flukes to the anus)	142	25.96	164	26.41	168	26.5	11.7	27.9
13	Distance from recess of tail flukes to the navel	253	46.25	300	48.31	290	45.7	19.0	45.2
14	Perineum (the distance from the anus to the center of the genital opening)	48	8.78	22	3.54	21	3.3	3.0	7.1
15	Distance from tail flukes notch to dorsal fin notch	145	26.51	201	32.37	180	28.3	12.3	29.3
16	The width of the tail blades	43	7.86	49	7.89	48	7.6	4.5	10.7
17	Length of tail blades	90	16.45	100	16.10	90	14.2	4.8	11.4
18	Tail blades spread	172	31.44	–	–	175	27.6	9.5	22.6
19	Distance from the end of the upper jaw to the dorsal fin	365	66.73	416	66.99	415	65.4	27.0	64.3
20	Distance from the end of the upper jaw to the pectoral fin	181	33.09	198	31.88	181	28.5	13.0	31.0
21	Semi girth at pectorals	182	33.27	190	30.60	–	–	–	–
22	Distance from the end of the upper jaw to the occipital condyle	164	29.98	163	26.25	146	23.0	10.5	25.0
23	Distance from eye to eye	–	–	98	15.78	–	–	–	–
24	Distance from the end of the upper jaw to the end of the lower jaw	–	–	–	–	20	3.1	0.3	0.7
25	Maximum height of caudal peduncle	–	–	–	–	75	11.8	5.0	11.9
26	Maximum height of the body at the pectoral fin	–	–	–	–	135	21.3	9.5	22.9
27	Girth at the base of the pectoral fins	–	–	–	–	380	59.8	26.5	63.1

Unfortunately, under whaling conditions, it was impossible to keep the skeletons of whales caught by the Soviet fleet intact. Harpoons broke some vertebrae, cutting the whale bodies and damaging some bones, and, in weighing animals while the vessel was pitching, some maxillary bones were chipped. Two prepared skeletons were given to museums at the Chisinau and Odessa universities, where due to a lack of proper facilities and the necessary funds, the skeletons are still not on display (excluding the male skull) and are therefore difficult to study. The skull of the third whale was transferred to Kaliningrad Institute of Fisheries and Oceanography. Unfortunately, it has not survived. Let's consider the description of skeletons as recorded on the decks of the whaling factories.

Skeleton Skull The skull is relatively small, approximately one-quarter of the body. The cranium is also relatively small. The shape of the skull is arcuate as a result of heavily overgrown baleens, which led to the widening of the maxillary, premaxillary and opener (Figs. 5.33, 5.34 and 5.35).

The ratio of the circumference of the skull to its condylobasal length is about 105%. The rostral part is arcuate, but relatively short – about 60%. Nasal bones (*nasale*) are symmetrical, accreted for females, but not for males. Moreover, in the male, they are relatively smaller and more prominent. Their front ends protrude in the shape of a triangle. The supraoccipital bone (*supraoccipitale*) of the males is relatively longer than that of the females, it hangs over the frontal bone (*frontale*) and is in contact with the nasal bones (*nasale*). For the females, a narrow (1.5–2.0 cm) strip of the frontal bone (*frontale*) can be seen between these bones. The medial ridge is well defined, its height being 4 cm. In males, the rear wings of the occipital bone (*basioccipitale*) is bent backwards more than in females.

Occipital condyles (*condilus occipitalis*) on the ventral side are divided by a deep groove. In males, the groove is barely noticeable. Foramen magnum is shifted to the dorsal side of the skull, so that it coincides with the plane of the circuit supra-occipital bone (*supraoccipitale*). Drum bones (*bulla tympani*) are rhomboid, similar in shape to other right whales, but not to rorquals or gray whales. Their length for

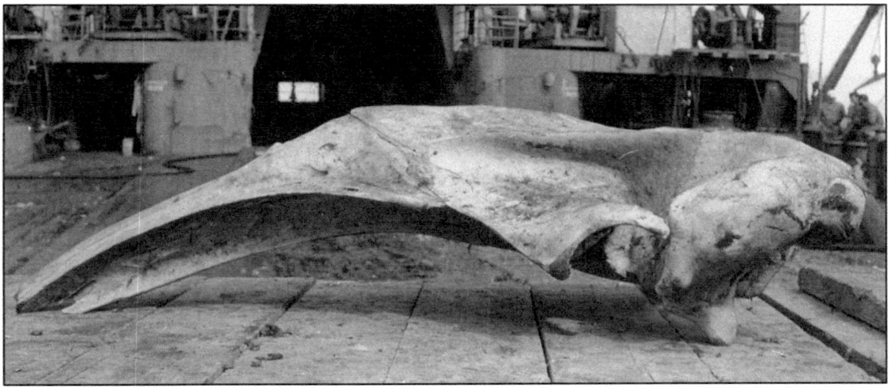

Fig. 5.33 The skull of the 6.35 m female pygmy whale (side view)

Fig. 5.34 The skull of the 6.35 m female pygmy whale (top view)

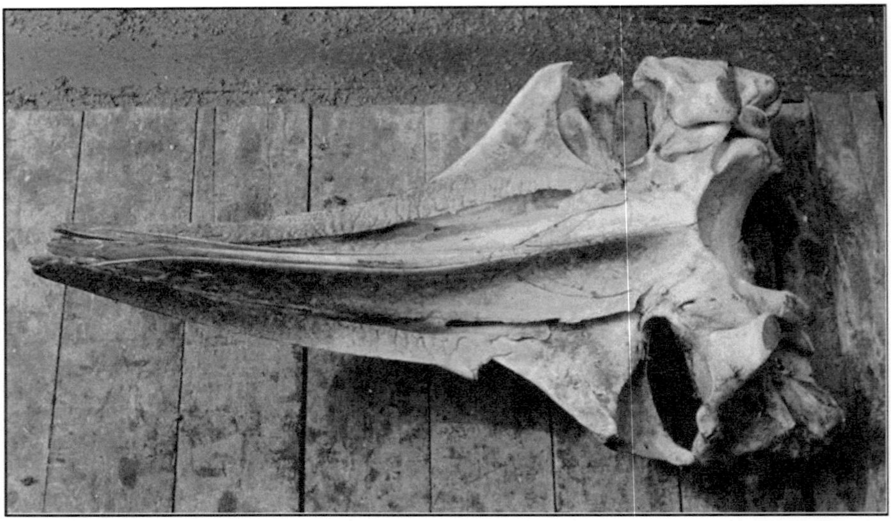

Fig. 5.35 The skull of the 6.35 m female pygmy whale (bottom view)

Table 5.3 Basic measurements of skulls of the 5.47 m male and the 6.21 m female

	Male 5.47 m		Female 6.21 m	
Measurement	length, cm	%	length, cm	%
Condylobasal length	139	100	157	100
Rostrum length	82	59	93	59.2
Rostrum width	36	25.9	41.5	26.1
Zygomatic width	79	56.8	79	50.3
Interorbital width	69	49.6	72	45.8
Jawbone length[a]	78.5	56.5	90	57
Nasal bone length	7	5.0	12.2	7.7
Supraoccipital bone length	56	40.3	61	38.9
Mandibular bone length	117	84.2	135	85
Mandibular bone width	21	15.1	19	12.1

Note: [a]From the frontal angle

males is 11.5 cm, with a width of 8.5 cm and for the 6.35 m female 12.5 cm and 7.0 cm respectively. The premaxillaries (*premaxillare*) protrude beyond the maxillary (*maxillare*). The mandibles (*mandibulare*) are symmetric and sigma shaped. Their width is relatively larger for males. Since the measurements of whales' skeletons caught by the flotillas *Soviet Ukraine* and *Yury Dolgoruky* were carried out by different schema, we present them separately in Tables 5.3 and 5.4.

Vertebral Column Both *Caperea* whales studied by the *Soviet Ukraine* flotilla had spines with 40 vertebrae. Beddard (Beddard 1901) counted 43 vertebrae for a large specimen from South Kensington, but the picture of the skeleton of the whale indicated fewer vertebrae! Having studied 24 skeletons from American, New Zealand and Australian museums, Buchholtz (Buchholtz 2010) came to the following formula for vertebral column: $C_7T_{18}L_1Cd_{16–17} = 42–43$ or $C_7T_{17}L_1Cd_{16–18} = 41–43$.

In our case, according to the structural features of the articulated surfaces of the vertebrae and ribs, we can assume that the first pair of ribs is attached not to the second, but to the first thoracic vertebra. Then there should be seven cervical vertebrae, 17 thoracic, three lumbar, and both male and female have 13 tail vertebrae. If our estimate of the number of thoracic and lumbar vertebrae is true, then the formula for the spinal column of the studied whales is S7D17L3Ca13. If, however, the first pair of ribs is attached to the second vertebra, then there should be 18 thoracic vertebrae and 2 lumbar ones: S7D18L2Ca13.

The relatively small number of vertebrae for *Caperea*, Beddard (Beddard 1901) is indicative of a distinctive feature in comparison with other species of baleen whales. We traced a constant relationship between the length of the body of different species of whales and the number of vertebrae in the spine, as was the case for the total number of vertebrae, and separately for thoracic (including lumbar), and tail vertebrae (Fig. 5.36).

When logarithms of the lengths of whales were used instead of absolute values, the semi-log plot proved to be more convincing. In this case, the regression line for all parts of the spine were straight (Fig. 5.37).

Table 5.4 Skull measurements for the 6.35 m female

	Female 6.35 m	
Measurement	length, cm	%
Condylobasal length	136	100
Rostrum length	75.0	55.1
Rostrum width at its base	35	25.7
Rostrum width in the middle of its length	20	14.7
Length of upper jawbone by curve (top)	117	86
Upper jaw bone length	105.0	77.2
Intermaxillary bone length	73	53.7
Jawbone fronto-nasal ridge length	30.0	22.1
Jawbone orbital ridge length	39	28.7
Distance between jawbone fronto-nasal and orbital ridges	21	15.4
Zygomatic width between the edges of squamosals	82	60.2
Interorbital width	74	54.4
Supraoccipital bone length	60	44.2
Occipital bone width	66	48.5
Width between the outer edges of the occipital condyles	20	14.7
Foramen magnum height	8	5.9
Foramen magnum width	6.0	4.4
Nasal bones length	8	5.9
Nasal bones maximum width	3	2.2
Mandibular bone length in straight line	140	103
Mandibular bone length on outer edge	145	107
Mandibular condyle height	17	12.5

This relationship is quite satisfactorily described by the following formulae: for the entire spine Nv = 17.03 + 34.29 log (L); for the thoracic and lumbar vertebrae: ND + L = 8.65 + 16.03 log (L); for the tail spine: NCa = 0.27 + 19.19 log (L). That is, there are no fundamental differences for this feature with other species of baleen whales. Reduction of the number of vertebrae with a reduction of the size of different species of whales is determined by an allometric relationship. For the same reason, body proportions change for different cetaceans (Mikhalev 1970a, 1984, 2007, 2008). It is likely that these proportional changes are determined by hydrodynamic laws. Maybe this is the reason why the relatively short *Caperea* has a distinct dorsal fin, which is absent for the much larger (three times) regular right whales? And maybe due to the same reason, the big-headed sperm whale's (*Physeter macrocephalus*) dorsal fin is virtually absent, while it is clearly expressed for the pygmy sperm whale (*Cogia breviceps*)? At the moment we can only conclude that the dependence of cetaceans' body shape on hydrodynamic laws is poorly understood.

The cervical vertebrae of *Caperea* we studied were fused into a single unit by bodies, spinous ridges and neural arches. For the 6.21 m female the merger line was

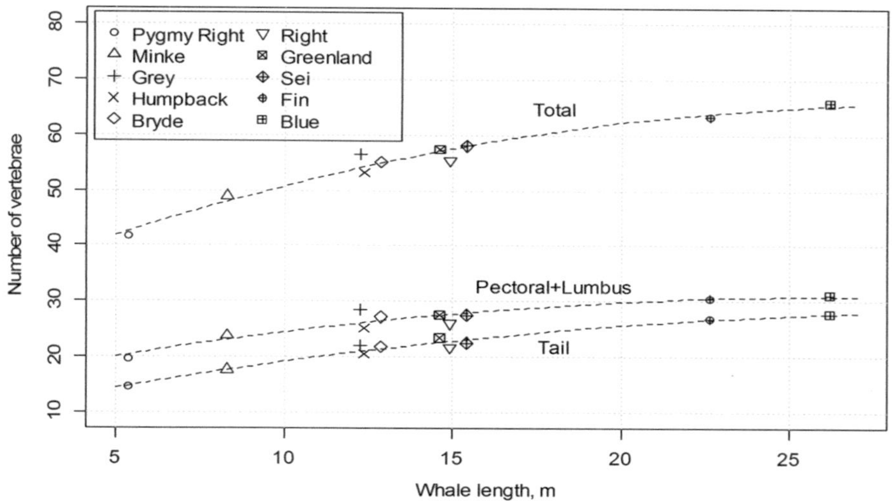

Fig. 5.36 Relationship between pygmy right whale body length and number of vertebrae in spine (direct coordinate system)

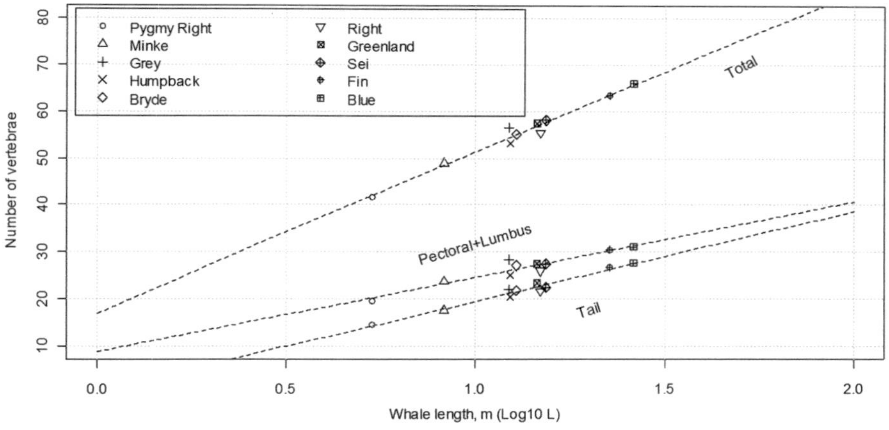

Fig. 5.37 Allometric relationship between the number of vertebrae in the spine and the length of baleen whales

evident amongst all the arcs, and for the 5.47 m male starting from fourth vertebra. The degree of epiphyseal ossification in thoracic vertebrae was greater for the females than for the male. The epiphyses of the thoracic vertebrae for the male were not fused. In the females, the first and second vertebrae were completely fused. From third to sixth – the merger was not complete. The rest of the thoracic vertebrae epiphyses were free. There were five hemal arcs, wherein the first and sixth arcs consist of two non-fused plates. The largest was the third chevron bone.

Ribs All *Caperea* we studied were physically mature animals with 17 pairs of ribs. The first pair had a single head, connected to the chest by its broad distal end. From the second to the eighth pairs in the female, and to the sixth in the male, the ribs had two heads. The longest neck was for the third pair. An articulated surface is well marked from the first to the ninth pair of ribs. From the tenth to the twelfth pairs, it increases in size, but becomes less expressed, and further it becomes invisible. From the second to the eighth pairs, the ribs are arcuate curved and have a rounded cross-sectional shape. The ninth pair differs sharply from the previous ribs, having extensions at both ends. Subsequent pairs of ribs are broad and flat, and form, together with the powerful development of the transverse ridges of the vertebrae, an almost continuous bone armor. The functional significance of such a structure of ribs is difficult to explain. Beddard (Beddard 1901) believes that the role of "armor" is to protect internal organs from the strong water pressure at depth. We don't think that this explanation is convincing enough. All *Caperea* observed by Soviet whalers immersed only in shallow depths. *Caperea* found in coastal waters were also in the shallows. The fact that these whales are not specialized divers is shown by the relatively small size of their heart and lungs, as well as the fact that the muscles (evidenced by their color) had little myoglobin.

Size and Weight of the Internal Organs The internal organs of the captured *Caperea* were extracted, measured and weighed (Table 5.5). Weighing results have mainly cumulative characteristics. Of some interest is a comparison with the characteristics of the Japanese right whales published by Klumov (1962). The *Caperea* liver index was significantly higher than those of right whales, the kidney index was nearly equal, while heart and lung were markedly lower.

Chemical Analysis of the Raw Material: Muscle and Organs Chemical analysis of muscle tissue and internal organs was carried out for *Caperea* caught by the *Soviet Ukraine* flotilla. The degree of fat was found to be 58.7%, and the moisture 21.1%. A more detailed chemical analysis of organs and tissues was performed onboard *Yury Dolgoruky* (Budylenko et al. 1973). According to their data the contents of fat, moisture and dry residue were as follows:

Raw material	Fat (%)	Moisture (%)	Residue (%)
Meat (muscle)	1.22	75.03	23.75
Fat coverslip	49.13	27.17	23.70
Bones	17.76	24.20	58.04
Tongue	38.70	6.80	54.50
Heart	1.52	79.17	19.35
Liver	5.93	69.60	24.47

Additionally, for *Caperea*, 1 g of raw liver contained 250 international units of vitamin A, and 1 g of liver fat 4170.

Table 5.5 Weight of internal organs of *Caperea* caught by Soviet whaling fleets[a]

							Index of the internal organs of whales in ppm					
	Caperea length, cm			Caperea weight, kg			Caperea			Japanese right whale		
	♂	♀	♀	♂	♀	♀	♂	♀	♀			
				547 cm	621 cm	635 cm						
Internal organ	547	621	635	2850 kg	3200 kg	3430 kg	547 cm	621 cm	635 cm	1630 cm	1660 cm	1740 cm
Heart	–	–	–	8.6	7.8	6.65	3.01	2.44	1.94	8.04	6.79	3.79
Lungs	90 × 23	112 × 33	–	5.5	9.0	–	–	–	–	–	–	–
	90 × 23	116 × 26	–	6.0	7.7	–	–	–	–	–	–	–
				Σ = 9.5	Σ = 16.7	Σ = 12.6	4.04	5.22	3.67	9.85	7.28	9.47
Kidney	55 × 18	59 × 20	–	4.2	4.7	–	–	–	–	–	–	–
	55 × 18	65 × 19	–	4.3	5.2	–	–	–	–	–	–	–
				Σ = 8.5	Σ = 9.9	9.13	2.91	3.08	2.66	4.99	5.21	3.15
Liver	–	88 × 91	–	34.85	42		12.23	13.12		9.22	6.16	9.89
Spleen	–	–	–	4	–	–	0.14	–	–	–	–	–
Pancreas	–	–	–	2.05	2.70	2.30	0.79	0.84	0.67	–	–	–
Thyroid	–	–	–	0.37	–	–	0.13	–	–	–	–	–
Ovaries: (left)	–	–	–	–	–	0.85[b]	–	–	0.24	–	–	–
(right)	–	–	–	–	–	0.95	–	–	0.28	–	–	–
Adrenals	–	–	–	–	–	0.03	–	–	0.01	–	–	–

Note: [a]By Klumov (1962); [b]Ovary weight without pregnancy corpus luteum trace

Biological Condition All three *Caperea* caught by Soviet whaling fleets were sexually mature individuals. The male's left testis weighed 900 g, and the right one 970 g. Both females were pregnant. The right horn of the 6.21 m female's uterus had a length of 60 cm, and a width of 11 cm, the left one was 45 cm and 8 cm respectively. The ovaries were almost the same size: length 11 cm, width 5 cm. On the right ovary the functioning corpus luteum of pregnancy and five traces of corpora lutea of ovulation were noted. On the left one there were three tracks of corpora lutea of ovulation. For the 6.35 m female, the ovaries had 11 lutea traces. If we assume that all the traces on the ovaries were signs of pregnancy (the difference between ovulation traces and signs of pregnancy have not been studied yet for *Caperea*), these females could already have produced eight–ten calves.

We know the length of the two pregnant females, caught by the Soviet whaling fleets, and one pregnant female (embryo length 60 cm) from the work by Guiler (Guiler 1961). The average size of these three females is 6.31 m. The relationship between the size of female baleen whales and their newborns is expressed by the formula: $l = 0.617\ L^{0,75}$ (Mikhalev 1972). Thus the average size newborn *Caperea* should approximately equal 2.21 m, which coincides with the assessment of Kemper (Kemper 2002a) – 2.20 m, who directly measured some newborns. This average size was close to the size of the known two prenatal fetuses – 200 cm and 212 cm (Ross et al. 1975; Kemper 2002a). Based on the prenatal growth rate for baleen whales (Mikhalev 1970a, 1975, 2007), *Caperea* newborn should achieve their average midsize in about 11 months after conception. For a more reliable estimate of the duration of *Caperea* pregnancy, much more material is certainly needed.

Embryos size allows us to judge the time of conception. Unfortunately, such information is very linited: besides a 6.21 m female caught on 29 November 1970, which had an embryo with a length of 24.2 cm, and a 6.36 m female, caught on December 4th, which had an embryo 42 cm in length (Fig. 5.38), we are aware of a 60 cm embryo found in a 6.37 m female caught near Tasmania on 26 June 1961 (Guiler 1961). Based on the prenatal growth rate for cetaceans (Mikhalev 1970a, 1975, 1980, 1984, 2007), the age of a 24 cm embryo is approximately equal to

Fig. 5.38 The 42 cm *Caperea* male embryo

3.2 months, 42 cm – 4.3 months, 60 centimeter – 5.1 months. So this means that the conception of the first embryo occurred in late August, the second in the second half of July, and the third in early February. That is, *Caperea* females caught in open waters of the South Atlantic were mating during the Southern hemisphere summer, and the female caught near Tasmania, almost 6 months earlier. Which implies that either *Caperea* mating season is really highly extended (Pavey 1992), or they, like many other baleen whales (Tomilin 1957; Laws 1959; Naaktgeboren et al. 1960: Mikhalyev 1972, 2008), also have an additional mating, with a time-shift of about 6 months. However, the embryos for our analysis are not from one regions, but from widely separated regions, so we cannot exclude population differences for this feature. The problem can be solved only if a larger number of embryos is collected from different regions of the Southern Hemisphere.

***Caperea* Migration and Population** So, an additional area of *Caperea* distribution is now known – the thirties latitude of the South Atlantic from the coast of Africa to the islands of Tristan da Cunha. It turned out that *Caperea* can be found not only in coastal waters, as previously thought, but also in open water. And they form quite numerous clusters. Even without a targeted observation, at least 25–30 individuals were reported in 1 week. If its range is so wide then, consequently, the number of species should be significantly higher than was previously assumed (Davies and Guiler 1957; Ross et al. 1975; Kemper 2002a, b). Moreover, recent studies have shown (Matsuoka et al. 1996; Gill et al. 2008), that *Caperea* can be found not only in small groups, but also in fairly large clusters of no less than 80–100 individuals (Fig. 5.39). However, it is still absolutely impossible to give any definite estimate of the total number of *Caperea* using casual observations.

There is no reason to consider *Caperea* to be only a coastal species. In recent years they have been recorded in open waters, not only in the Atlantic, but in the Indian and Pacific oceans. According to Matsuoka, Pitman and Marques (Matsuoka et al. 2005), in 2001 a group of pygmy whales comprising 14 individuals was found in open waters at a distance of 450 km from New Zealand. Apparently, they are able to make long migrations rather quickly, as evidenced by a fairly streamlined body shape and the presence of the dorsal fin. Otherwise it is difficult to explain the occurrence of pigmy right whales near islands like Tristan da Cunha, and Crozet, so far from the continents (Kemper 2009). However, the migration routes of this species are unknown. As for the south-eastern Atlantic, we can only assume that *Caperea* feeds there, and migrating from Africa to Tristan da Cunha and back, probably forms a single population.

At the moment, while there is no whaling and whale watching is limited, it is imperative to conduct a thorough inventory of logbooks from scientific research vessels and whale inspection logs recorded by biologists' research groups and preserved in both public and private archives.

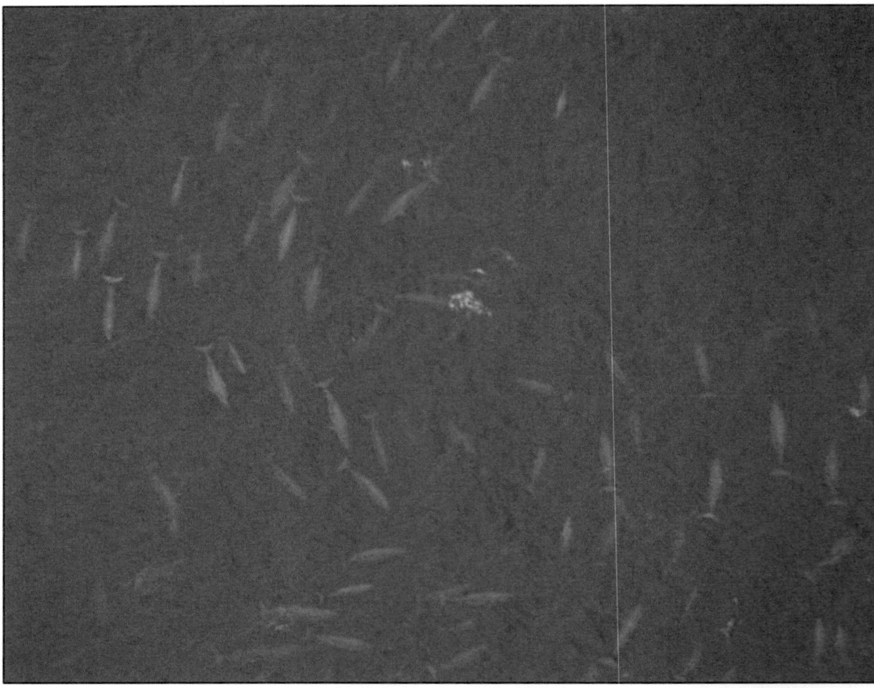

Fig. 5.39 Cluster of pygmy right whales off the coast of Australia (Photo by Peter Gill)

5.4.7 Occurrence of Ambergris in Sperm Whales

Brief Historical Overview Ambergris (*Ambra ambrosiaca grisea L.*) – is a very specific substance. It cannot be compared with anything, except, to some extent, "bezoar stone" (*Pierda del Porco*), which is most often found in ruminants, primarily in bezoar ibex (*Capra aegagrus aegagrus, Erxleben, 1777*). Since ancient times, ambergris found at sea, or on the coast, was known in China, India, Central Asia, Greece and Rome (Tomilin 1936, 1957; Abu Rayhan Biruni 1973; Schafer 1981; Avicenna 1982; Amasiatsi 1990). The Arabs and Chinese naively believed ambergris to be "dragon saliva" or "whale vomit" (Schaefer 1981). Equally naive were assumptions by Greeks and Romans that ambergris is formed by bitumen flowing into sea, or even that is deposited by some giant bird which eats fragrant plants and even the gum of coastal trees (Tomilin 1957). In the Middle Ages, in the tenth century, Amasiatsi suggested that hot sun melts honeycomb wax on coastal cliffs and it flows into sea, where it is swallowed by "sea cows", in whose stomachs the wax is allegedly digested and transformed into ambergris. During the same period, Abu Rayhan Biruni wrote that "… ambergris comes from an animal that lives in the sea", but he did not specify whales.

Fig. 5.40 Dana octopus squid

These mythical representations and guesses have been discarded by modernity. It has become clear that ambergris is formed in the intestines of sperm whales due to feeding on cephalopods (Dudley 1725; Schwedtawer 1783; Clarke 1949; Hardy 1949; Tomilin 1936, 1957; Zenkovich 1951; Ivashin 1963; Berzin 1971, etc.). And according to data from Soviet whaling fleets, the main food of sperm whales is squid. To a much smaller degree, their stomachs also contained toothfish, sharks, anglerfish and crustaceans. Among the squid, Dana octopus squid (*Taningia danae*) (Fig. 5.40), was most often found, along with the occasional deep-ocean giant squid (genus *Architeuthis*) (Fig. 5.41).

On broken pieces of ambergris the ruins of squid beaks and other impurities can be seen (Fig. 5.42). According to Tomilin (1957), impurities account for up to 25% of the weight and according to Ivashin (1966), 20%. It is assumed that squid beaks provoke the formation of ambergris in sperm whales, and that the process is somewhat similar to the process of formation of kidney stones, or formation of stones in the bile ducts of the liver of animals (Pouchet and Beauregard 1892; Clarke 1954; 2006; Ivashin 1966).

Yet there are many uncertainties concerning the origin of ambergris! At first glance it seems to be logical that ambergris found in sperm whales, which feed mainly on cephalopods, mostly squid, and indeed there are many remnants of squid beaks in ambergris. Additionally, some squid and other cephalopods have ink sacks

Fig. 5.41 Giant squid

Fig. 5.42 Grey ambergris

Fig. 5.43 Black ambergris, extracted from the intestines of the sperm whale

containing black inky liquid. The fecal masses inside the intestines of sperm whales are black, as is ambergris formed in the gut (Fig. 5.43), which again speaks in favor of the hypothesis of the formation of ambergris due to sperm whales' eating cephalopods. According to R. Clarke (Clarke 1954), it is by drying the semi-liquid feces of sperm whales that a product similar to ambergris was obtained.

But if we recognize that eaten cephalopods provoke the formation of ambergris in sperm whales, then ambergris would have to be formed in other whales that eat cephalopods: pilot whales, goose-beaked whales, mesoplodont whales, bottlenose whales and killer whales. Such evidence exists, but there is very little of it and its reliability is low. According to Cleyerus (Cleyerus 1689), Chevalier (Chevalier 1700), Kaempfer (Kaempfer 1727), Ratcliffe (Ratcliffe 1942) in Japanese waters ambergris was found in whales, which local whalers called "mako", "makko kouzira" or "ouki kouzira". According to Pouchet (Pouchet 1893) they were pygmy sperm whales (*Kogia breviceps* and *Kogia simus*). Yamada (Yamada 1954) confirmed that dwarf forms of sperm whales were called so in Japan in the nineteenth century. But, in fact, pygmy sperm whales are sperm whales too, or in any case, close relatives of the big-head sperm whale (*Physeter macrocephalus*). And what about other whale species? Jenkins (Jenkins 1932) reports the occurrence of ambergris in the northern bottlenose whales (*Nureroodon ampullatus*), however, there is no other evidence of this fact yet.

The belief that ambergris can be formed not only in toothed, but also in baleen whales (Sibbaldus 1773; Guibourt 1876), at first glance seems absurd, since they do not feed on squid. Nevertheless, Ommaney (Ommaney 1938) reported a surprising fact. About 30 nodules ranging in size from 5 cm to 25 cm were found in faeces in the rectum of a male blue whale, caught in the Durban area. Judging by the whaling area in which it was found, it was most a pygmy blue whale, Inside these lumps, many baleen bristles were discovered. According to Clarke (Clarke 2006), this confirms the hypothesis that the formation of ambergris in sperm whales is provoked by squid beaks, because both baleen bristles and squid beaks are chitin derivatives. But in fairness it should be noted that not only baleen bristles, but shells of crustaceans and plankton, which baleen whales feed on, are also chitin derivatives. However, we do not yet know of any indications of ambergris in other baleen whales.

Detection of Ambergris on the *Slava* and *Soviet Ukraine* Fleets While dissecting whales on these flotillas, ambergris, usually black (Fig. 5.43), less frequently gray (whalers sometimes call it golden), was found only in sperm whales and only in the intestine,. During the Soviet period of active whaling, ambergris was procured. For example, in the 1964/1965 season, the Slava whaling fleet alone procured 465 kg of ambergris. After a period of storage, "ripe" ambergris was transferred to the perfume industry, where it was used as an aroma fixative. At that time, using natural ambergris, the Soviet Union produced the perfumes "Moskva Belokamennaya", "Kometa", "Yaroslavna" (Tomilin 1957).

However, ambergris harvesting was not the main purpose of hunting sperm whale, so it was not always noticed. Unfortunately, the biological research groups of the fleets also paid insufficient attention to this subject and not all cases of ambergris detections were recorded in "whale inspection logs". Because of the poaching nature of whaling in those years, the material recorded by the research groups was kept secret, and only some of the data has been published (Tomilin 1936, 1957; Kiesewetter 1953; Zemsky and Berzin 1959; Ivashin 1963, 1966, 1968; Berzin 1971). After declassification and publication of actual Soviet whaling statistics, which revealed large-scale poaching and falsification of whaling data (Yablokov 1994; Zemsky et al. 1994), more complete publication of scientific papers, including those on the ambergris, became possible. Unfortunately, not all of the primary material has survived. We preserved and processed "whale inspection logs" from the *Slava* flotilla for the period 1961–1966, and from *Soviet Ukraine* for the period 1961–1978. i.e. from 22 commercial cruises.

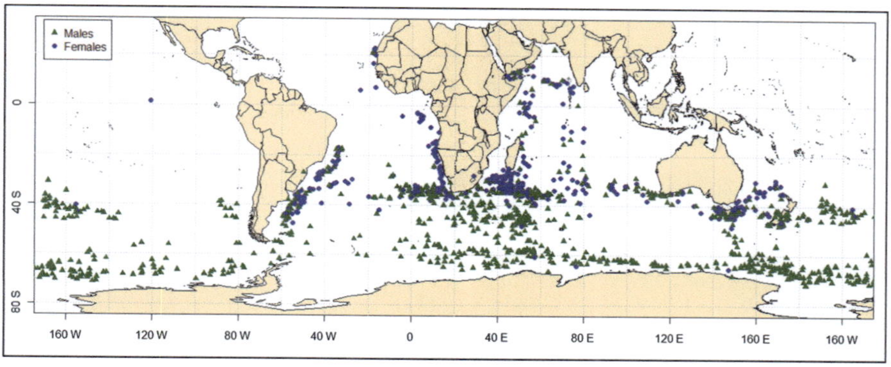

Fig. 5.44 Distribution of sperm whales caught by the Slava and Soviet Ukraine fleets for the period 1961–1978

For this period, the two fleets caught 43,211 sperm whales in total, 25,893 of which were males and 17,318 females. Fig. 5.36 illustrates the distribution map of the captured sperm whales (Fig. 5.44).

Ambergris was found only 150 times: – 108 times in males and 42 times in females (Table 5.6).

Distribution of sperm whales with ambergris (Fig. 5.45).

In 110 cases, ambergris was weighed, 81 times – for males, 30 times – for females. The ambergris weight distribution by sperm whale size range turned out as follows (Table 5.7).

As can be seen, there is a wide distribution of weights. The ambergris often consisted of several pieces, and these formations were weighed separately. Here are a few examples. Thus, in the 1962\63 season aboard the *Soviet Ukraine* flotilla, a 15.0 m male had ambergris which consisted of two pieces: 0.95 kg and 1.75 kg, and in a 14.7 m sperm whale, six pieces of ambergris weighing 0.675 kg were found. A 10.7 m male had ambergris which consisted of two pieces of approximately 6 kg each, and several smaller entities. Their total weight was 21 kg Ambergris from the 11.1 m male consisted of two pieces weighing 0.3 kg and 0.35 kg. And for the 11.2 m male, the ambergris was represented by three relatively large pieces (boulders) and many small pieces (Fig. 5.46). Ambergris from the 11.6 m male consisted of a large boulder weighing 80 kg and another four small pieces weighing about 5 kg in total. The ambergris found in 10.4 m male, weighed about 10 kg and consisted of many small pieces. Sometimes it was reported that ambergris consisted of several parts without specifying their size and weight. Thus, three pieces of ambergris were recovered from the 10.8 m female; from the 10.6 m female – two pieces; and from the 10.8 m female – about 20 pieces.

The linear dimensions of ambergris were only recorded a few times. The ambergris that weighed 2.4 kg had a length of 24 cm and a width of 12 cm. The ambergris that weighed 5 kg had the following transverse and longitudinal measurements: L = 30 cm, l = 15 cm and h (height) = 15 cm. A piece of grey ambergris from my

Table 5.6 Number of pieces of ambergris found in sperm whales caught by the fleets Slava and Soviet Ukraine

Fleet	Cruise	Number of sperm whales			Found ambergris	
		Total	Male	Female	Male	Female
Slava	1961\62	388	126	262	2	–
Slava	1962\63	865	546	319	1	–
Slava	1963\64	1656	941	715	21	5
Slava	1964\65	1767	610	1157	19	12
Slava	1965\66	888	321	567	6	5
Soviet Ukraine	1961\62	188	182	6	–	–
Soviet Ukraine	1962\63	883	854	29	8	–
Soviet Ukraine	1963\64	1534	997	537	2	–
Soviet Ukraine	1964\65	2570	871	1699	14	5
Soviet Ukraine	1965\66	1401	1062	339	3	–
Soviet Ukraine	1966\67	1735	1497	238	2	–
Soviet Ukraine	1967\68	1867	928	939	5	–
Soviet Ukraine	1968\69	1509	995	514	14	1
Soviet Ukraine	1969\70	3289	1632	1657	2	1
Soviet Ukraine	1970\71	2553	1532	1021	2	6
Soviet Ukraine	1971\72	4088	2183	1905	1	1
Soviet Ukraine	1972\73	2855	1807	1048	1	1
Soviet Ukraine	1973\74	2856	1808	1048	–	2
Soviet Ukraine	1974\75	2841	1804	1037	–	–
Soviet Ukraine	1975\76	3343	1945	1398	–	1
Soviet Ukraine	1976\77	1980	1573	407	3	2
Soviet Ukraine	1977\78	2155	1679	476	2	–
TOTAL:	22 cruises	43,211	25,893	17,318	108	42

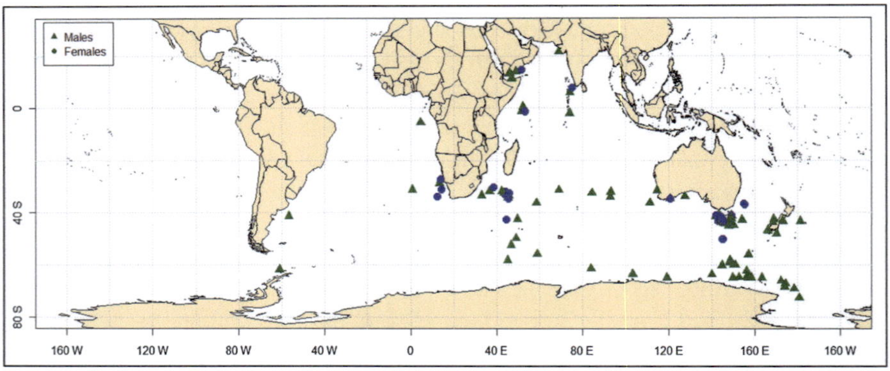

Fig. 5.45 Distribution of sperm whales with ambergris

Table 5.7 Ambergris weight distribution by sperm whale size range

Sperm whale length, L, m	8.1–9.0	9.1–10.0	10.1–11.0	11.1–12.0	12.1–13.0	13.1–14.0	14.1–15.0	15.1–16.0	16.1–17.0	*Total:*
Number of males ♂♂	–	2	10	13	13	14	20	7	2	*81*
Number of females ♀♀	1	7	15	7	–	–	–	–	–	*30*
Number of males and females ♂♀	1	9	25	20	13	14	20	7	2	*111*
Total ambergris weight, kg ♂♀	1.0	175.0	453.1	247.0	284.7	233.2	380.8	50.3	36.2	*1861.3*
Average ambergris weight, kg ♂♀	1.0	19.4	18.1	12.4	21.9	16.7	19.0	7.2	18.1	*16.8*
Average ambergris weight, kg ♂♂	–	6.5	14.3	15.2	21.9	16.7	19.0	7.2	18.1	*13*
Average ambergris weight, kg ♀♀	1.0	23.1	20.7	6.9	–	–	–	–	–	*17.4*

Fig. 5.46 Several boulders of ambergris found in the sperm whale male with length 11.2 m

Fig. 5.47 Grey ambergris weighing 110 g

Fig. 5.48 Black ambergris weighing 15 g

collection (Fig. 5.47) with a weight of 110 g has the following linear dimensions: L = 6.0 cm, l = 5.5 cm, h = 5.0 cm. And a piece of black ambergris from my collection (Fig. 5.48) with a weight of 15 g has a length of 4.0 cm, width of 2.5 cm and height of 2.5 cm.

Two pieces of grey ambergris weighing about 10 kg, found in a 10.9 m female sperm whale were nearly spherical (Fig. 5.49). The bigger one has the following linear measurements: L = 45 cm and l = 40 cm. Measurements of the smaller piece are respectively L = 22 cm and l = 20 cm. Unfortunately this is all we have at our disposal. Perhaps in the future we'll be able to find other "whale inspection logs," from Ukrainian and Russian whalers.

In the "whale inspection logs" the detection of black ambergris,, with an average weight of 6.9 kg, was reported 12 times. Grey ambergris was recorded half the number of times – 6 cases, with an average weight of 11.8 kg. That is, the weight of grey ambergris was 1.7 times higher than that of black ambergris. This is understandable, as grey ambergris is formed from black ambergris and is formed over a longer period so it is older and bigger. For males, the most significant pieces (boulders) of ambergris weighed 50 kg, 68 kg, 80 kg, 85 kg, 90 kg, 110 kg, and for the female – 50 kg,

Fig. 5.49 Boulders of ambergris found in the female sperm whale with length 10.9 m

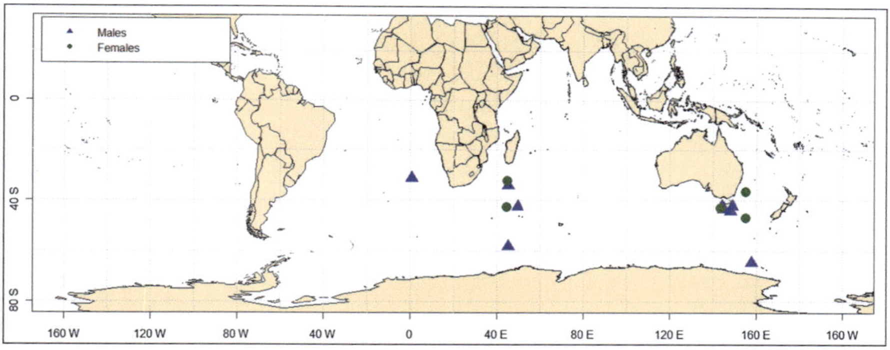

Fig. 5.50 Distribution of sperm whales with the largest ambergris boulders

53 kg, 70 kg and 94 kg. The geographical distribution of sperm whales with the largest single pieces of ambergris is shown in the following map (Fig. 5.50).

As can be seen (Fig. 5.50), sperm whales with large ambergris boulders were found south of 35°S. In addition, two aggregations (possibly two populations) of sperm whales are identified: one is migrating in the meridional direction to the south of Madagascar and the second is migrating from Australia and Tasmania to the Balleny Islands.

In the literature there are reports of even larger ambergris boulders than ones found by the *Slava* and *Soviet Ukraine* fleets. R. Clarke (Clarke 1954) writes about an ambergris boulder weighing 924 pounds (419 kg) with a cross section of 66 cm extracted from a sperm whale,. In another case an ambergris boulder weighing 926 pounds (420 kg), length 165 cm, with a cross-section of more than half a meter was extracted from a sperm whale caught in Antarctic waters.

What is the fate of sperm whales with such large ambergris formations? According to M.V. Ivashin (1968), the during digestion of food, in the intestines of sperm whales a soft black mass develops first, which in the process of moving through the intestines, adheres to other such formations, developing into larger ambergris formations. We need to agree with this assumption: in the above photograph (Fig. 5.46) it is shown that black ambergris, just extracted from the intestines of the sperm whale, is a lump of some smaller formations glued together. While moving further through the intestines, a lump of ambergris becomes larger and denser, gradually turning into solid grey ambergris (Fig. 5.49). While smaller formations of black ambergris are pliable and can be evacuated with faeces through the anus, such evacuation of large formations of solid grey ambergris is hardly possible, and the fate of these whales apparently abysmal. The ambergris extraction process from a dead sperm whale is colorfully described in *Moby Dick* (Melville 1851). However, in our practice, thin, or sick sperm whales with large ambergris boulders were not observed. Other investigators have reported emaciated sperm whales, burdened with large ambergris (Schwedtiawer 1783; Atwood 1869; Peelle 1932).

Incidence of Ambergris For a long time it was believed that ambergris is not found in female sperm whales, or that it was extremely rare and in any case smaller than for males (Schwedtiawer 1783; Hardy 1949). Our data indicate that ambergris occurs in both males and females (Table 5.6), but the percentage of occurrence in males is almost twice as high (0.42% of cases) than in females (0.24% cases). At the same time, for the same sized animals, the average weight of ambergris for male sperm whales was lower than for females (Table 5.7). And this is understandable – with the same body length, females are older than males, and, consequently, ambergris in their gut has taken longer to form.

As for the analysis of all captured sperm whales, according to the data by H. Thomas (Thomas 1955) ambergris was found in one of 100. M.V. Ivashin (1966) believes that the occurrence of ambergris is – three to four%. Of the data from 22 cruises we analyzed (Table 5.6), ambergris was not found at all in two cruises. For another two cruises only one occasion was registered for each. Between two and five cases of ambergris detection were noted in eleven cruises. And only for four cruises ambergris was observed relatively often – from 15 to 25 times. In general, the frequency of occurrence of ambergris in sperm whales according to our data was only 0.35% of cases.

Even taking into account the fact that ambergris was not always noticed, and that not every case when it was found was recorded by the biologists' research groups, it must be noted that ambergris in sperm whales is very rare. But, more importantly, it is not clear why the incidence of ambergris in different regions has such a different appearance. Thus, on neither our map (see Fig. 5.50), nor on the map published by M. Ivashin (1966), is ambergris found (except for individual cases) in the Atlantic or in the Pacific Ocean east of New Zealand. Almost all cases of detection of ambergris in sperm whales were in the Indian Ocean. And, more surprisingly, for some reason only in an area (see Fig. 5.51), which is very similar to the area in which baleen

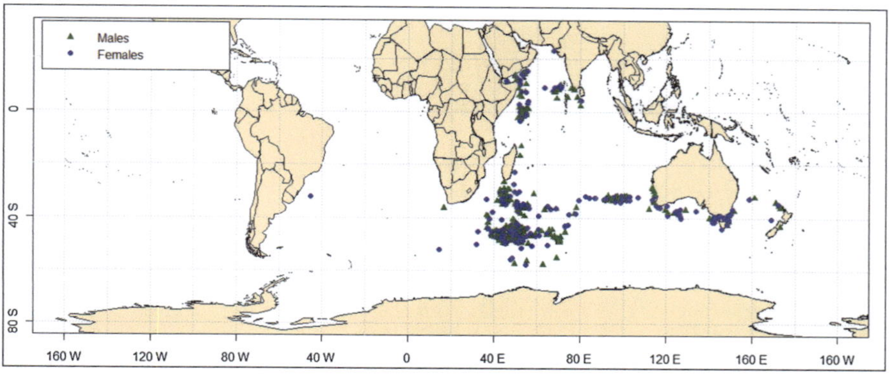

Fig. 5.51 Distribution of pygmy blue whales (Balaenoptera brevicaudis)

whales' representative – pygmy blue whales – are found. (Mikhalev 1998, 2008). Apparently, this area has some environmental characteristics that are essentially different from the neighboring Atlantic and Pacific areas!

Pygmy blue whales are thermophilic, they do not migrate south of the 50th latitude, and for that matter, neither do female sperm whales. Therefore, distribution areas for female sperm whales with ambergris and pygmy blue whales almost completely overlap. Some difference is observed for large sperm whale males, whose populations migrate by different southerly paths as far as the Antarctic coast. However, it appears that ambergris is formed in them in the warm waters of the lower latitudes.

In fairness it should be noted that apart from the main area where pygmy blue whales are found, these baleen whales, or a similar species, were observed in small numbers outside the Indian Ocean, including along the east coast of South America east of Brazil (Secchi and Della Rosa 1997; Pinedo and Barreto 1996). And it is precisely in this area according to records from the *Slava* and *Soviet Ukraine* that a few samples of sperm whale ambergris were detected (see Fig. 5.44). Sperm whales with ambergris were noted near Durban, where Ommaney (Ommaney 1938) found a similarity to ambergris in a baleen whales' representative – a blue whale. Let's recall that in the analyzed period the two fleets from Odessa caught more than 43,000 sperm whales. The sample is not small and this is not a random occurrence of such a coincidence of areas for sperm whale ambergris and pygmy blue whales. But what environmental causes – which biotic and abiotic conditions of the Indian Ocean – define this phenomenon? To answer this question, future researchers will first need to analyze the trophic relations.

Chemical Composition of Ambergris The chemical basis of ambergris is polyols, most of which are ambrein (25–45%) and free or bound epikoprosterol (30–40%), which give ambergris the properties of a scent retainer (Lederer 1953; Jegou 1977; Mookherjee and Patel 1979; Kiesewetter 1953). However, it is not clear what processes in the origin of ambergris lead to the formation of these substances.

Fig. 5.52 Ambergris coated with crystals

During my shift, working as a biologist on the *Slava* flotilla as shift biologist on 12th April 1966, ambergris was extracted from a 15.1 m male sperm whale, caught at 46°49′S 165°59′E. The top of the ambergris was covered with large, shiny, transparent, elongated crystals, which oxidized in the air and were covered with powdery white scurf. This scurf was easily washed off with water and the crystals again became transparent and bright. A piece of this ambergris is preserved in my collection (Figs. 5.44 and 5.52).

In the "whale inspection logs" there are two entries about ambergris crystals: a 15.8 m male sperm whale caught on 19th January 1966 at 66°04′S 172°10′E, and a 15.1 m male sperm whale caught on 8th January 1967 at 72°40′S 179°15′W. That is, only three such cases out of 150 cases of detection of ambergris were found. There is no other information in the national literature. In international literature, crystals on the surface of ambergris were described for a sperm whale caught in the area of British Columbia (Carter and Elsey 1954). According to R. Clarke (Clarke 1954) crystals were never observed in ambergris found on the shore, or at sea, but was occasionally seen in ambergris extracted from captured sperm whales. In particular, Clarke noted crystals on the surface of an ambergris boulder weighing 421 kg, extracted from a 14.9 m male sperm whale, caught on 21 December 1953, at 58°23′S 14°13′W, by the whaler *Southern Harvester*. The chief chemist onboard this whaler found that the crystals were magnesium ammonium phosphate (MgNHPO). The presence of magnesium ammonium phosphate in fecal masses of whales feeding on abundant, protein-rich prey had been pointed out by Pouchet

(Pouchet 1893). It is not clear just why the crystals formed on ambergris are so rare. There are more questions than answers.

To summarize:,of the 43,211 sperm whales (25,893 males and 17,318 females) caught from 1961 to 1966 by the Slava flotilla and from 1961 to 1978, by the Soviet Ukraine flotilla (22 commercial cruises), ambergris was found 150 times, which accounted for only 0.35% of the total sperm whales caught. Ambergris incidence in males (0.42% of cases) is almost twice as high as in females (0.24% of cases).

Black ambergris is more frequently found than gray ambergris. There were three incidents of ambergris covered by large transparent crystals, having the chemical composition of magnesium ammonium phosphate MgNH4PO4.

Ambergris was presented as individual pieces weighing a few grams and as boulders weighing up to 100 kg or more. The average weight of ambergris found in females was 17.4 kg and was higher than that in males – 13.0 kg, probably because at the same length females are older than males, and, consequently, the ambergris has more time to grow in females.

While pliable, soft and relatively small pieces of black ambergris can be freely evacuated from the intestines of sperm whales with faecal matter, the sperm whales burdened with large boulders of solid grey ambergris apparently died.

The nature of the distribution of sperm whales with the largest ambergris boulders (more than 50 kg) revealed the presence of two fairly isolated populations of sperm whales in the Antarctic region of the Indian Ocean – one south of Madagascar and the other south of Tasmania.

If we agree that the formation of ambergris is provoked by chitin from squid beaks, we cannot exclude the possibility of its formation in other whales that eat squid. So far, except for bighead sperm whales (*Physeter makrocephalus*) detection of ambergris is known for pygmy sperm whales (*Sogia breviceps* and *Kogia simus*) and northern bottlenoses (*Nureroodon ampullatus*), but the reliability of these data is low and requires confirmation.

The area of the most frequent occurrence of ambergris in sperm whales coincided almost exactly with the area frequented by the endemic Indian Ocean, representative of baleen whales, the pygmy blue whale. It is not clear what environmental features of the Indian Ocean are responsible for this coincidence.

5.5 Summary

We presented maps of whale distribution in the Southern Ocean and adjacent waters, taking into account tagging data, which were previously kept in secret, and poaching whaling (more than 100.000 whales). We produced more accurate data about whale distribution and their migrations. The maps differ significantly from published before by other researchers, and additional biological information is added to them. The highest concentrations of baleen and toothed whales are noticed in highly productive waters of the Southern Ocean. Also low productive ("dead") areas were determined, where whales are not present even during migrations.

We determined correlation between diatom coating of female whales and mean sizes of their embryos. Based on this we developed an original method to determine the speed of diatom coating. The method allowed to reveal complicated migrations of Minke whales: males arrive to the feeding areas first, then – mature females, then – pregnant and immature ones. Large pregnant females leave feeding areas first, then – males and mature females. A part of Minke whales stay in Antarctic waters even during winter.

We found differences of whale populations using position of "pair formations", extent and type of skin damage, peculiarities of sperm whale tooth structure, analysis of mean embryo sixes, peculiarities of body coloration and other features. According to these features fin whales of the Chilean-Peruvian area differ from fin whales of the area of the island South Georgia. Whales of the Bellingshausen Sea differ from whales of the area of islands Bouvet and Gough and from whales of the island area of the Indian Ocean. Sei whales of the southern Atlantic Ocean differ from whales of the Indian Ocean and Chilean area. There are sharp differences between Minke whales of the Amundsen, Bellingshausen, Weddell, Mawson and Cooperation Seas. There are noticeable differences between sperm whales of all three oceans.

Based on the analysis of body proportions, mean sizes of animals of the same age and rate of accumulation of ovulation and pregnancy traces, as well as isolation of their habitation areas, we concluded that in the Southern Ocean, in addition to regular killer whale (*Orcinus orca*, Linnaeus 1758) there is also a short-bodied killer whale – *Orcinus nana*, Mikhalev 1981.

Chapter 6
Southern Ocean Ecosystem Status and Perspectives on Whale Population Recovery

Y. Mikhalev, *Whales of the Southern Ocean*, Advances in Polar Ecology 5,
https://doi.org/10.1007/978-3-030-29252-2_6

6.1 Structure and Productivity of Communities in the Southern Ocean

Cetaceans, being one of the animals at the top of trophic pyramids, play an important role in marine (rarely in fresh-water) ecosystems. Their viability considerably defines the speed and efficiency of the cycle of organic matter and energy in biogeocenosis and substantially affects the biological efficiency of ecosystems. As a result of extended transzonal migrations, the value of cetaceans in the realization of transboundary transportation of substance and energy, both vertical and horizontal, is great. The influence of whales on various communities of the Southern Ocean ecosystems is so diverse and ambiguous, that with the current level of knowledge, it can hardly be estimated quantitatively or even qualitatively. Such estimations can only be of a rough nature.

The dramatic reduction in number of some whale species, caused by intensive catching during a very short period of time from the evolutionary point of view, should result in significant changes in the functioning of the ecosystems of the world's oceans. Researchers pay the greatest attention to the question of the biological consequences of mass destruction of southern hemisphere baleen whales. According to our calculations, from 1.5 up to 2.0 millions tons of biomass of whales per season were withdrawn from the Southern Ocean and adjacent waters. How did this affect the pelagic communities of the Southern Ocean? The definition of size of the "surplus" of Antarctic krill, which had been consumed by whales (the so-called "surplus of krill"), and the opportunities for its withdrawal and use by people, is also connected with this question.

These questions are rather complicated and difficult to solve. The difficulties are connected with a lack of information about the current condition and productivity of the Southern Ocean pelagic ecosystem, theh nature of trophic links, the importance of separate food chains in the transfer of matter and energy, and the abundance and biomass of organisms on different trophic levels. The available estimations, recorded by different researchers using a variety of methods and approaches, essentially differ (Bushuev and Mikhalev 2000). An even more difficult task is to reconstruct the conditions of the Antarctic marine ecosystem, which existed before the destruction of big baleen and toothed whales, and to define the importance and scale of the ensuing changes.

Retrospective estimations of the productivity of the pelagic ecosystem in the Southern Ocean region is based mainly on calculations of possible consumption of prey by whale populations at the original level. The publication of real Soviet whaling data from 1947 to 1972 (Yablokov 1994; Zemsky et al. 1994a, b, 1995, 1996, 1997; Tormosov et al. 1998; Yablokov, etc., 1998; Kasuya 1999; Mikhalev 2000, 2000a, 2002b, c; Zemsky and Mikhalev 2000a, b, etc.) should help to specify real volumes of initial whales stocks and give a more reliable estimation of their influence on marine communities.

Nowadays it is accepted that the Southern Ocean is not a single large-scale ecosystem. In its communities all trophic chains somehow include Antarctic krill (Euphausia superba Dana) and represent a number of ecosystems that differ in the structure and character of their functional links (Lubimova 1987). The opinion that the trophic system is based on a three-linked chain "phytoplankton-krill-toothed whales" has not been the prevailing one in the Southern Ocean (Naumov 1983; Voronina 1987).

G. Naumov (1983) distinguished four systems of food chains that have between three and seven trophic levels, with warm-blooded (hematothermal) vertebrates at the top: baleen whales, toothed whale's pinnipeds and birds. The nature of matter and energy streams in these systems essentially differs, whenever the efficiency of formation (power cost) of consumers' biomass of the highest level differs by some order in relation to primary production.

The main production part of consumers of the first order in the Southern Ocean trophic net is related to copepods. The production of copepods is estimated to be more than two billion t/year (Voronina 1987). Besides, they are mostly consumed by planktonic predators (copepods, chaetognatha, hyperiidea) and small fish (mainly myctophids). The importance of a short food chain, for instance "phytoplankton – copepods – sei whales", is rather insignificant.

As to estimations of krill stock, they vary widely depending on the applied method of calculation, such as direct calculation – trawling and hydro-acoustic samplings; or estimations based on the definition of volumes of primary production, abundance and consumption of consumers of various orders, and also on the probable area of mass development of *E. superba* and on the average value of krill's density of distribution in the Southern Ocean waters used as a basis. Therefore a range of existing estimations of krill's biomass is very large – from 60 million tons up to 25 billion tons, and forecasts about the volumes of its allowable withdrawal vary from 5 million tons up to 1.8 billion tons (Pequegnat 1958; McQuillan 1962; Mackintosh 1970; Gulland 1970; Allen 1971; Nemoto and Nasu 1975; Everson 1977; Klumov 1963; Bogdanov and Lubimova 1978; Naumov 1978; Lubimova and Shevtsov 1980; Lubimova and Shust 1980, 1987; Naumov and Chekunova 1983; Laws 1985). It is obvious, that with the current level of knowledge about the processes of circulation of substances and energy (cycle of matter) in the Southern Ocean ecosystems, it is difficult to determine the volumes of production and total biomass of krill. That is why many researchers, having determined the volumes of possible withdrawal of krill, abstain from recording estimations of the general biomass. Thus, the volume of allowable commercial withdrawal of krill is usually connected with amount of so-called "surplus" of krill at the expense of the decrease in its consumption by baleen whales.

6.2 Place and Role of Cetaceans in Ecosystems of the Southern Ocean

Originally it was considered that baleen whales were the main consumers of krill in the the Southern Ocean and that the amount of krill eaten by all other groups of animals was much less than that consumed by whales (Pequegnat 1958; McQuillan 1962), but subsequently this assumption has been challenged. A. G. Naumov (1978), A. G. Naumov and V. I. Chekunova (1983) considered that whales accounted for about 20% of krill consumption earlier, now it is believed to be 10–15%, with the majority being eaten by fish and cephalopods. Nevertheless, the majority of estimations about original amounts of krill consumption by baleen whales before their numbers were reduced were characterized by rather big sizes with a rather small range of values – 120–288 million t/year (Zenkovich 1969, 1970; Mackintosh 1970; Nemoto and Nasu 1975; Naumov and Chekunova 1983; Laws 1977). However, the actual volume of withdrawal was even bigger, because on the basis of the data on large-scale poaching of whales their initial population should be reconsidered towards substantial growth.

R. Laws (Laws 1977) made an attempt to analyze zonal distribution of all whale species in the Southern Ocean and to determine a degree of availability of krill and other prey, such as fish and squid, for every species separately. According to his calculations, the baleen whale population originally consumed about 190 million tons of krill per year, whereas contemporary consumption is 43 million t/year. The difference of 147 million tons corresponds to the "surplus" of krill biomass. As a result of seasonal whale migrations in low latitudes, about 42% of the energy which has been accumulated by whales in the form of fat during the feeding period, is removed from the Antarctic.

In recalculating the amount of krill consumed, the energy removed from the Southern Ocean is reduced from 60–70 million t/year as an initial level, up to 9.7 million t/year in 1970. Certainly, these figures are only approximate. A.G. Naumov and V.I. Chekunova (1983), using the same data based on the number and biology of whales as R. Laws (Laws 1977), but other energy expense coefficients for the animals' ability to live, have defined the size of krill consumption by baleen whales at an initial level of their number, which equaled 288 million t/year, and size of the krill "surplus" – as 218 million t/year.

Now, following the publication of data about large-scale poaching, the initial data demand an essential updating (Bushuev and Mikhalev 2000). For example, R. Laws (Laws 1977) used the initial number of minke whales as 200,000. Thus, it was calculated that this species has been consuming 20 million t. of Euphausidae annually. However, by later estimations of the SC IWC, the contemporary minke whale population in the Southern hemisphere was determined to be 760,000 (Borodin 1986), a number which is probably underestimated.

R. Laws (Laws 1977) has proceeded from the fact that the minke whale is a permanent inhabitant of the Antarctic region and continues to feed on krill all year round. This statement doesn't agree with the accepted opinion about the annual cycle of minke whales, which, similar to large baleen whales, has well determined feeding periods in the Antarctic region and winter feeding periods in low latitudes, which has been proved by visual observations, catches and the results of tagging (Ivashin et al. 1972; Best 1982; Mikhalev and Tormosov 1997; Kato 1987; Williamson 1975).

Only a small part of the population can remain in high latitudes all year round. In winter, krill aggregations stay mainly in a dissipated condition (Bogdanov and Lubimova 1978). In connection with the significant distribution of floating ice to the north, in winter practically all areas of *E. superba* are inaccessible to whales, and they are compelled to eat other, less abundant species of *Euphausidae*. Therefore, it is energetically inexpedient for the majority of minke whales to stay in the Antarctic (under condition of normal summer feeding period) for winter.

In view of the aforesaid, S.G. Bushuev (2000) calculated the amount of krill consumption by the initial population of Minke whales (382,000), which made 14.4 million t/year. By S.G. Bushuev's and Y.A. Mikhalev's (2000) calculations, krill consumption by the current population of Minke whales is 22–28 million t/year. A. Armstrong and V. Siegfried (Armstrong and Siegfried 1991) estimated annual krill consumption by these species to be even higher – 35 million t.

Despite the optimistic views expressed earlier about an opportunity to restore fin whale, humpback and blue whale populations through the long-term cessation of catch (Mackintosh 1970), the destruction of whales by poaching has not resulted in an essential increase of big baleen whale populations from the 1970s till now, and a decrease for some species. Anyway, the number of large baleen whales feeding in the areal of *E. superba* is much lower than that estimated by R. Laws' calculations (Laws 1977). Accordingly, the amount of krill consumed by whales should also be lower.

Of the total amount of krill formerly consumed by baleen whales, the majority – up to 95% – is eaten now only by minke whales, who maintained their numbers (Armstrong and Siegfried 1991), whereas initially their share did not exceed 8–10%.

6.3 Changes in the Sothern Ocean Ecosystems Related to Catches of Large Species of Baleen Whales

How has the "release" of significant amounts of krill following the extermination of big whale species affected the marine communities in the Antarctic region? Which groups of consumers has this "surplus" been used by? Unfortunately, there are still more questions than satisfactory answers.

In the second part of the twentieth century a significant number of researchers began to pay attention to changes in Antarctic communities. In most of them the intensity of reproduction, increase in number of populations and expansion of habitats of some species which consume Antarctic krill was registered. Directly or indirectly, these changes are caused by an improvement in feeding conditions.

Undoubtedly, these processes take place, but not all of the arguments submitted are convincing. In particular, changes supposedly occurred in the reproductive cycles of big species of baleen whales (fin whale, blue whale, sei whale) expressed by an increase in the number of pregnant females and decrease in the age of sexual maturity by 5–6 years (Sigurjonson 1980; Lockyer 1972, 1974; Laws 1977; Doi et al. 1970; Gambell 1968, 1973), have more likely occurred not from an improvement of feeding conditions but from methodical mistakes of researchers. In our opinion, the earlier seasonal recording of sei whales in the Antarctic and their detection in more southerly locations than usual (Gambell 1968) can be explained by an insufficient level of knowledge of this species' migrations, rather than by an expansion of its areal. In any case, these changes have not resulted in an appreciable increase either in numbers of sei whale or other baleen whale species.

Pinguins are eating krill too

Several works refer to a five- to eight-year decrease in minke whales' age of sexual maturity from the 1940s till the 1970s (Best 1982; Kato 1983, 1987; Kato and Sakuramoto 1991) and to a decrease in middle-aged females that are pregnant for the first time, along with an increase in growth rate and body length from the moment of physical maturity (Kato 1983, 1987). However, the proof on which these claims are based raises doubts. A change in biological parameters can be caused by changes of catch locations and numbers. Thus, M. V. Ivashin (1984) states that there was a 0.2–0.3 m increase in the average sizes of minke whales caught from the beginning of the 1970s up to the mid-1980s. However, while making his calculations, he used forged statistical data instead of actual catch results. Ivashin's conclusion that there had been a decrease in the age of onset of sexual and physical maturity on the basis of the analysis of the location of a so-called "transitional phase" in whale ear plugs is also very doubtful. This methodology is arguable and is not accepted by most researchers. J. Cook and V. De La Mar (Cook and De La Mar 1983) have expressed doubts about the reliability of these data, having specified the problems posed by defining age exactly and sample representation. We have also shown an inaccuracy in the method of defining baleen whale sexual maturity on the basis of the "transitional phase" position in ear plugs (Mikhalev 1990a, c, 2000; Mikhalev 1991).

The statement that the extermination of big species of baleen whales increases the number of minke whales and, hence, their consumption of Antarctic krill (Borodin 1996), does not stand up to criticism. To doubt the conclusion, it is enough to note that the estimations of minke whale numbers in the Antarctic region have improbably increased – from 70,000 (Ohsumi et al. 1970) up to 200,000 (Laws 1977), then up to 382,000 (IWC., Rep.int. Whal. Commn. 1985), and even up to 760,000 (IWC. Sci. Rep. int. Whal. Commn. 1991) in two decades. In other words, during two decades the number of minke whales has increased by ten times, which obviously contradicts their reproductive ability. But even if the consumption of krill by minke whales (whose numbers increased so incredibly for the last decades) has increased twice or three times (Kato 1987; Bushuev 2000; Armstrong and Siegfried 1991), the figure appears lower by ten times than the calculated surplus of krill. It is obvious that with a catastrophic reduction in numbers of big species of baleen whales, minke whales could have used only a small part of the "surplus" of krill.

Other research testifies to the increase in population of some species of penguin (Sladen 1964) and pinnipeds (Laws 1973, 1997, 1985; Payne 1977; Bengston and Laws 1985; IWC. Rep. Int. Whal. Commn. 1994) from the 1940s till the 1970s. By R. Laws's estimations (Laws 1985), for the last 20–30 years the population of Antarctic seals has been growing by 7.5% per year, Adélie penguins – 2–3% per year, Antarctic and Macaroni penguins – 6–10% per year. He believes that due to the growth in number of birds and pinnipeds, all "superfluous" krill has been consumed by them, and the upper trophic level of the Antarctic community changed from "whale" to mainly penguin and seal. The total consumption of krill by these groups of consumers has increased from 100 million to 260 million t/year (Laws 1985).

Unfortunately, there are many reasons to doubt the reliability of these conclusions. Firstly, because the observations of populations of marine animal growth, which were carried out mainly in one of the specified areas – in the Atlantic part of the Antarctic – were extrapolated through its whole areal. Secondly, there are no other works except for research by R. Laws (Laws 1977) in this area. Only rough data about the increase in the number of crabeaters from 15 million to 17 million (Borodin 1996), which is within the margin of error of the method used, are available. The initial population size of pinnipeds and birds is not reliable because more or less regular recording began only in the 1970s and the volume of data accumulated thus far is still insignificant.

It is also necessary to note that the "surplus" of krill, which had been eaten earlier by big species of baleen whales, could hardly be used completely or even to a greater extent by birds and seals, in view of their species-specific distinctions in distribution and food consumption. While baleen whale feeding areas and habitats of Antarctic seals and penguins overlap, zones of the largest mass concentration of these consumer groups do not coincide.

Thus, during the fattening period, seals and penguins gravitate towards either the zone of pack-ice, where they can access numerous cracks in the ice, ice-holes and fractures or to the open coast. In clear water far from the coast or ice edges, seals and penguins are rather rare. A characteristic feature of minke whale distribution during the feeding period is their concentration near the pack-ice edge, and when there is the least amount of ice – directly near the continental coast (Ichii T. 1990). Minke whales are very rarely seen in open water of significant distance from the ice edge, except during migrations. According to our observations, minke whales usually do not stay in open water for a long time, even where well expressed krill stains occur.

On the other hand, the main regions of concentration of big species of baleen whales are located appreciably to the north of the pack-ice edge (with the exception of blue whales in the IV–VI trade sectors of the Antarctic). Using to advantage their big size and the specificity of their baleen, large baleen whales can successfully feed on more rarefied and less stable concentrations of krill than minke whales, and even more so than seals and penguins. Oceanic birds, widely distributed through the whole Antarctic region, can take krill only from the water surface and, hence, the main concentrations of krill in the water column are completely inaccessible to these birds.

The main feeding regions of seals, penguins and minke whales may coincide with the reproductive area of *E. superba* populations in the system of closed circulations in the neritic zone of the Antarctic region. Big species of baleen whales to a greater extent feed on that part of the krill population which is carried away to the north by western branches of cyclonic circulations (Samyshev 1983). Thus, seals and penguins could most probably use only a part of the "surplus of krill" accessible to them.

According to the aforesaid, we can assume that with the slaughter of whales, the greatest part of the "surplus of krill" was consumed by fish and cephalopods. Krill is the source of nutrition of a significant number of Antarctic fish species of the bathypelagic complex (*Myctophidae* family) and the demersal complex (mainly families *Channichthyidae* and *Nototheniidae*), and also of southern blue whiting during their seasonal migration (Lubimova and Shust 1980). Other mass consumers of krill are Antarctic squid, first of all *Brachioteutis riisei*, which feeds only crustaceans, and also some species of the mesopelagic zone (Filippova and Stygar 1987).

An estimation of the total fish and squid biomass of the Antarctic region and the amount of krill they consume represents an even greater challenge than for warm-blooded animals. Furthermore, several estimations of the general biomass of squid has been executed on the basis of calculating the possible consumption of cephalopods by whales (Cetaceans), pinnipeds and birds, which was based on rather rough estimates of their numbers (Everson 1977; Filippova and Stygar 1987).

I. Everson (Everson 1977) estimated the quantity of krill eaten by squid as 80–100 million t/year. T. G. Lubimova and K. V. Shust (1980) defined the rough amount of consumption of krill by fish as 24–28 million t/year, squids – 50–54 million t/year. However, possible consumption of krill by *Myctophidae* was not taken into account. Although krill is not the basic feeding target for *Myctophidae*, keeping in mind the latest data about their plentiful numbers in the Antarctic region

(Lubimova and Shust 1987), the amount of krill they consume could be significant.

It is necessary to remember that, unlike warm-blooded vertebrates, fish and squid belong to intermediate levels of trophic chains. Therefore recycling of the "surplus of krill" could be accounted for by a displacement of emphasis from short to long food chain systems. A.G. Naumov (1983) calculated that a displacement of emphasis by 10–20% from a four-level food chain (primary production – krill – mesopelagic fish – squid – sperm whale) to longer food chains (for example, primary production – krill or mesoplankton – mesopelagic fish – giant squid – sperm whale) might have resulted in the full assimilation of the so-called "surplus of krill" in the feeding area of male sperm whales. A similar mechanism of the stabilization of energy balance could have allowed the complete or part utilization of the superfluous amount of krill without visible dramatic changes in the structure or number of basic consumer groups (Naumov 1983).

The most unpleasant development of events is not excluded when, owing to reduction of large whale number and to decrease of importance of trophic chain "phytoplankton – krill – baleen whales" in the ecosystem, reduction of production and biomass of *E. superba*. As a result, the hypothesized "surplus of krill" could appear to be not so high. N. M. Voronina (1987) considered the possibility of three variants of after-effects, resulting in a reduction in the role of *E. superba* in the ecosystem according to which it will be replaced by other groups of consumers of the first level – by copepods, finer *Euphausidae*, which do not swarm, and by salps (Samyshev 2006). In all these cases the production and biomass of the most valuable trade items of the Antarctic region – krill and whales – can be considerably reduced.

6.4 An Intraspecific Competition and Problems of Population Recovery in Large Species of Baleen Whales

A number of researchers adhere to the hypothesis, according to which the population recovery of large baleen whale species is restrained as a result of intraspecific competition with smaller and more numerous whale species. In particular, the southern right whale population is supposedly being restrained by sei whales (Mitchell 1974), and that of the blue whale by minke whales (Kawamura 1994). This hypothesis was considered by the Scientific Committee of the IWC (IWC. Rep. Int. Whal. Commn. 1994, 1995). Having analyzed the available data on this question, F. Clapham and R. Brownell, Jr. (Clapham and Brownell Jr 1995) have come to the conclusion that there is no authentic proof of the existence of tough intraspecific competition between these species.

In our opinion, from the ecological point of view there is every reason to believe that the intraspecific competition among baleen whales is minimal. Even in the case of limited feeding resources, whales neither establish territoriality nor protect certain areas of water, as they consume diversely distributed and mobile living resources. B. R. Tershy (Tershy 1992) while analyzing the data on feeding, character of distribution and social behavior of four baleen whale species in the California Gulf – blue, fin, Bryde's and minke whales), established for them precise ecological differentiation. Smaller Bryde's and minke whales feed in confined areas with a high density of prey, mainly fish. Fin whales and blue whales feed in spacious areas with a rather low concentration of *Euphausidae*. As mentioned above, a similar division in the development of feeding areas is also observed in the Antarctic region. Minke whales prefer to feed on *E. superba* swarms that are more dense, stable and close to the surface along the ice edge, while fin whale and blue whales feed on more rarefied and mobile aggregations of krill in more northerly areas in open water.

In the feeding zone preferred by large baleen whales, other consumers of the highest level – seals, penguins and oceanic birds - cannot compete with them.

Some researchers (Murphy et al. 1988; Fraser et al. 1992) believe that competition between whales and other smaller consumers of krill is rather low. A lack of food in feeding areas of large whales could be connected with the probable increased consumption of krill by fish and squid or with partial replacement of *E. superba* in the food chain of other zooplankton organisms. However there is no evidence of a reduction of krill biomass in this zone.

Apparently, the slow recovery of the population of large whales can be explained neither by competition with other species, nor by lack of food, but in the first place by rather low rates of reproduction, which is typical for them. The factor of disturbance and increasing anthropogenic pollution of the world's oceans may have a negative influence on the growth of whale populations, which is, fortunately, insignificant at present. The process of recovery of populations of large whale species is very long. The required condition is a long prohibition of commercial whaling, ecological monitoring and restriction of anthropogenic impact on marine ecosystems.

6.5 Perspective of Whale Population Recovery

What are the prospects of restoring whale pods? Mostly they are negative. Firstly it is necessary to say that stocks of all commercial whale species (except for minke whales) have catastrophically decreased, and some of them are almost extinct.

It is obvious that the prospective population recovery of some whale species cannot be determined only on the basis of data about their reproductive ability. It is also necessary to know the size of the initial population and that of the so-called current population. But there are significant difficulties for researchers here. Unfortunately, the methodology is unreliable, and the answers to these questions are too approximate for many whale species.

Minke Whales

Nowadays only minke whales are caught commercially in the Southern hemisphere, therefore we shall start with this species. For a long time minke whales had been extracted only as bycatch, they became an object of active catch only in the 1971–1972 season. Since then, the catch quota was controlled by the International Inspection (the latter was not absolutely innocent, indeed!) and lasted for less than 20 years and consequently, the stocks have remained close to the initial figures. That is why the estimation of their initial population could be rather demonstrative as to the level and reliability of the methodology which gave underestimated figures, and the influence of political conjuncture.

Remember that in the 1970s there were 70,000 minke whales in the Southern hemisphere, when these animals were almost never hunted (Ohsumi et al. 1970). Further estimations of their population began to grow uncontrollably. In 1981, after only a 10-year period of minimal hunting, the experts of the IWC estimated stocks of this species to be 150,000 animals, and in 5 years of more active catching – as 350,000 (Borodin 1985), later 380,000 animals (Borodin 1986). By 1988 the scientific committee of the IWC estimated the initial stock of minke whales to have been 690,000 animals. By the 1990s, based on visual observations, the estimation of minke whale stocks reached 700,000 animals (IWC.Rep. Int. Whal. Commn. 1994, 1995). During the same period an expert from VNIRO, R.G. Borodin, defined the number of minke whales within the limits of 500,000–700,000 animals, and then 760,000 (Nikonorov and Borodin 1997).

A paradoxical situation results – the more whales caught, the larger their initial population was! The fantastic assumption was even suggested that “the sharp rise” in minke whale numbers happened because of the lack of feeding competition with blue whales, which were practically completely destroyed by this time (Borodin 1996, p. 115).

Thus, if we admit that for approximately 20 years the population of minke whales in the Southern hemisphere had increased by ten times, it would contradict the knowledge about reproductive ability and the biology of reproduction of cetaceans in general. It is absolutely clear that the main reason for such estimations is hidden in imperfection and a lack of authenticity of methodology in defining whale populations based on visual observations and biological characteristics. Frankly speaking, the real reason is the adjustment of results in order to justify catch numbers. These circumstances justify the necessity for a more accurate establishment of the terms of the moratorium and recommendations for a renewal of commercial whaling, including the definition of catch quotas, even for scientific purposes, for not only minke whales, but also all other trade species of whales. At present, when the estimation of stocks of minke whales is so dubious, the possibility of their commercial catch is out of the question. Minke whales taken by Japanese whalers who, screened by the slogan "for scientific purposes" have come to catch minke whales commercially, should be under stricter IWC and public control.

Blue Whales

Blue whale observations have been conducted since 1931. Since that time it is known that 180,646 blue whales have been taken. Of those, according to IWC data, 57,330 animals were caught north of 60° South. Undoubtedly, there were also pygmy blue whales among them, which started to be identified only after 1963. With an extreme degree of approximation an initial stock of blue whales in the

Southern hemisphere is considered to have been 220,000 (Chapman 1964). By 1954 the population of blue whales was already estimated at 10,000–15,000 animals. Probably, it is a little bit underestimated, because since 1959 10,944 blue whales have been taken by soviet whale poachers alone (Zemsky et al. 1995) and exactly how many blue whales were caught by other countries is unknown. R. Laws (Laws 1977) has estimated the current population of blue whales at 10,000 animals. This figure is obviously overestimated, probably because pygmy blue whales were included in it. Otherwise it is impossible to understand how, by the 1980s, the stock of blue whales had been estimated only as 1000–1600 animals when commercial catching was not conducted (Butterworth et al. 1984). By the 1990s R. G. Borodin (1996) estimated the blue whale population as 2000–5000. This number is obviously overestimated. SC IWC (IWC. Rep Int. Whal. Commn. 1990) published the results of survey voyages in 1978/79–1983/84, when, during five seasons, only 453 animals were recorded and there is no firm evidence that the same animal was not recorded twice.

Time has shown that the prospect of increasing blue whale stocks for a 50-year period to 150,000 (IWC. Rep Int. Whal. Commn. 1990) proved to be completely unreal. The reproductive ability of this species was obviously overestimated, and various other factors influencing their recovery had not been taken into account. As we can see from the applied data, today there is no reason to speak about the recovery of the real blue whale population. If such a process exists, it is happening extremely slowly. With great regret, there are big fears that mankind could lose blue whales as a species, this being the largest animal that has ever existed on our planet. The strictest monitoring of this species is necessary for a long-term period (not less than 100 years), with prolongation of the term depending on the results of observations and calculations of its population. Any catches, even for the scientific purposes, are out of question. All scientific research of this species should be carried out only by non-lethal methods.

Pygmy Blue Whales

As to pygmy blue whales, their initial stock in the "island" area of the Indian sector of the Southern Ocean was estimated at 7600 animals (Ichihara and Doi 1964). E. G. Sazhinov (1980) defines their population within the limits of 12,500–13,000 animals. Meanwhile, in the 1960s, more than 9000 whales of this species were taken by Soviet flotillas alone (Zemsky et al. 1995). The number of whales taken by other countries is not clear, and how many pygmy blue whales were wrongly recorded as real blue whales is unknown. The number of pygmy blue whales in other areas of the world's oceans where this species was observed and whaling did not take place (for example, the Banda sea) is also not clear. It is not clear whether blue whales taken off the Brazilian coast are "real" or pygmy blue whales. Most likely, the initial stock of pygmy blue whales was much bigger – more than 50,000 animals.

In recent years they have more often been recorded in the northwest sector of the Indian Ocean, in Australian coastal waters and in the "island" area of the Indian sector of the Southern Ocean (Kasuya and Wada 1991; Robineau 1991). Recovery of all known populations of pygmy blue whales up to the level of stock stability in the course of 50–70 years is probable. However there must be strict monitoring with regular controlled estimations of their population.

Fin Whales

Fin whales, whose initial stock is believed to have been within the limits of 400,000 animals, after blue whales and right whales were almost completely exterminated, became the basic commercial species for more than six decades. As a result their population has dramatically reduced. By the 1970s R. Laws (Laws 1977) estimated their current population at 84,000 animals. According to other researchers (Borodin 1996; Masaki and Yamamura 1978) it was a little bit lower – up to 70,000. Note that as a result of the survey voyages in 1978/79–1983/84 only 2096 fin whales were recorded (IWC. Rep Int. Whal. Commn. 1990). Furthermore, it is possible that there was a repeated recording of the same whale.

Brazilian, Madagascan and East Australian populations have better prospects for recovery if we take into account the catch results and peculiarities of their distribution. However, it is difficult to agree that about 15 to 25 years will be required for this purpose, and for the recovery of all fin whale populations of the Southern Ocean – 10 to 40 years (Borodin 1996). Undoubtedly, this estimation is too optimistic. In our opinion, taking into account the facts of large-scale poaching, more than 50 years will be required to recover the numbers of the healthiest populations. And for the recovery of all fin whale populations in the southern hemisphere a much greater term – no less than 50–70 years – is required, with regular monitoring surveys.

Sei Whales and Bryde's Whales

Sei whales were actively caught rather late – in the 1960s. Their initial stock is estimated to have been approximately 170,000 animals (Borodin 1996), partly including Bryde's whales, which were not distinguished from sei whales for a long time. Moreover, according to recent investigations, it is possible that one more species of minke whale – *Balaenoptera olseni* – which is similar to Bryde's whale, exists in the area of Solomon Islands and Indian Ocean (Wada 1998), and was recorded as sei whales. For this reason we have to examine the stocks of these species together, although the greater part (no less than 150,000 animals) were undoubtedly sei whales.

In the 1970s the population of sei whales was estimated at 40,500 animals by R. Laws (Laws 1977). In two decades R. G. Borodin (1996) determined their number as 48,300 animals. If we accept these estimations, by the end of the catch the number of sei whales had decreased by three to five times. We shall note that as a result of survey voyages in 1978/79–1983/84 only 1498 sei whales were recorded (IWC. Rep Int. Whal. Commn. 1990).

Brazilian-Chilean, West African and East Australian populations suffered less. According to R. G. Borodin (1996), prohibition of their catch should be for no less than 25 years. Other populations of sei whales had been greatly damaged and, judging by their biological characteristics at the end of the catch, no less than 30–50 more years will be needed for their population to recover.

Humpback Whales

The initial number of humpback whales has been not established even approximately. For the entire period of whaling in waters of the Southern Ocean, according to the data of BIWS, no less than 300,000 whales of this species were taken. We can judge the low reliability of methods of calculation of the current population of whales by the example of the humpback whale. The population of Australian humpback whales in 1955 was estimated at 9800 animals, and in 1962 – at 2800 whales. But in two voyages from 1959–1961, more than 25,000 humpback whales were caught by Soviet whaling flotillas in this area (Soviet Antarctic Whaling Data (1947–1972, 1995). Hence, the population was three times underestimated! R. Laws (Laws 1977) believed the population of humpback whales to be 3000 animals. Very approximately, the humpback whale population was estimated at 6000 animals (Clapham., Brownell Jr. 1995), or a little bit lower – 5600 animals by the VNIRO expert R. G. Borodin (1996). In special surveys in 1978/79–1983/84, 4047 humpback whales were counted, a little higher than other species. According to the IWC scientific committee estimation (IWC. Rep Int. Whal. Commn. 1990), no less than 60 years is required for the recovery of the Southern Ocean herds of humpback whales. It is clear that not all the Antarctic populations can recover to the same extent. Time has shown that this forecast was too optimistic. But recently an increase of West Australian and East African populations has been reported. There are encouraging observations, which testify to an increase in the unique population of humpback whales of the northwest area of the Indian Ocean (Reeves et al. 1991).

However, in view of the data of large-scale poaching and hiding of the catch data in order to show growth of the humpback whale population in the Southern Ocean, no less than 50–70 years with constant monitoring will be required for the recovery of the population and, if necessary, prolongation of commercial catch prohibition.

Southern Right Whales

Among right whales the southern right whales were almost completely destroyed in the nineteenth century, that is why their initial population has not been determined even approximately. According to retrospective estimations, the South African southern right whale population could have been no more than 200 animals in the 1940s (Tormosov et al. 1998, 2000a, b). On the basis of Japanese experts' visual observations between 1965 and 1977 their population for the whole Southern hemisphere was estimated within the limits of 3500–3700 animals (Masaki and Yamamura 1978). The prospects of recovery of the population of southern right whales (especially those of Chilea, West Africa and New Zealand) are illusive, because in the 1960s, Soviet whalers took 3200 whales from their scant population. There are great fears that this species could go extinct.

As is the case for real blue whales, a strict moratorium on commercial hunting, in addition to scientific aerial and ship-based surveys, is necessary for southern right whales. All arguments about the current population of southern right whales and opportunities for their population recovery should be postponed until reliable information is received.

Sperm Whales

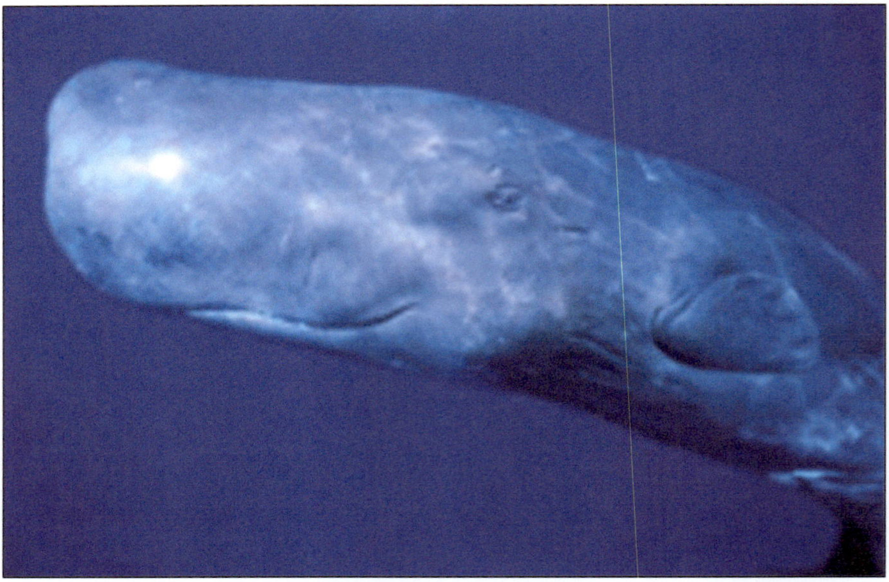

In comparison with baleen whales the estimation of sperm whale stock is complicated with several features of their biology, which have not been not entirely cleared up. Sperm whales have strongly expressed sexual dimorphism. By the period of sexual and physical maturity the males appear to be twice as big as the females. On this basis it is considered that sperm whales are polygamous animals. However we think that this question is disputable for the following reasons. That fact that large males migrate to high latitudes of the Southern Ocean, and females remain in the subtropics and tropics, moving only occasionally almost up to the sixtieth parallel (Budylenko et al. 1970), does not tell us anything.

According to K. Allen (Allen 1976), the initial stock of male sperm whales was about 230,000, females – near 360,000. R. G. Borodin (Borodin 1977) estimated the stock of male sperm whale in the Southern hemisphere at 210,000, females – 220,000. On the basis of visual observations, the stock of sperm whales of both sexes was estimated to be 180,000 to 250–000 from 1965 to 1977.

Bearing in mind disputable questions about the biological characteristics and unreliable methodology of population calculation and definition of stocks of sperm whales in the Southern Ocean, it is difficult to predict the terms for the recovery of their original population. But recent frequent sightings of sperm whales and the overview of all the biological parameters of the animals give cause for optimism regarding the restoration of their numbers in some areas. The current stocks are safest in the 2nd, third and eighth IWC management areas. Commercial whaling must be prohibited for at least 30–50 more years with the likelihood of further prolongation.

* * *

Summing up this chapter, it is necessary to emphasize once again, that all the forecasts of the previous researchers appeared to be too optimistic. The calculations made at the end of the 1960s (Mackintosh 1970) were unreal, hopes for the restoration of populations of blue whales, fin whales and humpback whales in 20–50 years was not justified. Hopes for the restoration of up to 150,000 blue whales in 50 years, fin whales in 15–25 years, and humpback whales in 60 years (IWC. Rep Int. Whal. Commn. 1990) also appears to have been completely unrealistic. The expectation that sei whales herds could be restored in 25–30 years also seems naive (Borodin 1996).

The experience of previous researchers shows that it is necessary to define the terms of recovery of whale populations with extreme caution and accuracy. Present estimations of whale populations are at such a low level, that our corrections of previous forecasts made on the basis of whaling data and more exact biological characteristics can only be accepted tentatively. The conditions for the period of prohibition of whaling should be regular monitoring of the occurrence of whales in the Southern hemisphere. There is a need to create testing areas to develop a method of calculating cetacean populations. In particular, such areas could be conveniently located in the most accessible areas – the Black Sea for dolphins and the Arabian Sea for large whale species of almost all taxonomical groups. As long as the size of whale populations raises doubts about their well-being, it will be necessary to prolong the terms of a whale sanctuary, i.e. all of the Southern hemisphere.

6.6 Summary

Whales of the Southern Ocean and adjacent waters have experienced the cruelest process of whaling. More than 2,500,000 whales were taken in that region during 80–90 years, up to 2,000,000 t of biomass were extracted from the ecosystem of the Southern hemisphere. The aftermaths of the catch have undoubtedly influenced the ecosystems of those waters.

The actual consumption of krill by whales was greater than has been believed. Taking into account large-scale poaching, the amount came to 200–300 million/t per year, or 20–25% of the total stock. After the number of whales had been reduced, the krill surplus came to 150–200 million t. The greater part of this surplus, which had previously been eaten by whales, was consumed by fish, cephalopods, pinniped and birds. But present methods do not allow us to determine precisely whether the biomass of those groups increased nor how much krill they consumed.

The main feeding grounds of seals, penguins and minke whales coincide with the reproduction areal of krill. Large baleen whale species mostly feed on krill carried to the north by the western arms of cyclones, thereby reducing competition for food between these groups of animals.

As a result of the dramatic decrease in the population of large baleen whales, the smallest representative of *Balaenopteridae* – the minke whale – has become the biggest krill consumer among cetaceans.

Some researchers declare that an improvement in minke whale nutrition has led to earlier sexual maturity and an increase in productivity, to an expansion of their areal and a dramatic growth in population, but the proof is not credible and does not withstand criticism.

Forecasts of some researchers (Borodin 1996; Mackintosh 1970; IWC.Rep Int. Whal. Commn. 1990) about the possibility of recovery in population of the large whale species in 15–50 years appears to have been too optimistic and is not justified. The reasons are an underestimation of the catch volume, unreliability of the methods used to calculate initial and current populations and an incorrect estimation of whale reproductive capacity.

Our research let us assume that more than 50 or 100 years will be necessary for the recovery of fin whale, sei whale, humpback whale and sperm whale populations. As blue whales and southern right whales are on the verge of extinction, their preservation and recovery is problematic.

The main research methods to be used during the prohibition of whaling are the following: genetic on the basis of biopsy, acoustic radio-marking, and visual observation methods using still- and motion picture cameras. The taking of a restricted number of whales for scientific purposes can be practiced only by a specialized, international small tonnage flotilla and under strict monitoring by the IWC and the public.

At a session of the IWC it is necessary to create a whale sanctuary for a long period for all waters of the Southern Ocean and the whole Southern Hemisphere, and a special scientific research and experimental area for the unique regions of the Black and Arabian seas with their isolated cetacean populations. Economic losses from prohibition of whaling can be compensated by the organization of ecological tourism to the preservation areas.

6.7 Conclusion

The necessity for writing this work was defined, firstly, by the absence of a general report about whales of the Southern Ocean and, secondly, by the facts of large-scale whale poaching and data falsification, which radically changed notions of the numbers, distribution and biology of whales. Without such data revision it is impossible to create a scientific concept for the preservation and restoration of the Southern Ocean whale populations. The following basic problems are being investigated: the growth of the Southern Ocean whales that were targeted by commercial whaling in all ontogenesis; biology of reproduction and female reproductive ability; location of reproductive zones; distribution and migration patterns of various whale populations; features of morphology and revealing distinctions in whale populations.

The constant functional dependence between cetacean length and body weight has been found, with a high degree of correlation, both in prenatal and postnatal development periods. It means that when studying whale growth in all ontogenesis, length, a more accessible measure, should be used instead of weight. Definition tables for weights, and amendments to them, have been created. These tables can be used to determine the live weight of whales caught at sea. It is important that the functional length-weight correlation is species-specific. Among rorqual whales the proportionality factor of smaller *kinds* appears higher, and the regress factor is higher for larger ones. It has been proved that the growth of skeleton, muscles and integumentary fat weight can also be described by power function. Accumulation of integumentary fat and liver weight is faster in females than in males. The intensity of increase in muscles, lung and liver mass of carnivorous killer whales is higher than that of sperm whales and minke whales. The rate of accumulation of weight in minke whales of the Herd and Prince Edward islands area is higher than that of minke whales of other areas of the Southern Ocean. In the same way killer whales of the Scotia Sea differ from killer whales of the Balleny Islands area.

In the embryonal development period (except for the earliest stages of exponential growth, and a prenatal phase, when growth is a little slower) growth of cetaceans, as for land mammals, appears a conservative enough sign to be described by monomial parabola dependence. In the course of growth, geometrical similarity of the body form does not remain.

Sexual dimorphism, expressed in the bigger size of female baleen whale and, on the contrary, of male toothed whales, can already be seen during the fetus development period. The average size of the same aged whale embryos of different species depends on the size of the mothers. By the neonatal period these distinctions reach 8–10% in length and 30–50% in weight.

The intensity of the relative growth (allometry) of individual parts of a whale body changes intermittently, and corresponds to the pre-fetal and fetal periods. In relation to the general embryogenesis duration, the germinal period accounts for approximately 17%, pre-fetal – 11% and fetal –72%, which corresponds to the *extent* of similar periods for terrestrial mammals.

Optical density of stratifications in toothed whale teeth is not usually defined by the maintenance of carbonic salts, but by the features of albuminous tooth stroma structure. The maintenance *of Ca* on a tooth axis of sperm whale males (2452) is higher than that on dentine periphery (2195). There is less Ca in females' dentine than in that of males'.

When defining age, quality standard of stratifications in registering structures of baleen and toothed whales is too subjective, and does not give authentic results. Quantitative characteristics of stratifications (recording structure graphs), recorded by means of registering devices, are a graphic representation of layering *on a certain scale*. On recording structure graphs, repeating zones of groups of layers and "labels" can be seen, which correlate with female reproduction periods. By means of such graphs the width of prenatal dentine and its annual gain/increase can be defined, and its tendency to decelerate with age is noted.

Not only endogenous, but also exogenous factors influence the nature of layering in registering structures, especially such powerful factors as lean years, pan-epizootic, solar activity, changing seasons and regular migrations of whales to the Antarctic *feeding grounds*. An original method of defining sperm whale age by comparing recording structure graphs of teeth has been offered. The result of the "original" method mentioned above is close to the result obtained by determining age using growth of the pre-pulp dentine layer.

Curves of postnatal growth of baleen and toothed whales have been created, along with an attempt at their mathematical description. The age of sexual maturity in whales was defined by comparing several methods: the percentage of mature and immature individuals; the presence of one trace of pregnancy on females' ovaries; the average age of primigravida females; and the line of regression of the rate of accumulation of traces of pregnancy and ovulation. For different species it varies within 2–6 years. When some authors define this indicator for baleen whales on the basis of the "transphase" position in ear plugs, it does not withstand criticism.

The connection between the average size of females and their newborns in a logarithmic system of co-ordinates is described by a straight line. This dependence is a systematic *sign*. It is specific not only for suborders, but also for *families and genera*. The newborn of smaller whale species appear to be rather large.

Pregnancy in whales defined on the basis of parabolic growth of embryos and the size of newborns, lasts from 9.5–11.5 months for baleen and toothed whales, and more than a year for sperm whales.

Pairing and whelping of 70% of baleen whales occurs within 2.5 months, and of 95% – within 5.5 months. Most species have an additional reproductive season with a six-month phase lag. Reproductive zones of whales (pairing, whelping and milking) coincide and are essentially located in the thirtieth-fortieth latitudes. Blue whales, fin whales, minke whales and humpbacks can also couple in Antarctic waters. When the reproductive zones coincide, the timing is different, which reduces the interspecific competition.

The rate of accumulation of ovulation traces for fin whales is 0.52 traces a year, Bryde's whales – 0.55, minke whales – 0.87, sperm whales – 0.34. Thus, one trace of pregnancy on minke whales is formed during approximately 18 months; fin whales and Bryde's whales– 2 years; sperm whales –2–3 years. For many species, lactation does not interfere with a new pregnancy. Over a lifetime, female baleen and toothed whales are capable of reproducing from five to seven calves. This high reproductive ability in whales, should effective security measures be provided, gives hope that their stocks may be restored, with the exception of blue whales, right whales and some humpback populations, whose numbers are too low.

It has been established that there are specific "pair formations" on both top and bottom jaws, and not only on baleen, but also on toothed whales. Their form differs for various species, and can be used as a phene for revealing population distinctions. A histologic analysis of the structure of "formations" reveals a considerable quantity of bodies of Vater-Pacini, but olfactory epithelium, bipolar neurons and mucous glands have not been found. It has been assumed that pair formations play their part in searching for and capturing a mother's nipple at the time of breast feeding.

The excessive nutrition of whale milk is probably an adaptation to feeding in a marine environment, when milk mixes with sea water. If this is so, the percentage of fat content of the diluted milk does not contradict "Aron's Rule". Such a hypothesis could explain why whale embryos at the last stage of development and young animals have excessively big kidneys, given that kidneys play an active role in the regulation of water and salt exchanges.

For the first time whale distribution charts for the Southern Ocean have been produced, taking into account the large-scale poaching of whales by the Soviet flotillas and hidden data from tagging. Areas of the greatest concentration of whales are the highly productive zones of the world's oceans. Unproductive "dead" zones where whales are practically never seen have also been detected.

On the basis of phenes, distinctions between various population of whales have been established. Pygmy blue whales of the Amsterdam Island aggregation differ from other affregations; the Chilean-Peruvian fin whales – from fin whales of the Atlantic and Indian oceans; sei whales of Gough and Bouvet islands waters – from sei whales found in Chilean waters and populations of the subantarctic islands zone of the Indian Ocean; sei whales of southwest Australian waters – from sei whales of Tasmania; minke whales of Commonwealth Bay – from Minke whales of the Amundsen, Bellingshausen, Weddell and Mawson seas; sperm whales of the Atlantic – from sperm whales of the Chilean coast and Indian Ocean.

From information recorded by tagging, dates and migration patterns of various whale populations have been detected. The speed with which whales are fouled by diatoms has been calculated on the basis of the correlation between the degree of fouling of females and the average sizes of their fetuses. On the basis of this dependence, the nature of migrations of various biological groups of whales to their feeding grounds and back has been discovered.

It is recommended that a scientific-experimental ground be reserved in the area of the Arabian sea, in order to determine methods of estimating whale numbers.

Distinctions in the average size, body proportions, terms of maturity, relative weight and quantity of traces of pregnancy and ovulation on the ovaries, and also isolated habitats, lead to the conclusion that in the Southern hemisphere there exists, in addition to the ordinary killer whale– *Orcinus orca* (Linnaeus 1758), a short-bodied killer whale – *Orcinus nana* (Mikhalev 1981).

Chapter 7
Photo Selection

1. Whales:

1e Southern right whale
2b, 2c 2e Humpbacks
5, 5c, 5d Sperm whales
6a, 6e Killer Whales

Y. Mikhalev, *Whales of the Southern Ocean*, Advances in Polar Ecology 5,
https://doi.org/10.1007/978-3-030-29252-2_7

Photo 1, 1a, 1b, 1c, 1d, 1e – Southern right whales

Photo 1f – Pygmy right whale

Photo 2, 2a, 2b, 2c, 2d, 2e, 2f, 2g – Humpbacks

Photo 3 – Pygmy blue whale

Photo 4a – Minke whale wounded by large mark

Photo 5, 5a, 5b, 5c, 5d – Sperm whales

Photo 6a, 6b, 6c, 6d, 6e – Killer whales

Photo 7 – Bottlenose whale

2. Whaling ships:

1a Slava and whaling steamboat after repair in Liverpool
2a The whaler "Soviet Ukraine" in ice of the Southern ocean
4b-1 The liner Gamburg after rebuilding into the whaler "Yuriy Dolgorukiy"

Photo 1a – Slava and whaling steamboat after repair in Liverpool

Photo 1b – Slava while accepting whales

Photo 1c – The whaler "Slava" while towing two whales

Photo 1d – The whaler “Slava” met whaler “Soviet Ukraine” in the Southern ocean off the island of Balleny

Photo 2a – The whaler "Soviet Ukraine" in ice of the Southern ocean

Photo 2b – The whaler "Soviet Ukraine" off the island of Balleny

Photo 2c – The twin whalers "Soviet Ukraine" and "Soviet Russia"

Photo 2d – The whaler "Soviet Ukraine" returned to Odessa

Photo 2e – The whaler "Soviet Ukraine" – arrival to Odessa

Photo 2f – The whaler “Soviet Ukraine” sold to Turkey for cupan metal

Photo 3a – The whaler “Soviet Russia” in the Southern ocean

Photo 4a – The liner Gamburg – future whaler "Yuriy Dolgorukiy"

Photo 4b – The whaler "Yuriy Dolgorukiy" – full speed to the Southern ocean

3. Hunting:

1 Loading whalers before a mission
3h Delights of the Southern Ocean
4b Modern harpoon gun
7, 7d Hunting
7l Whale 'on flag'
8f Queue to take whales to base
9c Dragging whale through slip

Photo 1 – Bunkering of whalers before departure

Photo 1a – Whaler "Mirny-21". The first domestic whaler of this series

Photo 1b – March in a wake to the Southern Ocean

Photo 2a, 2b, 2c, 2d, 2e, 2f, 2g, 2h, 2i, 2j – Storm Fortieth Latitudes

Photo 3, 3a, 3b, 3c, 3d, 3e – The delights of the Southern Ocean

Photo 3f – Meeting a Japanese whaler

Photo 4a – Svend Foin, the creator of the harpoon gun

Photo 4b – A modern harpoon cannon

Photo 4c – Experimental three-barreled harpoon gun

Photo 5a, 5b, 5c – The helicopter of the flotilla is preparing to take off for the search of whales

Photo 7, 7a, 7b, 7c, 7d, 7e, 7f, 7g, 7h, 7i – Hunting

Photo 7j, 7k – Whale on a line

Photo 7l, 7m – Whale "on flag"

Photo 8b – Towing the whales to the whaling base

Photo 8, 8a, 8c – Towing the whales to the whaling base

Photo 8e, 8f, 8g, 8 h, 8i, 8j – The queue to take the whales to the base

Photo 8k, 8l, 8m, 8n – Delivery of whales to the base

Photo 9, 9a – Dragging whales through the slip to the whaling base

4. Cutting:

1b, 1f, 1j Cutting a Baleen whale
2 Cutter sharpens his flasher knife
2b, 2c Cutting a Sperm whale
2d Separated spermaceti bags of sperm whales

Photo 1 – On the aft deck, a blue whale pulled through the slip

Photo 1a – On such a large whale is not a sin and take pictures!

Photo 1b, 1c, 1d, 1e, 1f, 1g, 1h, 1i, 1j, 1k, 1l, 1m, 1n, 1o – Cutting the baleen whale

Photo 1p – Cutting of whales at night

Photo 2, 2a, 2b, 2c – Cutting of the sperm whale

Photo 2d – Separated spermaceti bags of sperm whales

5. Everyday life:

1 Volleyball on deck while travelling to the Southern Ocean
10a Paining whaling vessel before returning to Odessa
9 Boris Simakov – carving from a Sperm whale tooth
5 Docked in Sydney harbour

Photo 1 – On the transition to the Southern Ocean, you can play on the deck in volleyball

Photo 2 – And in the tropical zone, it's not a sin to sunbathe on the helipad

Photo 3 – Meeting with Neptune. Neptune's exit with his retinue

Photo 3a – The order of Neptune to "baptize" the recruits in the font

Photo 4 – "Baptism" in the font

Photo 5 – During the voyage docked in Sydney

Photo 6 – Visit to Sydney University. Meeting with authoritative whale researcher Dawbin W.H. On the photo: Mikhalev Y.A., professor Dawbin W.H., captain Solyanik A.N.

Photo 7 – Visiting the whalers, the outstanding cellist Vyacheslav Rostropovich

Photo 8 – Meeting with the outstanding pianist Van Cliburn (next to him captain Solyanik AN with his wife Svetlana)

Photo 9 – On a voyage at leisure: motorist Boris Simakov is engaged in bone carving (sperm whale teeth)

Photo 10, 10a – Painting the whaling base before returning to Odessa

6a. & 6b. Research & Science Groups:

1 Scientific research vessel 'Bodry 25' with harpoon gun not yet removed
12 The Drake Passage with the Cape Horn as backdrop and Y. A. Mikhalev, A. Parokhno (Radio Operator) and a Watchman
6 Taking a sample of plankton
8a, 8b The whale is measured by Research Associate, Y. A. Mikhalev and State Inspector, A. N. Kulikov
13 Whale shark
1 The Scientific Committee of the International Whaling Commission (DATE??)
6a. Budylenko Gennady Alexandrovich (right), Head of the scientific group of the flotilla "Yuri Dolgoruky" and Y. A. Mikhalev Yu.A. Moscow (hotel "Russia")during a seminar of scientific groups.

Photo 1 – The scientific-search whaling vessel "Bodry-25". The harpoon gun has not yet been removed

Photo 1 – Kirpichnikov Apollon Alekseevich – the first head of the scientific group on the flotilla "Slava"

Photo 2 – Zemsky Vyacheslav Alekseevich is a participant of the Finnish and Great Patriotic War. He was a member of the scientific group of the first seasons of the flotilla "Slava". He was the head of the scientific group of the first season of the flotilla "Yuri Dolgoruky"

Photo 3 – In the voyage search for whales and other objects

Photo 4 – Mikhalev Yu.A. using a shotgun to mark a whale

Photo 5 – Marks "Discovery" for large (top) and for small whales (below)

Photo 6 – Taking a sample of plankton

Photo 7 – During the stay near the island of Chatham, a large merow fish was caught

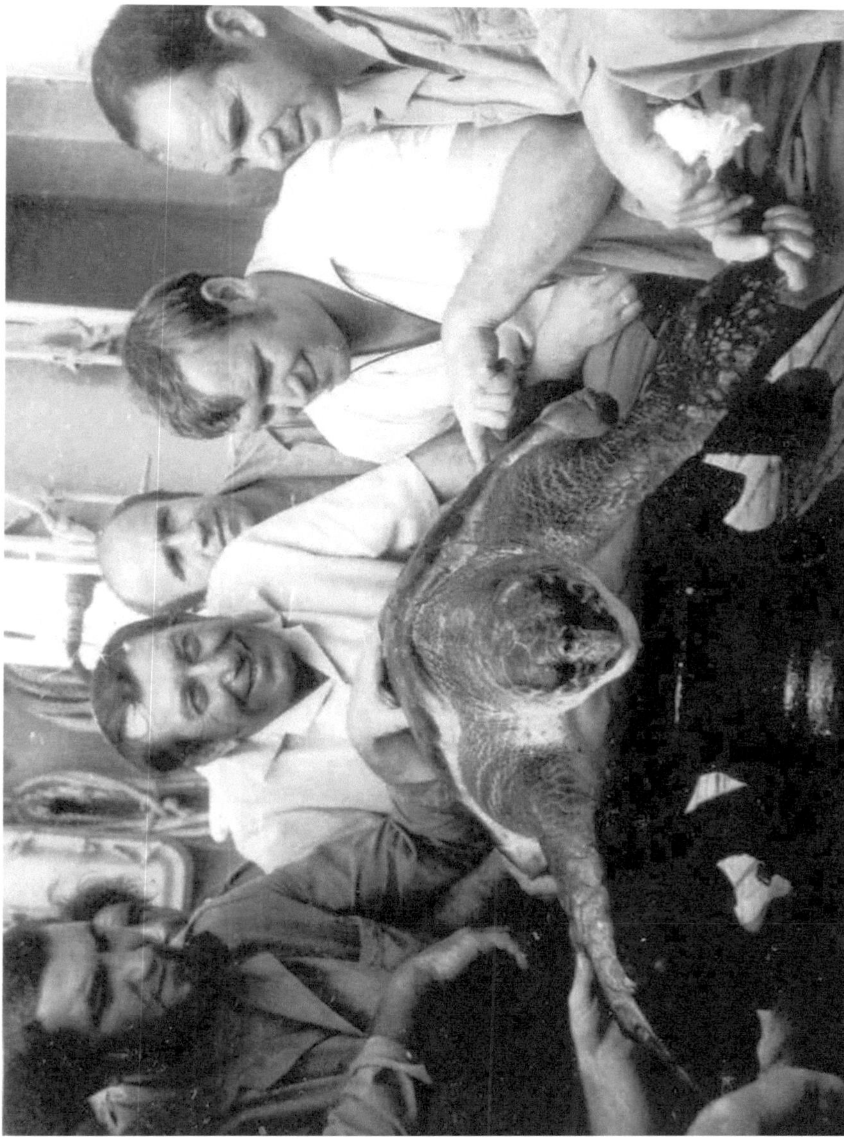

Photo 8 – A turtle is caught off the coast of Chile, there is a reason to be photographed

Photo 9 – Mikhalev Yu.A. on the background of the island of Scott at the entrance to the Ross Sea

Photo 10 – On the island of Kerguelen: sea elephants, royal penguins

Photo 11 – The radio operator Alexei Parokhno is trying to establish radio contact with Odessa

Photo 12 – The Drake Passage. Against the backdrop of Cape Horn: radio operator Parokhno A., Mikhalev Y.A. and a watchman

Photo 13 – Scientific conference in New Zealand. Report by Mikhalev Yu.A. (At the table, the far right)

Photo 14 – New Zealand, Wellington, 2010. Princess of the Kingdom of Tonga, whose father voluntarily decided to ban the whaling of the waters of the archipelago of Tonga, gives Mikhalev Y.A. A certificate "For the study and conservation of whales"

Photo 15 – In the same place in New Zealand, a friendly cartoon

Photo 1 – The staff of the Scientific Committee of the International Whaling Commission. In the center the chairman Gambell R., who headed the Scientific Committee for many years

Photo 2 – Ilyashenko Valentin is a representative of Russia in the International Whaling Commission

Photo N - Scientists Zemsky, Yablokov and Grigoriy Agarkov near a monastery on the Sovovetskie Islands

Photo 3 – Yablokov Alexey Vladimirovich – headed a group of former heads of laboratories for the study of cetaceans: Zemsky V.A. – Moscow; Mikhalev Y.A. – Odessa; Berzin A.A. – Vladivostok; Tormozov D.D. – Kaliningrad. On the basis of scientific reports, the group established the actual volume of whale extraction by the Soviet Union, and transferred this information to the Scientific Committee of the IWC

Photo 4 – Mukhamedov Lev Muharamovich is an outstanding scientist, sponsor of many studies and conferences on marine mammals

Photo 5 – Krotov Arkady Vasilievich – in the 1960s he headed the Laboratory of Ichthyology and Marine Mammals in Odessa, completed scientific groups on "Slava" and "Soviet Ukraine"

Photo 6 – Savusin Vladimir Pavlovich – in the 80s, the head of the scientific group on the flotilla "Soviet Ukraine" and Mikhalev Yuri Alekseevich, the head of scientific research seasons on the whaling ship "Bodry-25" in 1973–1975

Photo 6a – Budylenko Gennady Alexandrovich (right) – head of the scientific group of the flotilla “Yuri Dolgoruky” and Mikhalev Yu.A. – member of this group. In Moscow (hotel “Russia”) at a seminar of scientific groups

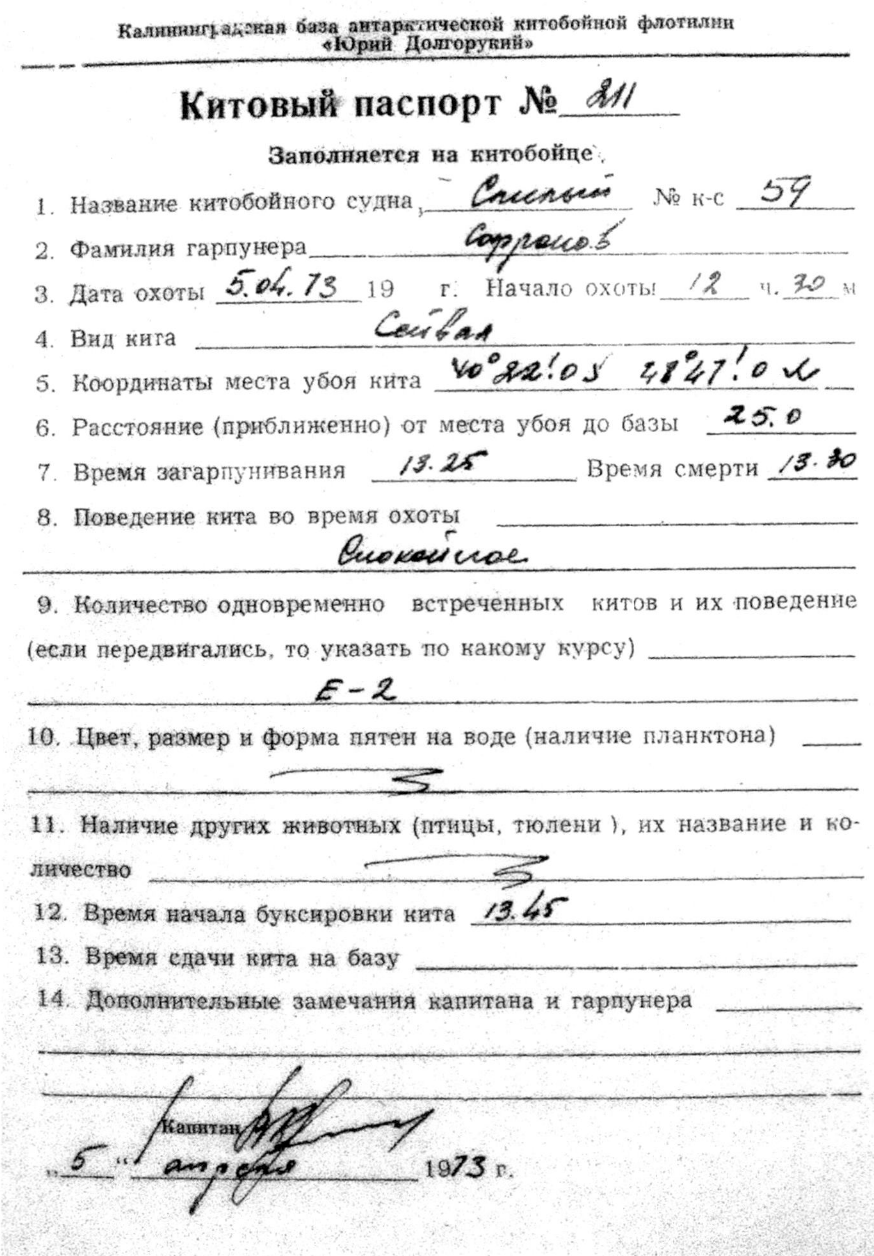

Калининградская база антарктической китобойной флотилии
«Юрий Долгорукий»

Китовый паспорт № 811

Заполняется на китобойце.

1. Название китобойного судна Слитный № к-с 59
2. Фамилия гарпунера Сафронов
3. Дата охоты 5.04.73 19 г. Начало охоты 12 ч. 30 м
4. Вид кита Сейвал
5. Координаты места убоя кита 40°22!0S 48°47!0 W
6. Расстояние (приближенно) от места убоя до базы 25.0
7. Время загарпунивания 13.25 Время смерти 13.30
8. Поведение кита во время охоты
Спокойное
9. Количество одновременно встреченных китов и их поведение (если передвигались, то указать по какому курсу)
E-2
10. Цвет, размер и форма пятен на воде (наличие планктона)
11. Наличие других животных (птицы, тюлени), их название и количество
12. Время начала буксировки кита 13.45
13. Время сдачи кита на базу
14. Дополнительные замечания капитана и гарпунера

Капитан
„5" апреля 1973 г.

Photo 7 – The whale passport. The passport is filled by the navigator of the whaling vessel that obtained the whale

Photo 8, 8a, 8b, 8c – The whale is measured by a series of standard measurements by research associate Mikhalev Y.A. and state inspector Kulikov A.N.

Photo 9 – Research associate Mikhalev Yu.A. is trying to extract the pituitary gland through the occipital opening of the whale. The process is supervised by the senior cutting engineer Yuri Saveliev

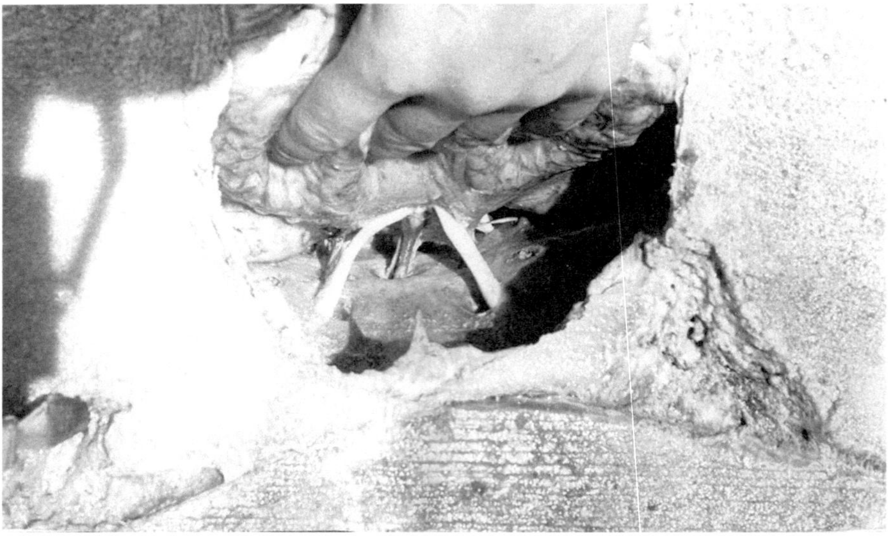

Photo 9a – On the cut of the skull in the brain box are visible the optic nerves and between them the pituitary leg

Photo 10 – The lower jaw of the sperm whale was overgrown with barnacles – Conchoderma

Photo 11 – Nipples of the nursing female of the Southern smooth whale

Photo 12 – Giant squid from the stomach of the sperm whale

Photo 13 – On the deck the largest fish is a whale shark

Photo 14 – Weighing the moon-fish

Photo 15 – In the laboratory on the whaling base “Soviet Ukraine”, research scientists Shevchenko Vladimir and Anatoly Korhov are studying samples selected from whales

7. Views:

1a, 1b Icebergs in Antarctica
3 Ukrainian Antarctic Station
5b Sunset, 5e morning, 5d Night

Photo 1a – Icebergs in Antarctica

Photo 1b – Icebergs in Antarctica

Photo 1c – Icebergs in Antarctica

Photo 2, 2a – The ice barrier at the mainland

Photo 2b – The shore is very close

Photo 2c, 2d – The shore is very close

Photo 2e, 2f – The shore is very close

Photo 3 – Ukrainian Antarctic Station named by Vernadsky

Photo 3a – Emblem of the Vernadsky Station

Photo 4, 4a, 4b – Penguins

Photo 4c, 4d, 4e – Boobies

Photo 4f, 4g – Gulls

Photo 5, 5a, 5b – Sunset

Photo 5c, 5d – Night

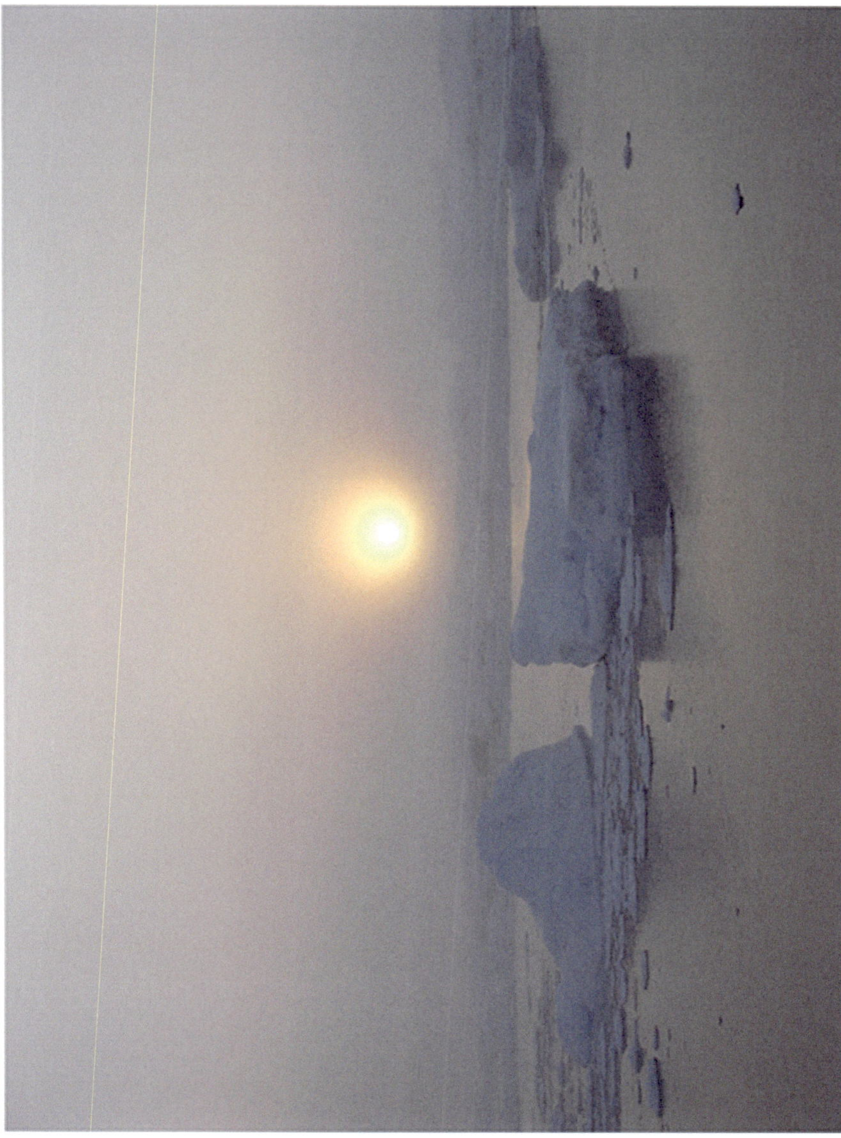

Photo 5e – Morning

Photo 6, 6a, 6b, 6c, 6d, 6e, 6f – Underwater world

8. Famous people, some with amazing destiny.

Photo 1; 1a; 1b. **Voronin Vladimir Ivanovich**. Captain of the Soviet icebreaking fleet, polar explorer, participant of many Soviet expeditions in the Arctic. In the First World War in the White Sea he worked as a navigator on the steamer "Fedor Chizhov". The ship was torpedoed by a German submarine. The wounded navigator Voronin directed the evacuation of people, left the ship last. In 1928, while commanding the icebreaker "Georgy Sedov", he took part in the search and rescue of the crew of the Umberto Nobile's second expedition, who was wrecked on his airship. In 1933–1934 he was the captain of the steamship "Chelyuskin" and took part in the famous tragic epic of the Chelyuskinites. During the Second World War he commanded the icebreaker "Siberia", conducted ships with cargo to the port of Murmansk. On the first voyage of the whaling flotilla "Slava" (1946–1947), was its captain. At the age of 62 years, he died suddenly from a stroke in the Arctic on the flagship icebreaker "Joseph Stalin". He was awarded two Orders of Lenin, the Order of Nakhimov 2nd degree (02.12.1945), medals.

In Antarctica, a cape on Victoria Land and the bay on Queen Maud Land were named by him.

Photo 1c; 1d. **Voronin V.I. and Schmidt O.Y.** A well-known scientist, Otto Schmidt, an astronomer, a mathematician, a polar explorer. In 1930–1934 he led the famous Arctic expeditions on the icebreakers "Sedov", "Sibiryakov" and "Chelyuskin". Organizer of the drifting station "North Pole". He was awarded the title of Hero of the Soviet Union.

In Antarctica, his name is one of underglaze plains.

Photo 1e. **Papanin I.D. and Voronin V.I.** Ivan Dmitrievich Papanin, Arctic researcher, Doctor of Geographical Sciences, Rear Admiral, twice Hero of the Soviet Union. In 1932–1933 he was the head of the polar station of the Bay of Tikhaya (Franz Josef Land), and in 1934–1935 – the station at Cape Chelyuskin. In 1937–1938 he headed the world's first drifting station "North Pole-1".

In the name of Papanin the mountains in Antarctica and the underwater mountain in the Pacific Ocean are named.

Photo 2a. **Solyanik A.N.** (1912–1984). Alexey Nikolayevich from November 1942 to October 1945 was a member of the Soviet government procurement commission in the United States. In October 1946, took in Liverpool the German whale base “Wikinger”, which after the repair was called “Slava”. On her first voyage to the Southern Ocean Solyanik served as captain's assistant in the field. In the second flight, he headed the flotilla and made 13 cruises on “Slava”. In 1958, he received from the Nikolaevsky Shipbuilding Plant the first domestic whale base “Soviet Ukraine” and became its captain-director. In October 1959 he was appointed the general captain of the United Flotilla “Slava” and “Soviet Ukraine”. Assigned the title of Hero of Socialist Labor. In 1965, on charges of corruption and a number of other violations, Solyanik was removed from the position.

Photo 2b. **Morgun B.M.** (1923–1970). Boris Makarovich had a tragic fate. While teenager in December 1941, during the Second World War, served as a sailor on an auxiliary boat, and participated in the famous landing of a military assault in Kamysh-Burun (Kerch, Crimea). He was injured. After recovering in February 1943 took part in an assault landing in the area of Myskhako (Novorossiysk). By direct hit of an artillery shell his tow ship was sunk. In April 1943 in the Gelendzhik area, his tug was blown up on a mine and sank. Morgun received a concussion. In 1952 in Odessa he graduated from navigator courses and was appointed as a senior assistant to the captain of the whaling ship "Slava-2", and then "Slava-8". In the 10th season of the flotilla he became the captain of this vessel. In 1958 he was appointed deputy captain-director of the flotilla "Slava", and in 1959 – captain-director of the flotilla. In 1965 he became the general captain-director of the united flotilla "Soviet Ukraine" and "Slava".

In January 1970, on a voyage on the whaling base "Soviet Ukraine" he fell under the obscure circumstances into the hold and died.

Photo 2c. **Kiryuhin G.L.** (1968–1969). German Lavrentievich was the last captain of the flotilla "Soviet Ukraine". In 1946 he graduated from the Astrakhan seaworthy school. Then (1946–1947) – navigator courses. In 1951, a one-year navigation course was held in Odessa. In 1952–1966 – 3rd, then 2nd and senior assistant captain of the whalers of the flotilla fleet "Slava" and "Soviet Ukraine". In 1967 he graduated in absentia from the navigational department of the Odessa Higher Naval School. In 1967–1968 – the captain of whalers "Divny", then "Dersky". In 1968–1969? – Deputy in whaling, captain of the flotilla "Soviet Ukraine". In 1970–1986 – captain of the flotilla "Soviet Ukraine".

Photo 3a, b. **Labunets A.S.** (1913–1984). Anatoly Stepanovich was one of the most respected captains in whaling flotillas. In 1928 he graduated from the cabin boy school in Tuapse. Since 1942 he was transferred to Vladivostok. A participant in the Japanese war as chief commandant of torpedo boats. From 1946 to 1965 years captain of whaling ships "Slava-1", "Slava-4", "Gumanny-44" of the flotilla "Slava". In 1965–1966, the captain of the flotilla "Slava". 1968–1970 – Captain-understudy, then captain-director of the flotilla "Soviet Ukraine". In 1971 – Deputy Director General of the Association "Antarctica". Awarded the Order of the Red Banner of Labor. In 1982 he retired. His name is the trawler "Captain Labunets".

Photo 3c. **Anatoly Stepanovich's** father, Labunets Stepan Andreevich – participant of the First World War, ensign. Three St. George's Cross and four medals.

Photo 4a. **Samolyanov F.I.** (1909). Fedor Ivanovich was graduated from the Marine Technical School (1931, Vladivostok). A participant of the Second World War. 1951–1954 – captain of the flotilla "Slava". In Warnemünde (Germany) he took part in the supervision and control over the conversion of "Hamburg" into the whale base "Yuri Dolgoruky", took part in its acceptance. The first captain-director of this flotilla (1959–1962).

Photo 4b. **Protsenko Fedor Trofimovich** (1926). He graduated from the Vladivostok Higher Naval School (1951). A captain of whaling flotillas in Vladivostok (1957–1959), a captain of the whaling ship "Skolzyaschy" of the whaling flotilla "Yuri Dogoruky", deputy captain-director (1960–1967), captain-director of this flotilla (1963–1966). Head of the Kaliningrad base of the Antarctic whaling flotilla (1967–1970). He was awarded the Order of Lenin.

Photo 4c. **Nebolyubov Vyacheslav Viktorovich** (1917). He graduated from the Leningrad Higher Naval School named by Frunze (1939). A participant of the war with Japan (1945). State Inspector of the whaling of the whaling flotilla "Yuri Dolgoruky" (1963–1965), captain of the flotilla "Yuri Dolgoruky" (1966–1975). He was awarded the Order of Lenin (1968). For military service – the Order of the Red Banner.

Photo 5. ***3 captains*****: Lulko Ivan Trofimovich** – captain-director of the flotilla "Soviet Russia"; **Solyanik Aleksey Nikolaevich** – captain-director of the flotilla "Soviet Ukraine"; **Protsenko Petr Trofimovich** – captain-director of the flotilla "Yuri Dolgoruky".

Photo 6. **Kravtsov Vilen Gavrilovich** – 1st assistant to the captain-director of the flotilla "Soviet Ukraine". He had great respect from the crew.

Photo 7. **Medvedev Stanislav Petrovich** – Assistant captain-director of the flotilla "Soviet Ukraine".

Photo 8. **Purgin Afanasy Nikolaevich** (1902–1971). Initially, he mastered the Arctic: in 1911, a nine-year-old boy on a two-masted sailboat "Saint Demetrius" went from Arkhangelsk to Norway; He sailed on the hunter schooners in the White and Barents seas; Helmsman on the expedition vessels "Rusanov" and "Georgy Sedov". In 1924 – drafted into the army and served in the Navy of the Northern Fleet in a submarine brigade. In 1931–1933 he studied at the Arkhangelsk seaworthy school and got a diploma of a navigator. On boats merchant as a boatswain went to England, France and ports of Africa. Then he transferred to the Far East. He worked as an assistant captain on fishing and whaling vessels. In 1935 he graduated from navigational courses and was appointed as a captain of the whaler "Avangard" of the flotilla "Aleut", simultaneously mastered the harpooner skills. In 1938 he was awarded the title of captain-harpooner. In 1946, he moved to the flotilla "Slava", took command of the whaler "Slava-4", then "Slava-8". In 1950 he was awarded the title of Hero of Socialist Labor. In 1956 he became captain-mentor of the flotilla "Slava". In 1962–1965 – captain mentor of the flotilla "Yuri Dolgoruky". Then he worked as a state inspector of whaling. In 1968 he retired.

Photo 9a. **Zarva Petr Andreevich** (1900–1976). Photo 9b (Solyanik A.N. and Zarva P.A. right). The first experience of whaling was obtained by Pyotr Andreyevich in the Far East in 1931. In 1932, he overtook the whaling ship “Trudfront”, built in Norway, and became its captain. From Germany, the “Aleut” whaling base and three whalers (including “Trudfront”), through the Panama Canal, traveled to Vladivostok. He showed himself as an insistent and skillful captain, and mastered the skills of harpooner. In 1938 he was awarded the title of captain-harpooner. Since 1946, he transferred to the flotilla “Slava”. He worked as an assistant captain-director of whaling, and in 1948 again returned to the whaling ship as captain-harpooner. He trained harpooners. He was awarded many orders and medals.

Photo 9b. **(Solyanik A.N. and Zarva P.A. right)** The first experience of whaling was obtained by Pyotr Andreyevich in the Far East in 1931. In 1932, he overtook the whaling ship "Trudfront", built in Norway, and became its captain. From Germany, the "Aleut" whaling base and three whalers (including "Trudfront"), through the Panama Canal, traveled to Vladivostok. He showed himself as an insistent and skillful captain, and mastered the skills of harpooner. In 1938 he was awarded the title of captain-harpooner. Since 1946, he transferred to the flotilla "Slava". He worked as an assistant captain-director of whaling, and in 1948 again returned to the whaling ship as captain-harpooner. He trained harpooners. He was awarded many orders and medals.

Photo 10. **Angelin Vladimir Vasilyevich** (right on the photo). In 1957 he graduated from the Kherson seaworthy school, navigator. In 1959-1971 he worked as a 3rd, 2nd, then a senior assistant to the captain of whaling ships of the flotilla "Slava" and "Soviet Ukraine". In 1964 he received a diploma of a sea captain. In 1971 to 1982 he was appointed captain of the whaler "Vyderzhany – 36".

Photo 11. **Gnilyak Nikolay Nikolaevich**. Harpooner of "Slava-3", Hero of Socialist Labor.

Photo 12. **Zolotov Aleksey Nikolaevich** (1917–2003) – Harpooner, Hero of Socialist Labor. Two Orders of Lenin. In the years 1943–1948. – sailor, harpooner assistant, harpooner of the whaling fleet "Aleut". Since 1948 – harpooner of whaling vessels of the flotilla "Slava" and "Soviet Ukraine". In 1973 he retired.

Photo 13. **Kobylnikov Anatoly Petrovich** (1917–2003). In the years 1950–1952. – harpooner of the Far Eastern flotilla. 1952 – Harpooner of the whaler "Avanguard" of the "Aleut" flotilla. Then the harpooner on the flotillas "Slava" and "Soviet Ukraine". In 1971 he was awarded the title of Hero of Socialist Labor.

Photo 14. **Panov Grigory Emelyanovich** (1914–?). Passed to the whaling fleet "Aleut" school of the seaman, learned skills harpooner. Then he worked as a harpooner in the flotilla "Slava". In 1950 he was awarded the title of Hero of Socialist Labor.

Photo 15. Three Heroes of Socialist Labor: **Gnilyak Nikolay Nikolaevich, Purgin Afanasy Nikolaevich, Panov Grigory Emelyanovich** (left to right).

Photo 16; 16a. **Tupikov Vasily Loginovich** (1913–?) – Harpooner, Hero of Socialist Labor. In 1938 – a whaler at the "Aleut" whaling base. In 1943 – assistant harpooner (harpooner Panov). Since 1948 – the harpooner of the whaler "Slava-5" (Captain A. Klarikyan) at the whaling flotilla "Slava". In 1960, the harpooner of the whaler, "Vyderzhanny".

Photo 17. **Mogilny Georgy Sergeevich**. Harpooner on the flotilla "Slava" and "Soviet Ukraine". He worked 30 years (!) without vacations, he needed to put his six children on their feet.

Photo 18. **Ishkov Aleksandr Akimovich** (1905–1988). Party and economic figure. In 1929–1930 – a student at the Practical Academy named by Andreev in Novorossiysk. In 1957 he graduated in absentia from the Pedagogical Institute. In the fishing industry since 1930, he has held various managerial positions, including the Minister of Fisheries of the USSR. In 1978, the investigative authorities of the USSR discovered a large-scale criminal organization (later called the "fish mafia"), engaged in large thefts in the catch and further sale (including export) of fish and black and red caviar. The scheme of thefts was closed at the All-Union chain of stores "Ocean", which was supervised by a deputy, A.A. Ishkov – V.I. Rytov. All the defendants at the interrogations pointed to Rytov, as the organizer and think tank of the mafia. Soon he was arrested and later sentenced to be shot. Despite the fact that the investigation had material on Ishkov, he, thanks to the personal intercession of L.I. Brezhnev, escaped criminal responsibility, only lost his post and was sent to retirement.

It is to the Minister of Fisheries, in the 60s, A.A. Ishkov, that representatives of scientific groups of whaling flotillas appealed with a statement about the necessity to adhere to the quota, and not to exceed it several times. They said that otherwise our grandchildren will not get anything. To which the Minister reasonably replied: "The plan is the law, and I must fulfill it. Grandchildren will not take me off work!"

Photo 19. **Kamentsev Vladimir Mikhaylovich.** (1928–2003). Soviet statesman. In 1950 he graduated from the Moscow Institute of Fisheries and Economics. He began his career in managing the trawling fleet of the MurmanRyba association. 1962 – Deputy Chairman of the State Committee for Fisheries of the Council of Ministers of the USSR. Further: the leading posts of the State Committee for Fisheries under the Council of National Economy of the USSR. 1965 – First Deputy Minister of Fisheries. In 1979–1986 he was Minister of Fisheries of the USSR. In 1986–1989 – Deputy Chairman of the Council of Ministers of the USSR. Since 1989 he has been retired.

To him, as to the 1st deputy minister at that time, the author of the book (being then the head of the Odessa Marine Mammal Laboratory) applied for the falsification of the scientific and industrial data of the whaling industry. A commission was established that confirmed these facts. The head and the composition of the scientific group were replaced, but the leadership of the flotilla was not touched. Two years later, by order of another deputy minister (V.I. Rytov, who was subsequently executed for major fraud with the all-Union chain of stores "Ocean"), my laboratory was abolished.

Photo 20. **Baranov Igor Alekseevich** (1930). A participant of the Second World War. In 1953 he graduated from the Kherson Naval School, navigational department. 1954–1957 – 3rd navigator of the trawler "Latvia"; Then the 3rd navigator of the scientific research vessel "Persey-2", where he participated in the first underwater nuclear explosion. 1957–1960 – the captain of the vessel "Academician Berg". 1960–1962 – Senior Assistant Whaler "Gnevny-40" and a number of other whalers. 1971–1976 – captain-director of the fishing base "Vostok". 1976–1994 – General Director of the Black Sea fishing association "Antarctica".

Literature

Abu Rayhan Biruni (1973) Selected works. T. 4 Saydali (Pharmacognosy). Tashkent: "FAN" of the Uzbek SSR

Aguayo LA (1974) Baleen whales of continental Chile. In: WE Schevill (ed) The whales problems: a status report. Harvard University Press, Cambridge, MA, x+419p, pp 209–217

Allen KR (1971) Relation between production and biomass. J Fish Res Board Can 28(10):1573–1581

Allen KR (1976) A more flexible model for baleen whale populations. Rep Int Whal Commn 26:247–263. Cambridge

Allen KR (1980) Size distribution of male sperm whales in the pelagic catches. Rep Int Whal Commn (Special issue) 2:51–56. Cambridge

Amasiatsi Amirdovlat (1990) Useless for ignoramuses. M.: Science

Armstrong AJ, Siegfried WR (1991) Consumption of Antarctic krill by minke whales. Antarct Sci 3(1):13–18

Arnold A (1987) Portland's chance encounter with a pygmy right whale. Aust Nat Hist 22:266–270

Arnold P (1997) Occurrence of dwar minke whales (Balaenoptera acutorostrata) on the northern great Barrier Reef, Australia. Rep Int Whal Commn 47:419–424. Cambridge

Arnold P, Marsh H, Heinsohn G (1987) The occurrence of two forms of minke whales in east Australian waters with a description of external characters and skeleton of the diminutive or dwarf form. Sci Rep Whal Res Inst 38:46. Tokyo

Aron H (1927) Biohemie des Wachstums des Menschen und der hoheren Tiere. In: Oppenheime's Handbuch der Biochemie, H, 7

Arsenyev VA (1958) Marine Mammal and Bird Observations. USSR Academy of Sciences. Description of expedition of the ship "Ob'" 1955–1956. Publishing house of USSR Academy of Sciences, Moscow

Arsenyev VA, Zemsky VA (1973) Studenetskaya I.S. Marine mammals. Food industry, Moscow, 232p

Arshavsky IA (1968) Biology of the neonatal period in mammals. Biological basis of the neonatal period. Works of MOIP, vol 29, pp 7–22

Ash CE (1952) The body weights of whales. Norsk Hvalf-Tid 41, Arg. (7):364–374

Ash CE (1953) Weight of Antarctic humpback whales. Norsk Hvalf-Tid 7:387–391

Ash CE (1957) Weight and oil yield of Antarctic humpback whales. Norsk Hvalf-Tid 10:569

Atwood N (1869) Ambergris. Am Nat 2

Avicenna (Abu Ali Ibn Sina) (1982) The canon of medicine, 2nd edn, vol 2. "FAN" of the Uzbek SSR, Tashkent

Backmann G (1931) Das Wachstumsproblem. Ergebnisse der Physiologie, bd.

Baer KM (1950) The history of the development of animals. In: Observations and thoughts, vol 1. Publishing house of USSR Academy of Sciences, Series «Classics of science», Moscow

Y. Mikhalev, *Whales of the Southern Ocean*, Advances in Polar Ecology 5,
https://doi.org/10.1007/978-3-030-29252-2

Baer KM (1952) The history of the development of animals. In: Observations and thoughts, vol 2. Publishing house of USSR Academy of Sciences, Series «Classics of science», Moscow
Bailey N (1970) Mathematics in biology and medicine. Publishing house Mir, Moscow, 326p
Balcomb KC, Goebel CA (1977) Some information on a Berardius bairdii in Japan. Rep Int Whal Commn 27:485–486. Cambridge
Beddard FE (1901) Contribution towards a knowledge of the osteology of the pygmy whale (Neobalaena marginata). Tran Zool Soc Lond 16:87–114
Beklemeshev KB (1959) Antarctic divergence and whale food fields. Publishing house of USSR Academy of Sciences, pp 90–93
Beklemeshev KB (1961) The effect of atmospheric cyclones on whale feeding fields in Antarctica. Proc USSR Acad Sci 51:122–141
Belkin AN (1964) A new species of seal from the Kuril Islands Phoca insularis sp. nova. Rep USSR Acad Sci 158:5
Beloborodov AG, Potelov VA (1969) Some features of the initial period of life of crested pups and characteristics of their hairline – Marine mammals. Publishing house Nauka, Moscow, pp 220–225
Bengston JL, Laws RM (1985) Trends in crabeater seal age and maturity: an insight into Antarctic marine interactions. Selected papers presented to the scientific committee of CCAMLR 1982–1984. Hobart, Australia. P. 2, pp 341–368
Bennet FD (1840) Narrative of a whaling voyage around the globe from the year 1833 to 1836. II, London, 395p
Bennett AG (1932) Whaling in the Antarctic. Henry Holt, New York
Berg LS (1920) Bipolar distribution of organisms and the Ice Age. Math. Academy of Sciences, Moscow, S. 273–302
Berzin AA (1961) About methods for determining the age of female sperm whales. Rep USSR Acad Sci 139:491–494
Berzin AA (1963) Methods for determining the age and age composition of Pacific sperm whale herds. Vladivostok
Berzin AA (1971) Sperm Whale. Food Industry, Moscow. p 368
Berzin AA, Vladimirov VL (1982) A new kind of killer whale from Antarctica. Priroda 6:31–32. Moscow
Berzin AA, Vladimirov VL (1983) A new kind of killer whale (Cetacea, Delphinidae) from the waters of the Antarctic. Zool Mag 62 (2):287–295. Moscow
Berzin AA, Lagerev SI, Isakov GI (1975) Materials on the population morphology of Pacific sperm whales. In: Marine mammals. Part. 1. Materials of the 4th All-Union meeting, Kiev, pp 28–30
Berzin AA, Volkov AF, Moroz IF (1976) Soviet-American Whale expedition in the Eastern Pacific. Mar Biol 4:74–75
Best PB (1969) The sperm whale (Physeter catodon) off the west coast South Africa. 4. Distribution and movements. Invest Rep Div Sea Fish 78:1–12. South Africa
Best PB (1977) Two allopatric forms of Bryde's whale of South Africa. Rep Int Whal Commn (Special issue 1):10–34. Cambridge
Best PB (1982) Seasonal abundance, feeding, reproduction, age and growth in minke whale off Durban (with incidental observations from the Antarctic). Rep Int Whal Commn 32:759–786. Cambridge
Best PB (1983) Sperm whale stock assessments the relevance of historical whaling records. Special issue on historical whaling records. IWC, HWR/1, pp 41–56. Cambridge
Best PB (1988) Right whales at Tristan da Cunha – a clue to the 'non-recovery of depleted species? – Biol Cons 46:23–51
Best PB (1989) Some comments on the BIWS catch record data base. Scient Repts Whales Res Inst 39:363–369
Bezikovich Y (1931) Approximate calculations. State Scientific and Technical Publishing House, Leningrad
Bloich D, Lockyer C, Zachariassen M (1993) Age and growth parameters of the long-finned pilot whale of the Faroe Islands. Rep Int Whal Commn (special issue 14):163–207. Cambridge

Bodrov VA, Grigoriev SN (1963) Processing whale raw at whalers. PischePromIzdat, Moscow, 363p
Bogdanov AS, Lyubimova TG (1978) Soviet studies of krill in the South Ocean. Fisheries 10:6–9. Moscow
Bogolyubsky SN (1967) Embryology and modern problems of animal husbandry. Embryonic development of farm animals. Publishing House Nauka. Proceedings of MOIPIYU, vol 26, Moscow, pp 7–14
Borodin RG (1977) Estimation of initial stock and catchability coefficient of male sperm whales in the Southern Hemisphere. IWC Scientific Committee SP/Doc.22, Cambridge, 368p
Borodin RG (1985) On the relationship between a stock estimate and a "net" recruiment rate for minke whales of the Southern hemisphere. Intr Whal Commn, SC/37/Mi2, pp 1–4
Borodin RG (1986) On the assessment of reserves and the possible degree of exploitation of minke whales of the Southern Hemisphere – Marine mammals. In: Abstracts of the 9th All-Union meeting on the study, protection and rational use of marine mammals. Arkhangelsk, pp 55–56
Borodin RG (1996) Whales: fishery control measures and stock status. Publishing House VNIRO, Moscow, 208p
Boylsten (1724) Ambergris found in whale. Philos Trans R Soc Lond 33
Bradis VM (1935) Theory and practice of computing. UchPedGiz, Moscow
Brand K (1940) Whale oil. An economic analysis. – Stanford University, 264p
Brinkmann A (1948) Studies on female fin and blue whales. Hvalradets Skr 31:1–38
Brodie PF (1982) The beluga (Delphinapterus leucas); growth and age based on a captive specimen and a discussion of factors affecting natural mortality estimates. Rep Int Whal Comm 32:445–447. Cover Financial Year 1980–1981. Cambridge
Brody S (1927) Growth and development. IV. Growth rates during the self accelerating phase of growth. University of Missouri Agricultural Experiment Station Bulletin, 98p
Brown R (1868) Notes on the history and geographical relation of the Cetacea frequenting Davis strait and Baffin's bay. Proc Zool Soc 35:533–556. London
Brown SG (1968) The results of sei whale marking in the Southern Ocean to 1967. Norsk Hvalf-Tid 57:77–83
Brown SG (1976) Whale marking process report, 1970. In: Twenty-first report of the international Whaling commision, pp 51–55
Brown SG (1995) Whale marking: a short review, from a voyage of Discovery. In: George Deacon 70th anniversary volume edited by Nartin Angel. Pergamon Press Ltd., Oxford
Brownell RL Jr, Yablokov AB, Zemsky VA (1998) USSR Pelagic catches of North Pacific sperm whales, 1949–1979: conservation implications. Materials Scientific Committee, IWC, SC/SO/CAWS27, Cambridge, pp 1–8
Buchholtz EA (2010) Vertebral and rib anatomy in Caperea marginata: implications for evolutionary patterning of the mammalian vertebral column. Marine Mammal Science, (in lit.), p 16
Budylenko GA (1970a) Some questions of the biology of the sei whale of the South Atlantic – Whales of the southern hemisphere (biology and morphology). In: Proceedings of AtlantNIRO, Kaliningrad, issue 29, pp 17–33
Budylenko GA (1970b) Observation of the behavior of a newborn humpback whale – Whales of the southern hemisphere (biology and morphology). In: Proceedings of AtlantNIRO, issue 29, pp 231–233. Kaliningrad
Budylenko GA, Panfilov BG, Pakhomov AA, Sazhinov EG (1973) New data on the dwarf whale smooth Neobalaena marginata (Gray, 1848). In: Proceedings of the AtlantNIRO, vol L1, S. 122–132
Budylenko GA(1978) Distribution of small cetaceans in the South Atlantic. In: Marine mammals (Abstracts of the 7th All-Union meeting), Moscow, pp 50–52
Budylenko GA, Mikhalev YA (1970) To the biology of reproduction of the Antarctic sei whale – Whales of the southern hemisphere (biology and morphology). In: Proceedings of AtlantNIRO, issue 29, pp 3–16. Kaliningrad

Budylenko GA, Pervushin AS, Naumov GA (1970) Why female sperm whales enter Antarctica – Whales of the southern hemisphere (biology and morphology). In: Proceedings of AtlantNIRO, Kaliningrad, issue 29, pp 203–215
Bunak VV (1946) The main type of mammalian growth curve. MOIP Bull Biol Div 51(4–5):13–43. Moscow
Bushuev SG (1986) Feeding of minke whales, Balaenoptera acutorostrata, in the Antarctic. Rep Int Whal Commn 36:241–245. Cambridge
Bushuev SG (1989) Population-morphological characteristics of the minke whale of the Southern hemisphere. In: Abstract of the candidate dissertation, Kiev, 20p
Bushuev SG (2000) Distribution and feeding of minke whale on feeding fields in the 1st sector of the Antarctic. In: Materials of the Soviet Whaling industry (1947–1979), Moscow, pp 206–232
Bushuev SG, Mikhalev YA (2000) The effect of whale extermination and its possible biological effects. In: Materials of the Soviet whaling industry (1947–1979), Moscow, pp 233–248
Butterworth DS, Best PB, Hembree D (1984) Analysis of experiments carried out during the 1981/2 IWC/IDCR Antarctic minke whale assessment cruise in Area II. Rep Int Whal Commn 34:356–392. Cambridge
Camerer (1910) Gewichts-und langenwachstum der kinder. Wurtemberg med. Korrespon. Blatt, 23
Carter NM (1935) Ambergris. Prog Rep Pac Boil Stn 25
Carter NM, Elsey CR (1954) Ambergris found in a British Columbia Sperm whale. Prog Rep Pac Coast Stn, Fishery Research Board of Canada 99:35–36
Chapman DG (1964) Reports of the committee of three Scientists on the special scientific investigations of Antarctic whale stocks. Rep Int Whal Commn 14:32–106. Cambridge
Chapsky KK (1965) Growth and onset of puberty in females of the Caspian seal. Marine mammals. Publishing House Nauka, Moscow, pp 122–137
Chevalier N (1700) Description de la pièce D'Ambre Gris que la Chambre D'Amsterdam a recenë des Indes Orientales, pesant 182 livres; avec un petit traité de son Origine et de sa Vertu. Chez l'Anteur, Amsterdam
Chittlbotough RG (1965) Dynamics of two populations of the humpback whale, Megaptera novaeangliae (Borowski). Aust J Mar Freshw Res 16(1):33–128
Chittleborough RG (1954) Studies on the ovaries of the humpback whale, Megaptera nodosa (Bonnaterre) in the Western Australian coast. Aust J Mar End Freshw Res 5(1):35–63
Chittleborough RG (1958) The breeding cycle of the female humpback whale, Megaptera nodosa (Bonnaterre). Aust J Mar End Freshw Res 6:3–29
Chugunkov DI (1969) Far Eastern Larga – hunting and hunting, vol 10. pp 18–20
Clapham PJ, Brownell RL Jr (1995) The potential for interspecific competition in baleen whales. Rep Int Whal Commn, SC/47/SH/27, Cambridge, 15p
Clark H, Florio B, Hurowitz R (1955) Embryonic growth of Thampophis sp. sirtalis in relation to fertilization date and placental function. Copeia 1:9–13
Clarke R (1949) Ambergris, soap parfumery and cosmetic, Tr. Rew., A Great Haul if Ambergris. Nature 174:155
Clarke R (1954) A great haul of ambergris. Nature. Norsk Hvalf.-tid 174(8):155–156
Clarke R (1956) The biology of sperm Whales captured in the Azores. Norsk Hvalf-Tid 8
Clarke R (1957) Migratons of marin mammals. Norsk Hvalf-Tid 11
Clarke R (2006) The origin of ambergris. LAJAM 5(1):7–21
Cleyerus A (1689) De Ceto minore ambrophago. Ephem Med Phys Germ Acad Nat Curios 8:69
Clusius C (1605) Caroli Clusii Atrebatis, Aulae Cesoreae quondam Familiaris, Exoticorum Libri Decem. Leyden. (De Ambaro, pp 147–149)
Cock AG (1966) Genetical aspects of metrical grows and form in animals. Quart Rev Biol 41(2):131–190
Collett R (1911–1912) Norges pattedyr, Cetacea. Christiania, 1911–1912. pp 543–722
Collin-Asselineau C, Lederer E, Mercier D, Polonsky J (1950) Sur quelques produits d'oxydation et de dégradation l'ambréine; obtention de substance à odeur ambrée. Bulletin Societé Chimique de France, année:720–728

Cook JG, De La Mar WK (1983) The effect of variability in age data on the estimation of biological parameters of minke whales (Balaenoptera acutorostrata). Rep Int Whal Commn 33:333–346. Cambridge
Crosier J (1927) On curves of growth, especially in relation to temperature J Gen Physiol v. 10. N 53
Davies IL, Guiler ER (1957) A note on the pygmy righ whale, Caperea marginata Gray. Proc Zool Soc Lond 129(4):579–589
Dawbin WH (1964) Movements of humpback whale marked in the south-west Pacific Ocean 1952 to 1962. Norsk Hvalf-Tid 48:213–238
Della Rosa L, Secchi ER (1997) Stranding of an ordinary blue whale (Balaenoptera musculus) in southern Brazil: 'true' or pygmy? Rep Int Whal Commn 47:425–430. Cambridge
Deméré TA, McGowen MR, Berta A, Gatesy J (2008) Morphological and molecular evidence for a stepwise evolutionary transition from teeth to baleen in mysticete whales. Syst Biol 57:15–37
Devis JS, Guiler ER (1958) A newborn piked whale in Tasmania. J Mammal 39(4):593–594
Doi T, Ohsumi S, Nasu K, Shimadzu Y (1970) Advanced assessment of the fin whale stock in the Antarctic. Rep Int Whal Commn 20:60–87. Cambridge
Donaldson H (1924) The rat data and reference tables. Philadelphia
Donovan GP (1991) Blue whales off Peru, December 1982, with special reference to pygmy blue whales. Rep Int Whal Commn 34:473–476. Cambridge
Donovan G, Briwick I, Best P (1981) A not on comparative readings of sperm whale teeth. Rep Int Whal Commn 31. Cambridge
Dorofeev SV, Arsenyev VA (1936) Reference indicators for the beluga whaling in the Far East. Marine mammals of the Far East. In: Proceedings of VNIRO,vol III, Moscow/Leningrad, pp 49–62
Doroshenko NV (1978) On the relationship of killer whale and minke whale (predator-prey) in the Antarctic. In: Marine mammals. Abstracts of the 7th All-Union conference, Moscow, pp 107–109
Doroshenko NV (1996) The results of a study of blue pygmy whales (Balaenoptera musculus brevicauda), caught by Soviet whaling fleets in the Indian Ocean in seasons 1962/63-1971/72. In: Proceedings of TINRO, vol 121. Vladivostok, pp 150–162
Doroshenko NV (2000) Soviet whaling for humpback Megaptera novaeangliae in the North Pacific – Materials of the Soviet whaling (1949–1979). Moscow, pp 48–95
Dowbin WH (1966) The seasonal migratory cycle of humpback whales. In: KS Norris (ed) Whales, Dolphins and Porpoises. University of California Press, pp 147–170
Drukker GF, Galichko SI (1936) Pacific beluga whale as an industrial raw material. In: Proceedings of VNIRO, 3
Dudley (1725) An essay on the natural history of whales, with a particular account of the ambergris found in the spermaceti whale. Phil Trans R Soc Lond Vjl 33
Dubin VN, Mikhalev YA, Savusin VP, Zinchenko VL, Ivashin MV (1982a) Characterization of the distribution of calcium in cement and dentin of sperm whale teeth. In: Materials of 8th All-Union conference on protection and rational use of marine mammals, Astrakhan, pp 116–118
Dubin VN, Mikhalev YA, Savusin VP, Zinchenko VL, Ivashin MV (1982b) Characterization of the distribution of calcium in cement and dentin of sperm whale teeth. In: Materials of the 8th All-Union conference on the study, conservation, and rational use of marine mammals. Astrakhan, pp 116–118
Dunker Q (1924) Die korrelation zwischen lange und gewicht bei fischen. Wissenschaftliche meeresuntersuchungen, B-XV, H-1, 4
Eckstein D, Bauch J (1969) Beitrag zur Rationalisierung eines dendrochronologischen Verfahrens. – Forstwissenschftliches Centralblatt, Bd. 88. H. 4., Hamburg
Elfsberg (1863) Information about the whale fishery, vol 2. Marine digest, Saint Petersburg
Eschricht DF, Reinhardt J (1866) On the greenland right-whale. Recent Memoirs on the Cetacea. Royal Society, London
Essapian FS (1953) The birth and growth of a porpoise. Nat Hist 62(9):92–399

Evans WE, Yablokov AV, Bowles AE (1982) Geographic variation in the color pattern of killer whales (Orcinus orca). Rep Int Whal Commn 32:687–698. Cambridge

Everson I (1977) The living resources of the Southern Ocean. Southern Ocean Fisheries Survey Programme, FAO, pp 1–156

Farris DA (1956) Diet-Juduced differens in the weight length. Relation-ship of aquorium fed sardines (sardinops cacrules). J Fish Res Board, Canadds 13:503–513

Fedoseev GA (1965) Puberty and reproduction rates of the Okhotsk ringed seal-- Marine mammals. Publishing House Nauka, Moscow, pp 113–117

Fergson CW (1968) Bristlecone Pine. Science and esthetics. Science v. 159, N 3817

Filippova YA, Strygar IE (1987) Features of the distribution and biology of squids of the Southern Hemisphere. Raw materials of the Southern Ocean and the problems of their rational use: extracts of 2nd All-Union meeting. Kerch, pp 13–15

Fraser JFD, Huggett ASG (1959a) The breeding season and length of pregnancy in four species of large whales. In: Proceedings of the XV, International Congress of Zoology, London, 311p

Fraser JFD, Huggett ASG (1959b) The growth rate of foetal whales. J Pohysiol 146(2):21–22

Fraser WR, Trivelpiece WZ, Ainley DG, Trivelpiece SG (1992) Increases in Antarctic penguin populations: reduced competition with whales or a loss of sea ice due to environmental warming? Polar Biol 11(8):525–531

Fritts H, Mosimann J, Bottorff C (1969) Revised computer programm for standartizing tree-ring series. Tree-Ring Bull 29(1–2). Arizona

Fujino K (1955) On the body weight of the sei whale located in the adjacend waters of Japan. Sci Rep Whal Res Inst 10:133–141. Tokyo

Gambell R (1964) Pygmy blue whale at Durban. Norsk Hvalf-Tid 3:66–68

Gambell R (1966) Foetal growth and the breeding season of sperm Whales. Norsk Hvalf-Tid 55:113–118

Gambell R (1967) Seasonal movement of Sperm Whales aspects. Academic Press, New York/London. Marine Zool

Gambell R (1968) Seasonal cycles and reproduction in sei whales of the Southern Ocean. Disc Rep 35:31–134

Gambell R (1973) Some effects of exploitation on reproduction in whales. J Reprod Fert (Suppl.) 19:533–553

Gambell R (1983) Outstanding whale assessment problems requiring analysis of historical data. Special issue on historical whaling records. IWC, HWR/R, Cambridge, pp 15–17

Gambell R, Grzegorzewska C (1967) The rate of lamina formation in sperm whale teeth. Norsk Hvalif-Tid 6

Gawn RWL (1948) Aspects of the locomotion of Whales. Nature 161(4080)

Gihr M, Pilleri G (1968) Relationship between body length end body weight of stenella gray and delphinus delphis Linne from the Western Mediterranean. Rev Suisse sool 75(4):813–817

Gill PC, Kemper CM, Matthew T, Susie AL (2008) Large group of pygmy right whales seen in a shelf upwelling region off Victoria, Australia. Mar Mamm Sci 24(4):962–968

Golovlev IF (2000) Mysteries of the Whales. Materials of the Soviet whaling industry (1947–1979). Center for Environmental Policy of Russia. Marine Mammal Council, Moscow, pp 11–24

Goncharov AI, Smetanin MM (1974) A method for objectively determining the age and studying the growth of fish by scales. Biology of inland waters. Inform Bull 22:63–67

Govorokov IV (1934) Beluga whale. Whaling and processing. SnabTekhIzdat

Gray JE (1846) The zoology of the voyage of H.M.S. Erebus and Terror, vol 3–5. Janson, London, pp 13–53

Gray JE (1936) Studies in animal locomotion. The propulsive powers of the dolphin. J Exptl Biol 13:170–191

Guibourt NJBG (1876) Histoire naturelle de Drogues Simples, Paris, 4 vols, 7th edn. (Ambergris 1876, 4:118–123)

Guiler ER (1961) A pregnant female pygmy right whale. Aust J Sci 24:297–298

Gulland JA (1970) The development of the resources of the Antarctic seas. In: MW Holgate (ed) Antarctic ecology, vol 1. Academic Press, London, pp 217–223
Hale HM (1964) The pygmy right whale (Caperea marginata) in South Australian waters, Part 2. Rec South Aust Mus 14:679–694
Hardy E (1949) The sperm whale. The perfume and essent. Oil Record, vol. 40? No.9
Harmer SF (1920) Memorandum on the present position of the southern whaling industry: report of the Interdepartmental Committee on Research and Development in the Dependencies of Falkland Island, London, pp 69–83
Harrison RJ (1969) Reproduction and reproductive organs. In: Andersen HT (ed) The biology of marine mammals. Academic Press, New York/London, pp 253–348
Hart T (1935) On the diatoms of the skin film of Whales and their possible bearing on problems of Whale movements. Discov Rep 10. Cambridge
Hector J (1875) Notes on New Zealand whales. Transactions and Proceedings of the New Zealand Institute 7:25 1–265.
Hershkovitz Ph (1966) Catalog of living whales. Bull US Nat Mus 246:259. Smithsonian Institution
Herzfeld P (1888) Uber das Jacobson'schen organ desMemschen der Saugethier. Zool Jahrb Bd 3:551–575
Hinton M (1925) Report on the paper left by the late Major Barret-Hamilton to the whales of S. Georgia. Crown Agents for Colonies, London, pp 57–209
Hoffmann PB (1938) Quantitative patterns of animal growth. Adv Mod Biol 9(1):39–67
Hohn A (1980) Age determination and age related factors in the teeth of Western North atlantic bottlenose dolphins. Sci Whales Res Inst 32:39–66. Tokyo
Huber B, Siebenlist V, Niess W (1965) Jahrringchronologie Hessischer Richen. Budinger Geschichtsblatter. Bd. V
Hugget ST, Widdas WF (1951) The relationship between mammalion foetal weight and conception age. J Physiol CXIV(3):306–317
Huxley J (1932) Problems of relative growth. Methuem & Co., London, 276p
Huxley J, Teisier G (1936) Terminology of relative growth. Nature 137(3471):780–781
Ichichara T (1961) Blue Whales in the water around Kergelen Island. Norsk Hvalf Tid 1:1–20
Ichichara T (1963) Identification of pygmy blue whale in the Antarctic. Norsk Hvalf-Tid 5:128–130
Ichichara T (1964) The pygmy blue whale, Balaenoptera musculus brevicauda, a new subspecies from the Antarctic. University of California press
Ichichara T (1966) Criterion for determining age of fin whale with reference to ear plug and baleen plate. Sci Rep Whal Res Inst 20:17–80. Tokyo
Ichihara T, Doi T (1964) Stock assessment of pygmy blue whales in the Antarctic. Norsk Hvalf-Tid 6:145–167
Irzhak PI (1967) Determining the age and growth rate of moose during fetal development. In: Proceedings of the Pechery-Ilych State Reserve, issue 12, pp 5–20
Ivanova EI (1961a) The proportions of the body and the nature of the growth of whales in the Far East. In: Proceedings of the meeting on the ecology and whaling of marine mammals. Publishing house of USSR Academy of Sciences, Moscow, issue 12, pp 72–78
Ivanova EI (1961b) The proportions of the body of the sei whale (Balaenoptera borealis L.), obtained from the Kuril Islands – Cetaceans of the Far Eastern Seas. Proceedings of the Institute of Animal Morphology named after A.N. Severtsov, USSR Academy of Sciences, Moscow, vol 34, pp 194–204
Ivanova EI (1961c) To the morphology of Japanese whale (Eubalaena sieboldy Grey) – Cetaceans of the Far Eastern Seas. Proceedings of the Institute of Animal Morphology named after A.N. Severtsov, vol 34. USSR Academy of Sciences, Moscow, pp 216–225
Ivanova EI (1961d) The proportions of the body of the minke whale (Balaenoptera acutorostrata L.) from the Kuril Islands – Cetaceans of the Far Eastern Seas. Proceedings of the Institute of Animal Morphology named after A.N. Severtsov, vol 34. USSR Academy of Sciences, Moscow, pp 225–231

Ivashin MV (1959) About the breeding of a humpback whale in the southern Atlantic Ocean. Information collection of VNIRO, vol 7. Moscow, pp 12–40
Ivashin MV (1963) Ambrayu. Zool J 7
Ivashin MV (1965) About the duration of pregnancy and the size of newborn humpback whale cubs in the southern Atlantic Ocean. In: Marine mammals. Publishing House Nauka, Moscow, pp 67–69
Ivashin MV (1966) Ambergris and fishing sperm whales. Nature 7
Ivashin MV (1968) Laboratory in sperm whales. Chem Life 2
Ivashin MV (1971) Some results of whale tagging conducted from Soviet vessels in the Southern Hemisphere. Zool J. 50(7):1063–1078
Ivashin MV (1973) Marking of whales in the Southern Hemisphere (Soviet materials). In: Twenty-third report of the international Whaling commission, pp 174–191
Ivashin MV (1976) About breeding of minke whale in the Indian sector of Antarctica. Zool J 55(6):893–903
Ivashin MV (1984) Large-scale pelagic whaling in the oceans in 1970–1980 – Marine mammals. Publishing house Nauka, Moscow, pp 40–60
Ivashin MV(1986) Progress report on cetacean research. June 1984 to May 1985. Rep Int Whal Commn 36:179–182. Cambridge
Ivashin MV (1990) Whaling and tagging of large species of baleen whales in the Southern Hemisphere. In: Marine mammals. Collection of scientific papers of VNIRO, pp 103–111
Ivashin MV, Mikhalev YA (1978) To the problem of the prenatal growth of Minke Whales Balaenoptera acutorostrata of the Southern Hemisphere and of the biology of their reproduction. Rep Int Commun 28:201–205
Ivashin MV, Popov LA, Tsapko AS (1972) Marine Mammals (Handbook). Food Industry, Moscow, 304p
Ivashin MV, Shevchenko VI, Yuchov VL (1972a) The pygmy right whale Caperea marginata (Cetacea). Zoologicheskii Zhurnal 51:1715–1723
Ivashin MV, Popov LA, Tsapko AS (1972b) Marine mammals (handbook). Edited by Prof. Moiseeva PA. Publishing House "Food Industry", Moscow, 304
Jackson JRS (1840) Dissection of a spermaceti whale and three other cetaceans. J Nat Hist Boston 5(2):137–171
Janisch E (1927) Das exponentialgesetz als Grundlage einer vergleichenden biologie. Springer, Berlin
Japha A (1905) Uber der Bau der Haut des Seiwal (Balaenoptera borealis). Zool Anz – Bd 29(14):422
Jegou E (1977) Nouveau journal de chimie. 1(6):529–531
Jenkins JT (1932) Whales and modern whaling, Wither by, London
Jones EC (1971) Isistius brasiliensis, a squaloid shark, the probable cause of crater wounds on fishes and cetaceans. Fish Bull 69(4):791–798
Jonsgard A (1951) Studios on the littie piked whale or minke whale (Balaenoptera acutorostrata). Norsk Hwalf-Tid 40(5):5–54
Kaempfer E (1727) The history of Japan …, translated by JG Scheuchzer from the High-Dutch. (Ambergris 1:133)
Kapralova LT (1967) On the comparative growth of the pancreas in fetuses of sheep and pigs. Embryonic development of farm animals. Publishing House Nauka, Moscow
Kasuya T (1977) Age determination and growth of Baird's beaked Whale with a comment on fetal growth rates. Sci Rep Whales Res Inst 29:1–29
Kasuya T (1999) Examination of the reliability of catch statistics in the Japanese coastal sperm whale fishery. J Cetacean Res Manag 1(1):109–122
Kasuya T (2002) Distortion of fishing statistics in the past of Japanese coastal whaling. Marine mammals of the Holarctic, Moscow, KMK, 128p
Kasuya T, Brownell RL (1979) Age determination, reproduction and growth of franciscana dolphin Pontoporia blainvillei. Sci Rep Whales Res Inst 31:45–67

Kasuya T, Wada S (1991) Distribution of large cetaceans in the Indian Ocean: data from Japanese sighting records, November-Marc. Cetaceans end cetacean research in the Indian Ocean sanctuary. Marin mammal. Technic Report N 3,UNEP, pp 139–170
Kasuya T, Brownell RL Jr, Balcomb KC (1998) Life history of Baird's beaked whales off the Pacific coast of Japan. Rep Int Whal Commn 47:969–979, Cambridge
Kato H (1979) Unusual Minke Whale With deformed jaw. Sci Rep Whal Res Inst 31:101–103. Tokyo
Kato H (1982) Some biological parameters for the Antarctic minke Whale. Rep Int Whal Commn 32:935–945. Cambridge
Kato H (1983) Some consideration on the decline in age at sexual maturity of the Antarctic minke whale. Rep Int Whal Commn 33:393–399. Cambridge
Kato H (1987) Density dependent changes in growth parameters of the southern minke whale. Sci Rep Whal Res Inst 38:47–73. Tokyo
Kato H, Sakuramoto K (1991) Age and sexual maturity of southern minke whales: a review and some additional analyses. Rep Int Whal Commn 41:331–337. Cambridge
Kawamura AA (1994) review of baleen whale feeding in the Southern Ocean. Rep Int Whal Commn 44:261–271. Cambridge
Kellog R (1928) The history of whales. Their adaptation to life in the water. Quart Rev Biol 3:29–76, 174–208
Kimura S (1957) The twinning in Southern fin whales. Sci Rep Whal Res Inst 12:103–125. Tokyo
Kemper CM (2002a) Distribution of the pygmy right whale in the Australasian region. Mar Mamm Sci 18:99–111
Kemper CM (2002b) Pygmy right whale. In: Perrin WF, ursig BW¨, Thewissen JGM (eds) Encyclopedia of marine mammals. Academic, San Diego, pp 1010–1012
Kemper CM (2008) Pygmy right whale. In: Perrin W, Wursig B, Thewissen J (eds) Encyclopedia of marine mammals. Academic, London, pp 939–941
Kemper CM (2009) Pygmy right whale Caperea marginata. Enciklop. Mar Mamm:939–941
Kiesewetter IV (1953) Marine mammals. Vladivostok
Kleinenberg SE, Yablokov AV (1960) Materials on the biology of breeding of belugas of the northern seas of the USSR. In: Proceedings of PINRO, issue 12, pp 165–173
Kleinenberg SE, Yablokov AV, Belkovich VM, Tarasevich MN (1964) Beluga whale. Nauka, Moscow, 456p
Kleinenberg SE, Yablokov AV, Klevezal GA, Belkovich VM, Etin VY (1965) Reference indicators on the characteristics of some pinnipeds and cetaceans. Publishing House Nauka, Moscow, pp 251–256
Klevezal GA (1963) On the causes of dentin stratification of marine mammalian teeth. In: The second All-Union conference for the study of marine mammals. Abstracts. Publishing House of the USSR Academy of Sciences, Leningrad
Klevezal GA (1966) Age-related changes in the structure of dentin, cement, and the periosteal zone of mammalian bones. PhD thesis, Moscow
Klevezal GA (1973) On the limitations and new possibilities of using layers in tooth and bone tissues to determine the age of mammals. Zool J 52:757–765
Klevezal GA (1975) On seasonal growth rhythms in hibernating mammals. Zool J (1):95–102
Klevezal GA (1988) The recording structures of mammals in zoological studies. Nauka, Moscow, pp 1–286
Klevezal GA, Klenenberg SE (1967) Determination of the age of mammals. Publishing House Nauka, Moscow
Klevezal GA, Myrick AC (1984) Marks in tooth dentine of female dolfphins (Genus Stenella) as indicators of parturition. J Mammal v. 65, N 1. – pp 103–110
Klevezal GA, Tormosov DD (1969) On the allocation of local groups of sperm whales by the nature of the layers in the dentin of the teeth. In: Abstracts of the meeting on the study of marine mammals, Moscow, pp 30–34

Klevezal GA, Tormosov DD (1971) On the allocation of local groups of sperm whales by the nature of the layers in the dentin of the teeth. In: Proceedings of AtlantNIRO, Kaliningrad, issue 39, pp 35–43
Klumov SK (1955) On the locality of whale herds. Proc USSR Acad Sci 18
Klumov SK (1962) Right Pacific whales. Trans Inst Oceanol ANSSSR 58:202–298
Klumov SK (1963) Food and helminth fauna of baleen whales (Mystacoceti) in the main whaling areas of the World Ocean. In: Proceedings of the Institute of Oceanology of the Academy of Sciences of the USSR, vol 71, Moscow, pp 94–194
Kolchin BA (1963) Dendrochronology of Novgorod. New methods in archeology, vol 117. Materials and Research on Archeology of the USSR, Moscow, pp 5–103
Kolchin BA, Rybina EA (1982) Excavation on Kirov Street – Novgorod collection. 50 years of excavation of Novgorod. Publishing House Nauka, Moscow, pp 178–238
Komatsu M, Misaki S (2005) Whales and the history of whaling: a look from Japan (Russian edition). BAM. BOOK, 142p, ISBN 5-87317-262-5
Krylov VI (1965) Determination of age, growth rate, and analysis of the age structure of pacifiers of the Pacific walrus – Marine mammals. Publishing House Nauka, Moscow, pp 201–211
Krylov VI (1969) Dates of mating and birth of Pacific Walrus – Marine mammals, Publishing House Nauka, Moscow, pp 275–285
Kuzin AE (1970) About the weight characteristic of the northern fur seal embryos. Izvestia TINRO 70:52–59. Vladivostok
Lakhtin LK (1922) Distribution curves and construction of interpolation formulas for them according to the methods of Pearson and Bruns. Moscow
Laurie AH (1937) The age of female blue whales and the effect of whaling on the stock, vol 15. Discovery Reports, Cambridge
Laws RM (1958) Recent investigation fin whale ovaries. Norsk Hvalf-Tid 5:225–229
Laws RM (1959) The foetal growht rate of whales with special reference to the fin whale, Balaenoptera physalus Linn. Discov Rep 29:281–308
Laws RM (1961) Reproduction growth and age of southern fin whales. Discov Rep 31:328–486
Laws RM (1973) Populations increases of fur seals at South Georgia. Polar Record 16:856–858
Laws RM (1977) The significance of vertebrates in the Antarctic marine ecosystem. In: GA Llano (ed) Adaptations within Antarctic ecosystem. Proceeding of the 3-rd SCAR Symposium. – On Antarctic Biology, Smithsonian Institute, Washington, XLIV, pp 411–438
Laws RM (1985) The ecology of the Southern Ocean. Amer Sci 1:36–40
Laws RM (1997) Seals and whales of the Southern ocean. Phil Trans R Soc London 279:81–96
Lederer E (1949) Chemistry and biochemistry of some mammalian secretions and excretions. J Chem Soc 3:2115–2125
Leduc RG, Duizon AE, Goto M, Lux CA, Pastene LA, Brownell RL (1998) Discrimination of the two forms of blue whales in the Southern Hemisphere. Rep Int Whal Commn, SC/49/SH25 48:122. Cambridge
Lederer E (1953) Chimie de l'ambréine. Industrie de la Parfumerie 8:189–197
Leimbach R (1951) Die ätherischen Öle
Lillie DG (1915) Cetacea. British Antarctic ("Terra nova") expedition, 1910. Zoology 1(3):85–124
Lindholm O (1888) About whaling in the Sea of Okhotsk. Russian shipping 33
Linnaeus K (1758) Systema naturæ per regna tria naturæ, secundum classes, ordines, genera, species, cum characteribus, differentiis, synonymis, locis. Editio decima, reformata. Holmiæ [Stockholm]: impensis direct. Laurentii Salvii. [4] Bl., S. 6-823
Lockyer C (1972) The age at sexual maturity of the southern fin whale (Balaenoptera physalus) using annual layer counts in the ear plug. ICES J Mar Sci 34(2):276–294
Lockyer C (1974) Investigation of the ear plug of Southern sei whale (Balaenoptera borealis) as a valid means of age determining. J Cons Int Explor Mer 36:71–81
Lockyer C (1977) A preliminary study of variations in age at sexual maturity of the fin whale with year class, in six areas of the Southern Hemisphere. Rep Int Whal Commn 27:141–147. Cambridge

Lockyer C (1979) Changes in a growth parameters associated with exploitation of southern fin and sei whales. Rep Int Whal Commn 29:191–241. Cambridge
Lockyer C (1980) Age determination studies on Physeter macrocephalus. In: Perrin WF, Myrick AC (eds) Age determination of toothed whales and sirenians. Rep Int Whal Comm (Spec. Issue N 3) IWC. Spec (1980) Issue 3. Cambridge
Lockyer C (1981) The age at sexsual maturity in fin whales off Iceland. Rep Int Whal Commn 31:389–393. Cambridge
Lockyer C (1993) Mortality rates for mature Southern sei whales. Rep Int Whal Commn (special issue 14), 53–57, Cambridge
Lockyer C (1995) Review of factors involved in zonation in odontocete teeth, and an investigation of the likely impact of environmental factors and major life events on harbour porpoise tooth structure. Rep Int Whal Commn (special issue) 16:511–529, Cambridge
Loird AK (1966) Dynamics of embryonic growth. Growth 30(2):263–275
Loird AK, Barton AD, Tybr SA (1968) Growth and time: an interpretation of allometry. Growth 32(4):347–354
Lotke A (1925) Elements of physical biology. Williams & Wilkins Co., Baltimore
Lubimova TG (1987) Research objectives and prospects for the use of Antarctic bioresources from the perspective of the international Antarctic Commission (CCAMLR). In: Raw materials of the Southern Ocean and the problems of their rational use: abstracts of reports of the 2nd All-Union Conference, Kerch, pp 18–19
Lubimova TG, Shevtsov VV (1980) Some data on the amount of krill in selected areas of the Southern Ocean. Fisheries 11: 21–25. Moscow
Lubimova TG, Shust KV (1980) Estimation of the level of consumption of Antarctic krill by the main groups of consumers – Biological resources of Antarctic krill. VNIRO, Moscow, pp 203–224
Lubimova TG, Shust KV (1987) The results of the study of biological resources of the Antarctic and the tasks of further fisheries research. Raw materials of the Southern Ocean and the problems of their rational use. Abstracts of the 2nd All-Union Conference, Kerch, pp 3–4
MacDowell EC (1928) The growth curve of the suckling mouse. Science
MacDowell EC, Allen E (1927) Proc Soc Exptl Biol Med 24:672
Mackintosh NA (1942a) The southern stocks of baleen whales. Discov Rep 22:197–300
Mackintosh NA (1942b) The southern stocks of whalebone whales. Discov Rep 22:197–300
Mackintosh NA (1965) The Stocks of whales
Mackintosh NA (1970) Whales and krill in the twentieth century. Antarct Ecol 1:195–212
Mackintosh NA, Wheeler JFG (1929) Southern blue and fin whales. Discov Rep 1:257–540
Magakyan YA (1967) Some issues of growth regulation in embryo development. Biol J Armenia 20(6):19–32
Masaki J (1973) Some notes on age estimation of baleen whales based upon the ear plug. In: Paper R I presented to the meeting on age determination in baleen whales. Reykjavik, April
Masaki J (1976) Biological Studies of the North Pacific sei whales. Bull Far Seas Fish Res Lab 14
Masaki Y, Yamamura K (1978) Japanese pelagic Whaling and whale sighting in the 1976/77 Antarctic season. Rep Int Whal Commn 28, SC/29. Doc. 31, pp 251–254. Cambridge
Matsuura Y (1940) Statistical study on whale fetuses. III. Sperm whales in the adjacent waters of Japan. Bull Jpn Soc Sci Fish 9(4)
Matsuoka K, Yoshihiro F, Pastene LA (1996) A sighting of a large school of the pygmy right whale, Caperea marginata, in the southeast Indian Ocean. Mar Mamm Sci 12:594–597
Matsuoka K, Pitman RL, Marquez FC (2005) A note on a pygmy right whale (caperea marginata) in the southwestern pacific ocean. J Cetacean Res Manag 7:71–73
Matthews LH (1937) Humpback whale, Megaptera nodosa. Discov Rep 17
Matthews LH (1938) The sperm whale, Physeter catodon. Discov Rep 7:93–168
Mattsuura I (1936) On the lesser rorqual found in the adjacent waters of Japan (in Japanese). Bull Jpn Sci Fish 4(5):325–330

McBrde AF, Krizler H (1951) Observations on pregnancy, parturition and postnatal behaviour in the bottlenose dolphin. J Mammal 32(3):251–266
McLaren IA (1958) The biology of the ringed seal (Phoca hispida Schreber) in the Estern Canadian Arctic. Bull Fish Res Board Canada 118:97
McQuillan H (1962) The Antarctic krill. West Fish 63(4):354–358
Melville H (1851) Moby-dick; or the Whale
Melville G (1962) Moby Dick, or white whale. State.izdat.geograf.lit, Moskva
Mikhalev YA (1966) Morphological features of the pituitary gland of cetaceans. In: Scientific conference on the basis of research work. Chisinau, 144p
Mikhalev YA (1969) On the embryonic development of some species of Antarctic whales. In: 4th All-Union meeting on the study of marine Mammals, Kaliningrad, 17p
Mikhalev YA (1970a) Change in weight of the adenohypophysis of some species of whales in connection with their biological state. In: Proceedings of young scientists. VNIRO, ONTI, issue III, pp 226–232
Mikhalev YA (1970b) Analysis of the growth of embryos of fin whales of the Antarctic (Balaenoptera physalus L). Protection of fish stocks and increased productivity of water bodies in the southern zone of the USSR. Proceedings of the inter-university meeting, Chisinau, pp 92–93
Mikhalev YA (1970c) Prenatal growth and some issues of the biology of reproduction of the Antarctic fin whale – Whales of the southern hemisphere (biology and morphology). Proceedings of AtlantNIRO, Kaliningrad, vol 29, pp 53–82
Mikhalev YA (1971a) To the question of determining the age and reproductive ability of Physeter macrocephalus (Cetacea). In: 1st conference of young biologists and doctors, Odessa
Mikhalev YA (1971b) Whale milk. Nature 3:74–76. Moscow
Mikhalev YA (1972) To the question of determining the average size of newborn fin whale. In: Abstracts of the 5th All-Union meeting on the study of marine mammals. Makhachkala, pp 206–209
Mikhalev YA (1973) Experience in improving the methodology for determining the age of marine mammals and fish. In: Conference of young scientists AzCherNIRO, Kerch
Mikhalev YA (1975) The growth of whales in the prenatal period. Marine mammals. 6th all-union conference, vol 2. Naukova Dumka, Kiev, pp 12–13
Mikhalev YA (1977) Method for graphical record of surface relief of decalcinated sections of sperm whale teeth with the aim to determine their age. Rep Int Whal Commn 27:356–360. Cambridge
Mikhalev YA (1978a) Graphic recording of stratifications in sperm whale teeth using a recording microphotometer. In: Marine mammals. Abstracts of the 7th All-Union Conference on Marine Mammals. Moscow, pp 223–225
Mikhalev YA (1978b) Occurrence and distribution of cetaceans in the Pacific sector of the Antarctic Regions to the results of observations, 1973/74 and 1974/75 seasons. Rep Int Whal Commn 28: 263–267. Cambridge
Mikhalev YA (1980) General regularities in prenatal growth in whales and some aspects of their reproductive biology. Cambridge: Rep. Int.Whal.Commn. No 30, pp 249–254
Mikhalev YA (1981) Observations from aircraft on the distribution of bottlenose dolphins in the Black Sea – Biological aspects of the protection of rare animals. All-Russian Research Institute of Nature Protection and Conservation, Moscow, pp 97–98
Mikhalev YA (1982) Subjective and Objective Assessments of the Laminations in Sperm Whale Teeth. Rep Int Whal Commn 32:227–233. Cambridge
Mikhalev YA (1984) Growth and some questions of the biology of breeding of the minke whale of the Southern hemisphere. Marine mammals. Nauka, Moscow, pp 61–80
Mikhalev YA (1986) The rate of formation of dentin layers in sperm whale teeth – Marine mammals. In: Abstracts of the 9th All-Union meeting on the study, protection and rational use of marine mammals, Arkhangelsk, pp 277–278
Mikhalev YA (1990a) Analysis of methods for determining the age of puberty in female cetaceans – Marine mammals. In: Abstracts of the 10th All-Union meeting on the study, protection and rational use of marine mammals. Moscow, pp 198–200

Mikhalev YA (1990b) On the question of "transphase" in the ear plugs of baleen whales – Marine mammals. In: Abstracts of the 10th All-Union meeting on the study, protection and rational use of marine mammals. Svetlogorsk, pp 196–198
Mikhalev YA (1990c) The nature of calcifications of stratifications in the teeth of sperm whales. In: Scientific-practical conference dedicated to issues of organization, prevention and treatment. Odessa, 88p
Mikhalev YA (1991) Pair formations at the ends of jaws of Cetaceans. Rep Int Whal Commn 41:591. Cambridge
Mikhalev YA (1995) Humpbacks of the Arabian Sea. In: International conference on the study and protection of marine mammals, Moscow, pp 59–61
Mikhalev YA (1997a) Additional information about the catches of Soviet whaling fleet Sovetskaya Ukraina. Rep Int Whal Commn 47:147–150. Cambridge
Mikhalev YA (1997b) Hampback whales (Megaptera novaeangliae) in the Arabian sea. G. Mar Ecol Prog Ser 149:13–21
Mikhalev YA (1998) Pygmy blue whales catch by Soviet Antarctic whaling fleets, 1962/63–1971/72. Cambridge: Rep. Int.Whal.Commn 48. R. 196–197
Mikhalev YA (2000a) Pregnant females caught by whaling fleet Sovetskaya Ukraina. Materials of the Soviet whaling industry (1949–1979): Center for Environmental Policy of Russia, Moscow, pp 141–181
Mikhalev YA (2000b) Whaling in the Arabian See by fleets Slava and Sovetskaya Ukraina. Materials of the Soviet whaling industry (1949–1979): Center for Environmental Policy of Russia, pp 141–181
Mikhalev YA (2000c) Structural formations at the ends of cetacean jaws. Marine mammals of the Holarctic. Arkhangelsk, pp 250–255
Mikhalev YA (2000d) Recently retrieved data on blue whale catches by the Slava, 1946–1957. G. Cetcean Res Manag 2(Suppl.), SC/51/CAWS41, pp 174. Cambridge
Mikhalev YA (2002a) Expert or subjective assessment of stratifications in the teeth of sperm whales Marine mammals of the Holarctic. Moscow, pp 190–191
Mikhalev YA (2002b) Pygmy blue whales of the Arabian Sea. Marine mammals (results of studies conducted in 1995–1998), Moscow, pp 357–387
Mikhalev YA (2002c) Analysis of whaling in Antarctica and adjacent waters. Marine mammals (results of studies conducted in 1995–1998), Moscow, pp 329–356
Mikhalev YA (2007) The overall pattern of prenatal growth cetaceans (Cetacea). Naukova visnyk Pivdennoukraïnskogo sovereign pedgogichnogo universitetu IM. KD Ushinskiy (Collection of scientific works). Yuvileyny Preview Issue. Type of PDPU IM. KD Ushinskiy, Odes, pp 43–56
Mikhalev YA (2008) Southern Hemisphere whales: biology, whaling, prospects for population recovery. LLC "INVATS", Odessa, 328p
Mikhalev YA, Ivashin MV (1978) About prenatal growth and breeding biology of minke whales In: Marine mammals. Abstracts of the 7th All-Union meeting, Moscow, pp 225–226
Mikhalev YA, Ivashin MV (1982) Qualitative and quantitative assessment of the recording structures of marine mammals. In: Materials of the 8th All-Union meeting on the study, protection and use of marine mammals. Astrakhan, pp 240–242
Mikhalev YA, Shevchenko VI (1972) To osteology pygmy right whale. Abstracts of the fifth All-Union conference on the study of marine mammals. Part 1. Makhachkala, pp 209–213
Mikhalev YA, Shevchenko VI (1973) Size of newborns and duration of pregnancy in sperm whales. In: Conference of young scientists of AzCHERNIRO. Kerch, pp 10–11
Mikhalev YA, Shevchenko VI (1986) Morphological and size-balanced characteristics of a newborn male sperm whale– Marine mammals. Abstracts of the 9th All-Union meeting on the study, protection and rational use of marine mammals. Arkhangelsk, pp 273–275
Mikhalev YA, Sventsitskaya GA (1970a) Sex ratio in fin whales and sei whales of the Southern Hemisphere. In: Proceedings of the University Scientific Conference of Young Scientists, Odessa State University, pp 156–158

Mikhalev YA, Sventsitskaya GA (1970b) Fat content of milk of some whales and growth of newborns. In: Proceedings of the University Scientific Conference of Young Scientists, Odessa State University, pp 159–162
Mikhalev YA, Tormosov DD (1997) Corrected data about non-Soviet whale marks recovered by Soviet whaling fleets. Rep Int Whal Commn 47:1019–1027. Cambridge
Mikhalev YA, Zelenaya FE (1986) Description of paired formations at the ends of the cetacean jaws – Marine mammals. In: Abstracts of the 9th All-Union meeting on the study, protection and rational use of marine mammals. Arkhangelsk, pp 275–277
Mikhalev YA, Shevchenko VI, Neizhko VI (1975) The location of the Jacobson organ as a population sign in some baleen whales. In: Marine mammals. 6th All-Union conference, Part 2. Publishing House Naukova Dumka, pp 14–15
Mikhalev YA, Ivashin MV, Savusin VP, Zelenaya FE (1981) The distribution and biology of killer Whales in the Southern Hemisphere. Rep Int Whal Commn 31:551–566. Cambridge
Millais JG (1906) The mammals of Great Britain and Ireland. London/New-York
Mina MV, Klevezal GA (1976) Animal growth: body level analysis. Nauka, Moscow, 291p
Mitchell E (1974) Trophic relationships and competition for food in northwest Atlantic whales. Proc Can Soc Zool 123–133
Mizue K, Jimbo H (1950) Statistic study of fetuses of whales. Sci Rep Whal Res Inst 3:119–131. Tokyo
Mookherjee BD, Patel RR (1979) VII International congress on essential oils, pp 479–482
Murphy EJ, Morris DJ, Watkins JL, Priddle J (1988) Scales of interaction between Antarctic krill and the environment. In: D Sahrhage (ed) Antarctic ocean and resources variability. Springer, Berlin/Heidelberg/New York, pp 120–130
Myrick AC (1980) Examination of layered tissue of odontocetes for age determination using polarized light microscopy. Rep Int Whal Comm (Spec. issue. N 3):105–112. Cambridge
Myrick AC (1984) Time significance of laering in some mammalian hard tissues and its application studies. Acta Zool Fenn 171:217–220
Naaktgeboren C, Slijper EJ, Utrecht WL (1960) Researches on the period of conception, duration of gestation and growth of the foetus in the fin whales, based on data from International Whaling Statistics. Horsk. Hvalf-Tid 49(3):113–119
Naumov VP (1933) Seals of the USSR. KOIZ, Moscow
Naumov AG (1978) The main features of the biology and distribution of Antarctic krill. Fisheries 10:9–12. Moscow
Naumov AG (1983) On the systems of trophic chains and their role in stabilizing the energy balance in the community of Antarctic waters. Raw materials of the Antarctic zone of the ocean and the problems of their rational use: abstracts of reports of the All-Union Scientific Conference, Kerch, pp 23–25
Naumov AG, Chekunova VI (1983) Analysis and revision of energy balance data for baleen whales and their role in the Antarctic community. Raw materials of the Antarctic Ocean and problems of their rational use: abstracts of the All-Union Scientific Conference, Kerch, pp 122–123
Naydenov (1927) Patterns in the growth of young cattle and pigs. Gorki
Nemoto T (1959) Food of baleen whales with reference to whale movements. Sci Rep Whal Res Inst 14:149–290. Tokyo
Nemoto T, Nasu K (1975) Present status of exploitation and biology of krill in the Antarctic. Oceanol Int 75(16/2):353–360
Nikonorov IV (1995) Mystery of the whales. VNIRO Publishing House, Moscow, 40p
Nikonorov IV, Borodin RG (1997) Whales and their fate. VNIRO Publishing House, Moscow, 60p
Nishiwaki M (1950) Age characteristics in baleen plates. Sci Rep Whal Res Inst 4:162–183. Tokyo
Nishiwaki M (1952) On the age determination of Mystacoceti chiefly blue and fin whales. Sci Rep Whal Res Inst 7:87–119. Tokyo
Nishiwaki M (1957) Age characteristic of ear plugs of whales. Sci Rep Whal Res Inst 12:23–32. Tokyo

Nishiwaki M (1966) Distribution and migration of marine mammals in the North Pacific area. In: Symposium No. 4 of the XIth Pacific Scientific Congress, 49p
Nishiwaki M, Ichichara T, Ohsumi S (1958) Age studies of fin whale based on ear plug. Sci Rep Whal Res Inst 13:155–169. Tokyo
Ogawa T, Shida T (1950) On the sensory tubercles of lips and of oral cavity in sei and fin whales. Sci Rep Whal Res Inst 3:1–16. Tokyo
Ognetov GA (1987) Ecological and morphological characteristics of the beluga whale of the western region of the Soviet Arctic and the rational use of its reserves: abstract of dissertation for the degree of candidate of biological sciences: 03.00.08, Moscow, 19p
Ognev SI (1935) Animals of the USSR and surrounding countries. III. Predatory and pinnipeds. BioMedGiz, Moscow
Ohno M, Fujino K (1952) Biological investigation on the whale caught by the Japanese Antarctic whaling fleets, season 1950/51. Sci Rep Whal Res Inst 7:125–188. Tokyo
Ohsumi S (1960) Relative growth of the fin whale, Balaenoptera physalus (Linn.). Sci Rep Whal Res Inst 15:17–84. Tokyo
Ohsumi S (1965) Reproduction of sperm whale in the North-West Pacific. Sci Rep Whal Res Inst 19:35. Tokyo
Ohsumi S (1966) Allomorphosis between body length at sexual maturity and body length at birth in the Cetacea. J Mammal Soc Jpn 3(1):3–7
Ohsumi S (1986) Ear plug transition phase as an indicator of sexual maturity in female Antarctic minke whales. Sci Rep Whal Res Inst 37:17–30. Tokyo
Ohsumi S Masaki V (1975) Biological parameters of the Antartic minke whale of the virgeinal population level. J Fish Res Board Can 32(7):995–1004
Ohsumi S, Nishiwaki M, Hibiya T (1958) Growth of fin whale in the Northern Pacific. Sci Rep Whal Res Inst 13:97–133. Tokyo
Ohsumi S, Kasuya T, Nishiwaki M (1963) Accumulation rate of dentinal growth layers the maxillary tooth of the sperm whale. Sci Rep Whal Res Inst 17:15–35. Tokyo
Ohsumi S, Masaki J, Kawamura A (1970) The stock of the Antarctic minke whale. The whales research institute, Tokyo, vol 22, pp 75–125
Ommaney FD (1938) South latitude. Longmans, Green & Co., London
Omura H (1950) On the body weight of sperm and sei whales located in the adjacent waters of Japan. Sci Rep Whal Res Inst 4. Tokio
Omura H, Sakiura H (1956) Studies on the little piked whale from the coast of Japan. Sci Rep Whal Res Inst 11:1–37. Tokyo
Omura H, Ohsumi S, Nemoto T, Nasu K, Kasuya T (1959) Black right whales in the North Pacific. Scient Repts Whales Res Inst 21:78. Tokyo
Ostwald W (1908) Uber die zeitlichen eigenschaften der entwicklungsvorgange, vortrage und aufsatze uber entwick ungasmechanik der organismen. Leipzig, 5p
Otha K, Watarai T, Oishi T, Ueshida Y, Hirose S, Yoshizawa T, Akikusa J, Sato M, Okaho H (1955) Composition of fin whale milk. Sci Rep Whal Res Inst 10:151–167. Tokyo
Ozharovskaya LV (1990) Delivery dates in captive bottlenose dolphins in captivity – Marine mammals (abstract of the 10th All-Union Conference on the Study, Protection and Rational Use of Marine Mammals). Moscow, pp 221–2
Pastukhov VD (1969) The onset of puberty in females of the Baikal seal. Nauka, Moscow, pp 127–135. Marine mammals
Pavey CR (1992) The occurrence of the pygmy right whale, Caperea marginata (Cetacea: Neobalaenidae), along the Australian coast. Aust Mamm 15:1–6
Payne RM (1977) Growth of fur seal population. Phil Trans R Soc L – Ser B 279:67–79
Payne PR, Wheeler EF (1967) Growth of the foetus. Nature 215(6103)
Pearl R, Reed L (1925) Skew-growth curves. Proc Nat Acad Sci USA v. 11, N 16
Peelle N (1932) Whaling in Northeastern Japanese waters. Science 75:666–667
Pequegnat WE (1958) Whales, plankton and man. Sci Amer 198(1):84–90
Pervushin AS (1966) Sperm whale birth monitoring. Zool J 45(12)

Peters N (1938) Der Neue Deutsche Walfang. Hamburg
Pike G (1966) Progress report on the study of sperm whales from British Columbia. Background Pap Spec Sci. Meeting International Whaling Commission, Honolulu, 9p
Pinedo MC, Barreto AS (1996) Identity of a blue whale stranded in southern Brazil. Cambridge: Paper SC/48/SH16 presented in the IWC Scientific Committee
Pirrin WF, Myrik AC (1980) Age determination of toothed whales and sirenians. Rep Int Whal Commn (special issue 3):229. Cambridge
Pitman RL, Ensor P (2003) Three forms of Killer whales (Orcinus orca) in Antarctic waters. J Cetacean Res Manag 5(2):131–139
Plokhinsky NA (1961) Biometrics. Publishing House Siberian Branch of the USSR Academy of Sciences, Novosibirsk, 364p
Plokhinsky NA (1967) Biometrics algorithms. Moscow University Press, Moscow
Plokhinsky NA (1978) Mathematical methods in biology. Moscow State University Publishing House, Moscow, 265p
Pogorelov AV (1968) Analytic geometry. Publishing House Nauka, Moscow, 176p
Popov LA, Timoshenko YK (1965) Some data on the Greenland seal of the Newfoundland herd. Marine Mammals, Publishing House Nauka, Moscow, pp 158–170
Potelov VA (1969) Periods of pregnancy, childbirth and areas of childbirth of a sea hare. In: Abstracts of the 4th All-Union meeting on the study of marine mammals, Moscow
Pouchet G (1893) Sur l'ambre gris. Extrait du volume commémoratif du centenaire de la fondation du Muséum de Histoire Naturelle, Paris, p 17
Pouchet G, Beauregard H (1892) Recherches sur le Cachalot (suite). Nouv. Arch.Du Mus. (3), Vol.4
Przibram H (1922) Form und formel in tierreich. Leipzig und Wien
Purgin AN (1953) "Slava" in Antarctic. Publishing House Molod', Kyiv
Quay WB, Mitchell E (1971) Structure and sensory apparatus of oral remnants of the nasopalatine canals in the fin whale (Blaenopter physalus). J Morphol 134(3):271–280
Ratcliffe HL (1942) Autopsy of a male pygmy sperm whale (Kogia breviceps). Notulae Naturae Philadelphia 112:4
Reeves RR, Leatherwood S, Papastavrou V (1991) Possible stock affinities of humpback whales in the Northern Indian Ocean. Cetaceans and Cetacean Research in the Indian Ocean Sanctuary. Marine mammal. Technical report, vol 3, pp 259–269
Rice DW (1977) A list of the marine mammals of the world. NOAA Tech Report NMFS SSRF-711, US Department of Commerce,15p
Rice DW, Scheffer VB (1968) A list of the marine mammals of the world. U. S. Fish and Wildlife Service. Spec Sci Rep Fish 579:1–16
Rice DV, Wolman AA (1971) The life history and ecology of the gray whale (Eschrichtius robustus). Am Soc Mammal. Special publication No. 3:142
Risting S (1928) Whales and whales fetuses. Cons. perm. intern., Pour. L'explor. de la mer. Rapp. Prac-Verb., Copengagen, v. L, 122p
Robertson TB (1925) Growth and development. Philadelphia/London
Robineau D (1991) Balaenopterid sighting in the western tropical Indian Ocean (Seychelles Area), 1982–1986. Cetaceans and Cetacean Research in the Indian Ocean Sanctuary. Marine Mammals, Technical Report Number 3, pp 171–178
Rokitsky PF (1961) Basics of variation statistics for biologists. Publishing House of the Belarusian State University, Minsk
Rokitsky PF (1964) Biological statistics. High School, Moscow
Ross GJB, Best PB, Donnelly BG (1975) New records of the pygmy right whale (caperea marginata) from South Africa, with comments on distribution, migration, appearance and behaviour. J Fish Res Board Can 32:1005–1017
Ruspoli M (1955) A la recherches du cachalot. Paris
Ruud JT (1940) The surface structure of the daleen plates as a possible clue to age determination in whales. Hvalrdets Skr 23:1–47

Ruud JT (1945) Further studies on the structure of the baleen plates and their application to age determination. Hvalradets Skr 29:1–69
Salnikov NE (1953) Fin whale and blue whale feeding in Antarctica. In: Proceedings of VNIRO, vol 25, pp 54–67
Salnikov NE (1956) Some issues of breeding biology of fin whale and blue whale of Antarctica. In: Proceedings of BaltNIRO, Kaliningrad, issue II
Samyshev EZ (1983) On the life cycle diagram of Antarctic krill Euphausia superba. Raw materials of the Antarctic zone of the ocean and the problems of their rational use: abstracts of reports of the All-Union scientific conference. Kerch, pp 32–33
Samyshev EZ (2006) On the life cycle diagram of Antarctic krill Euphausia superba. On the transformation of the Antarctic ecosystem – Problems of biological oceanography of the XXI century: International scientific conference dedicated to the 135th anniversary of InBYUM (Sevastopol, Ukraine, September 19 – 21, 2006). Sevastopol, 34p
Sasaki T, Nikaido M, Hamilton H, Goto M, Kato H, Kanda N, Pastene LA, Cao Y, Fordyce RE, Hasegawa M, Okada N (2005) Mitochondrial phylogenetics and evolution of mysticete whales. Syst Biol 54:77–99
Sazhinov EG (1970) The onset of puberty and physical maturity in blue pygmy whales (Balaenoptera musculus brevicauda) – Whales of the southern hemisphere (biology and morphology). Proceedings of AtlantNIRO, Kaliningrad, issue 29, pp 34–52
Sazhinov EG (1980) Blue whale pygmy: abstract of dissertation for the degree of Doctor of Biological Sciences: 03.00.08. USSR Academy of Sciences Institute of Zoology, Kaliningrad, 24p
Schafer E (1981) Golden Peaches of Samarkand. The book is about the foreign outlandish in the Tang Empire. M.: Science, 608 p
Scheffer VB, Slipp JW (1948) The whales and dolphins of Washington State a key to the cetaceans of the west coast of North America. Am Midl Nat 39(2):259–337
Schmalhausen II (1927) On laws of embryo growth – Works of department of physics and mathematics. VUAN 6(1)
Schmalhausen II (1935) Definition of basic concepts and methods of growth research. Animal growth. BioMedGiz, Moscow, pp 8–60
Schmidt-Nielsen K (1987) Animal sizes: why they are so important. Publishing House Mir, Moscow, 260p
Schulman E (1958) Bristlecone Pine, oldesi known living thing. National Geothapic Magazine, March
Schwedtiawer (1783) An account of ambergris. Philos Trans 73
Scoresby W (1820) An account of the Arctic region with a history and description of the Northern whale-fishery, vol 1. Edinburgh, 551p
Scott EOG (1948) Neonatal Length as a linear function of adujt length in Cetacea. In: Paper and proceedings of the Royal Society of Tasmania, 75p
Secchi ER, Dalla Rosa L (1997) Stranding of an ordinary blue whale (Balaenoptera musculus) in southern Brazil: "true" or pygmy? Rep IntWhal Commn 47:425–430. Cambridge
Sekiguchi K, Best PB, Kazcmaruk BZ (1992) New information on the feeding habits and baleen morphology of the pygmy right whale caperea marginata. Mar Mamm Sci 8:288–293
Semendyaev KA (1933) Empirical formulas. GTTI, Moscow/Leningrad
Sergeant DE (1959) Age determination in Odontocete whale from dentinal growth layers. Norsk Hvalingst-Tidende 6:273–288
Sergeant DE (1962) The biology of the pilot or pothead whale Globicephala melaena (Traill) in Newfoundland waters. Bull Fish Res Board Can 132:1–84
Sergeant DE (1963) Minke whales, Balaenoptera acuturostrata Lacepede of the western North Atlantic. J Fish Res Board Can 20(6):1489–1504
Shevchenko VI (1970) Riddle "white scars" on the body of whales. Nature, Academy of Science, Moscow, No 6, pp 72–73
Shevchenko VI (1971) On the origin of the "white scars" on the body of whales. Study of marine mammals. In: AtlantNIRO proceedings, Kaliningrad, vol 39, pp 67–74

Shevchenko VI (1975) The effects of small shark attacks on whales. In: Marine mammals. Part 2. Materials of the 6th All-Union conference, Kiev, Naukova Dumka, pp 175–177

Shevchenko VI (1977) Application of Whit Scars to the study of the location and migrations of Sei Whale populations in Area III of the Antarctic. Cambridge: Rep. Int.Whal.Comm (special issue I), pp 130–134.

Shevchenko VI, Yukhov VL, Golubovskii YP (1971) Pygmy right whale. Nature 6

Sibbaldus R (1773) Phalainologia Nova; sive Observationes de Rarioribus QuibusdemBalaenis in Scotiae Littus nuper ejectis … Edinburgh, Second Impression

Sidorov M (1867) Natural wealth of the north of Russia. – Mirskoe slovo, pp 2–4

Sigurjonson JA (1980) Preliminary not on ear plug from Icelandic minke whales. Rep Int Whale Commn 30:193–194. Cambridge

Siliman RP (1969) Comparison between Gompertz and von Bertallanffy curves for expressing growth in weight of fishes. – J Fesh Res Board Canada v. 26, N 1. pp 161–165

Sladen WJL (1964) The distribution of Adelie and Chinstrap penguins. In: R Carrick et al. (eds), Biologie antarctique. Hermann, Paris, pp 359–365

Sleptsov MM (1940) On the peculiarities of the birth and nutrition of pups of the Black Sea dolphin Delphinus delphis. Zool J 19(2):297–304. Moscow

Sleptsov MM (1941) About reproduction biology of the Black Sea dolphin. Zool J 20(4–5):632–653. Moscow

Sleptsov MM (1952) Cetaceans of the far eastern seas. Izvestia TINRO, Vladivostok, issue 38, 166p

Sleptsov MM (1955a) Features of the development of cetaceans in the early embryonic stages. In: Proceedings of the Institute of Oceanology of the USSR Academy of Sciences, Moscow, 18p

Sleptsov MM (1955b) Cetaceans of the Far Eastern Seas. Vladivostok, 161p

Sleptsov MM (1961) The weighing results of large and small cetaceans caught in the Far East – Moscow. Proceedings IMEZh USSR Academy of Sciences, – issue 34, pp 144–150

Sleptsov MM (1965a) Orca of the Southern Hemisphere. In: Marine Mammals, Publishing House Nauka, Moscow, pp 60–64

Sleptsov MM (1965b) Minke whale of the Southern Hemisphere. In: Marine mammals. Nauka, Moscow, pp 53–59

Slijper EJ (1962) Whales. Hutchinson, London, 475p

Smetanin MM, Kasyanov AN (1979) The use of the accumulation method in the analysis of the optical density of fish scales – biology of inland waters. Inform Bull (41):79–82

Spenser RP, CoulombeMJ, Wagenen G (1966) Interspecies comparison of fetal growth. – Growth. – v. 30, N 1. – pp 1-7

Starbuck A (1878) History of the American Whale fishery from its earliest inception to the year 1876. – Rep US Commn Fish and Fisheries 1875–1876/ 4, 779p

Stephenson W (1951) The lesser rorqual in British waters. Dove Mar Lab Rep III Ser 3(12):7–48

Sukhovskaya LI, Klevezal GA (1982) On the dentin structure of bottlenose teeth. In: Study, conservation and rational use of marine mammals (Abstracts of 8th All-Union Conference), Astrakhan

Symons HW, Weston RD (1958) Studies on the humpback Whale (Megaptera nodosa) in the Bellimghausen Sea. Norsk Hvalf-Tid 47:53–81

Tavolga MC, Essapian ES (1957) The behaviour of the bottlenose dolphin (Tursiops truncatus): mating, pregnancy, parturition and mother-infant behavior. Zoologica 42:11–31

Tershy BR (1992) Body size, diet, habitat use, and social behavior of Balaenoptera whales in the Gulf of California. J Mammal 73:477–486

Thomas H (1955) Ambra. Dragoco Berichte, Nr. 5

Tikhomirov V (1894) About the whaling industry in Russia. Russian shipping, St. Petersburg, pp 146 –147

Tomilin AG (1937) Whales of the Far East. Scientific notes of Moscow State University, Moscow, issue 13

Tomilin AG (1936) Ambergris and its origin. Nature, No 5

Tomilin AG (1940) Some issues from the ecology of cetaceans (adaptation and water temperature). Bull MOIP (Department of Biology) 49(5–6)
Tomilin AG (1945) Determination of age of whales by baleen. Doc USSR Acad Sci 49(6):472–476
Tomilin AG (1953) On the systematic relationship of northern and southern minke whales. Bull MOIP 58(6)
Tomilin AG (1957) Cetaceans. Animals of the USSR and adjacent countries, vol 9. Publishing House of the USSR Academy of Sciences, Moscow, 756p
Tomilin AG (1962) Cetacean fauna of the seas of the USSR. Qualifier to the fauna of the USSR. Publishing House of the USSR Academy of Sciences, Moscow, 212p
Tomilin AG (1965) On migrations, geographical races, thermoregulation and the influence of environmental temperature on the distribution of cetaceans In: Animal migrations. Publishing House of the USSR Academy of Sciences, issue 25, pp 4–26
Tomilin AG (1969) Dolphins serve man. Nauka, Moscow, 247p
Tomilin AG (1980) The results of labeling mammals. Cetaceans. - Ser. Theriology issues. Publishing House Nauka, Moscow, pp 272–281
Tomilin AG, Morozov DA (1969) Suction of feed – a previously unknown way of eating of Phocaena phocaena. In: Proceedings of the VSHIZO, vol 31. pp 201–202
Tomilin AG, Plavsky VI (1962) The chemical composition of milk and its calorie content in different groups of mammals. In: Proceedings of the VSHIZO, issue 10, pp 98–109
Tormosov DD (1970) Geographical distribution of various physiological and age groups of sperm whales in the Southern Hemisphere. Whales of the Southern Hemisphere (biology and morphology). In: Proceedings of AtlantNIRO, issue 29, pp 41–52
Tormosov DD (1982) Use of osteodentin variability as a population trait. In: Study, protection and rational use of marine mammals (Abstracts of 8th All-Union Conference). Astrakhan, pp 371–372
Tormosov DD (2002) On the reproductive potential of female sperm whales. In: Marine mammals of the holarctic: abstracts of the 2nd international conference, Moscow, pp 255–259
Tormosov DD, Mikhaliev YA, Best PB, Zemsky VA, Sekiguchi K, Brownell RLJ (1998) Soviet catches of southern right whales Eubalaena australis, 1951–1971. Biological data and conservation implications. Biol Conserv 86:185–197
Tormosov DD, Mikhaliev YA, Best PB, Zemsky VA, Brownell RL Jr. (2000a) Soviet catches of southern right whales Eubalaena australis, 1951–1971. Biological data and aspects of conservation. Center for Environmental Policy of Russia – pp 366–382
Tormosov DD, Mikhalev YA, Best PB, Zemsky VA, Brownell RL Jr (2000b) Soviet fishing of the southern Australian whale Eubalaena australis in 1951–1971. Biological data and aspects of protection. Materials of the Soviet Whaling industry (1949–1979). Center for Environmental Policy of Russia, Moscow, pp 366–382
Townsend CH (1931) Where the nineteenth century whales made his catch. Bull N Y Zool Soc 34(6)
Townsend CH (1935) The distribution of certain whales as shown by logbook records of American Whale ships. N Y Zool Soc.19(1):1–50
Turner W (1875) Further contribution to the placentation of the Cetacea (Monodon monoceros). Proc Roy Soc Edinb 9:103–110
Tyurin PV (1927) On the relationship between the length of the fish and its weight. Proc Sib Ichthyol Lab 2(3):3–21
Uralskaya IV (1957) The chemical composition of milk of dolphins of the Black Sea, vol 57. Scientific notes of Rostov State University, Rostov, issue 1
Urbakht VY (1963) Mathematical statistics for biologists and physicians. Publishing House of the USSR Academy of Sciences, Moscow, 205p
Urbakht VY (1964) Biometric methods. Nauka, Moscow
Utrecht-van WL (1965) On the growth of the baleen plate of the fin whale and the blue whale. Bijdr dierkunde 35:3–38
Utrecht-van WL, van Utrecht- Cock CN (1967) Comparison of records of baleen plates and of ear plugs in female fin Whales, Balaenoptera physalus (Linnaeus, 1758). Ibid Aft 39:81–97

Utreht-van WL (1968) Notes on some aspects of the mammary glands in the fin whale, Balaenoptera physalus (L.), with regard to the criterion Lactating. Norsk Hvalf-Tid 1:1–13
Veberman E (1914) Whaling in Russia. Part 1. History of whaling. News of Moscow Institute of Commerce. Department of commerce and technology, Moscow. book 2
Ventsel ES (1969) Probability theory. Nauka, Moscow, p 576
Veynger GM (1974) Identification of the population structure of sperm whales based on color analysis. In: Studies in fish biology and commercial oceanography. TINRO, Vladivostok, issue 5
Vidal O (1995) Population biology and incidental mortality of the vaquita. phocoena sinus. Rep Int Whal Commn (special issue 16):247–272. Cambridge
Vinberg GG (1966a) Growth rate and metabolic rate in animals. Success Mod Biol 61(2):274–293
Vinberg GG (1966b) The growth rate and metabolic rate in animals. Adv Mod Biol 61(2):274–293
Vinberg GG (1971a) Linear dimensions and body weight of animals. Adv Mod Biol 32(6):714–722
Vinberg GG (1971b) Linear dimensions and body weight of animals. Success Mod Biol 32(6):714–722
Voronina NM (1987) To the study of the trophic network of the Antarctic. In: Raw materials of the Southern Ocean and the problems of their rational use: abstracts of the 2nd All-Union meeting, Kerch, pp 12–13
Wada S (1998) A review of Brayde's whale taxonomy for the Indian Ocean and Solomon Sea. Rep Int Whal Commn, SC/49/NP7 48:123. Cambridge
Wandler A, Huber W (1969) Gewichtswachstum und jabreszeitliche Wewichtsschwankungen bei Reh und Gemes. Rev Suisse Zool 76(3):686–694
Wheeler JE (1933) Notes on a young sperm whale from the Bermuda Islands. Proc Zool Soc Lond 8
Wheeler JF (1934) On the stock of whales at South Georgia. Discov Rep 9
Wheeler JE (1945) Observation on whales the South Atlantic Ocean in 1943. In: Proceedings of Zoological Society of London, v.116, part 2
White JCD (1953) Composition of Whales milk. Nature 171:612–617
Williamson GR (1975) Minke whales off Brazil. Sci Rep Whal Res Inst 27:37–59
Yablokov AV (1959) Features of the reproduction of beluga whales as a representative of toothed cetaceans. In: Abstracts of the III scientific youth conference. IMZh USSR Academy of Sciences
Yablokov AV (1961) On the "sense of smell" of marine mammals. In: Proceedings of the conference on the ecology and fishing of marine mammals. Publishing House of the USSR Academy of Sciences, Moscow, issue 12, pp 87–93
Yablokov AV (1964) Convergence or parallelism in the evolution of cetaceans? Paleontol J 1:97–106
Yablokov AV (1966) Mammalian variability. Nauka, Moscow, 363p
Yablokov AV (1994) Validity of whaling data. Nature 367(6459):108. London
Yablokov AV, Belkovich VM, Borisov VI (1972) Whales and dolphins. Nauka, Moscow, 472p
Yablokov AV, Zemsky VA, Mikhalev YA, Tormosov DD, Berzin AA (1998) Materials of the Soviet whaling in Antarctica in 1947–1972 (Population aspects). Ecology 4:43–48
Yamada M (1954) Some remarks on the Pygmy Sperm Whale, Kogia. Sci Rep Whales Res Inst Tokyo 9:37–58
Yosipov VK (1929) To the question of the relationship between the body length of the fish and its weight. Proc Sib Sci Fish Stn 3(3):163–176
Yukhov VL (1971) Some data on the nutrition of sperm whales in the high latitudes of the Antarctic. Research of marine mammals. In: Proceedings of AtlantNIRO, Kaliningrad, issue 39, pp 54–59
Zaika VE (1970) The ratio of the maximum specific growth rates of warm-blooded animals. Zool J 49(2):181–192
Zasosov AV (1965) Methodology for determining the fatness of whales. Fisheries 10:62–64. Moscow
Zemsky VA (1950) Materials for the study of the embryonic development of the Antarctic fin whales. BMOIP (Department of Biology) 55(6):15–20. Moscow

Zemsky VA (1956) About the method for determining the traces of the corpus luteum of pregnancy and ovulation on the ovaries of female fin whale. In: Bulletin of the Moscow Society of Naturalists, Department of Biology. 61, 6, pp 5–13
Zemsky VA (1958) Determination of traces of corpus luteum at the fin whales of Antarctica. In: Proceedings of VNIRO, vol XXXIII, Moscow, pp 105–160
Zemsky VA, Berzin AA (1959) Gem ambergris. Nature:2
Zemsky VA (1960) Materials on the reproduction of Antarctic fin whale (B. physalus). BMOIP (Department of Biology), 65(1). Moscow
Zemsky VA (1961) On the lifespan of fin whales and periodicals of their reproduction. In: Proceedings of the meetings of the Ichthyological Commission of the USSR Academy of Sciences, issue 12
Zemsky VA (1965) To the distribution of baleen whales in the Southern Hemisphere. In: Marine Mammals. Nauka, Moscow, pp 20–26
Zemsky VA (1972) On the taxonomic position of the pygmy blue whale (B. Musculus brevicauda). Makhachkala, pp 163–167
Zemsky VA (1974) The main features of the distribution of marine mammals in the oceans. In: Results of science and technology gray Zoology of vertebrates, vol 6, Marine Mammals. Moscow, pp 21–39
Zemsky VA, Boronin VA (1964) On the question of the pygmy blue whale taxonomic position. Norsk Nvalfangst – Tedende, 11p
Zemsky VA, Budylenko GA (1995) The taxonomic position of some species of whales with bipolar distribution. In: Marine mammals. Part 1. Materials of the 4th All-Union meeting. Publishing House Naukova Dumka, Kiev, pp 116–118
Zemsky VA, Mikhalev YA (2000a) Data collected by the research group of the whaler "Slava". Materials of the Soviet whaling industry (1949–1979). Center for Environmental Policy of Russia, Moscow, pp 182–205
Zemsky VA, Mikhalev YA (2000b) Data collected by the scientific group of the whaling factory Sovetskaya Ukraina. Materials of the Soviet whaling industry (1949–1979). Center for Environmental Policy of Russia, Moscow, pp 366–382
Zemsky VA, Mikhalev YA (2000c) Data collected by the scientific group of the whaling factory Slava. Materials of the Soviet whaling industry (1949–1979). Center for Environmental Policy of Russia, Moscow, pp 182–205
Zemsky VA, Berzin AA, Mikhalev YA, Tormosov DD (1994a) Sovet Antarctic pelagic Whaling after WWII: review of actual catch data. IWC, SHWP1, Cambridge, pp 1–5
Zemsky VA, Mikhaliev YA, Tormosov DD (1994b) Humpback whale catches by area by the Soviet Antarctic whaling fleets. IWC, SHWP1, Cambridge, 151p
Zemsky VA, Berzin AA, Mikhalev YA, Tormosov DD (1995) Soviet Antarctic pelagic Whaling after WWII: review of actual catch data. Rep Int Whal Commn 45:131–135. Cambridge
Zemsky VA, Mikhalev YA, Berzin, AA (1996) Supplementary information about Soviet whaling in the southern hemisphere. Rep Int Whal Commn 45:131–138. Cambridge
Zenkovich BA (1935) On whale embryos. Reports of the USSR Academy of Sciences, vol II, pp 3–4
Zenkovich BA (1937) Weighing whales. DAN USSR, vol 16, issue 3. Moscow
Zenkovich BA (1938) Milk of large cetaceans. DAN USSR, nov ser, 20:2–3. Moscow
Zenkovich BA (1951) Ambergris. Fisheries No 4
Zenkovich BA (1952) Whales and whaling. PishchePromIzdat, Moscow
Zenkovich BA (1955) Brief history of whaling and its current state in USSR/Whaling in Soviet Union. Publishing House Rybnoe Hoziaystvo, Moscow, pp 5–29
Zenkovich BA (1966) Whales and whaling in the waters of the Antarctic and the North Pacific and stock status considerations. In: Abstracts of the III All-Union meeting on the study of marine mammals. Vladivostok
Zenkovich BA (1969) Whales and Plankton in Antarctica – Marine mammals. Nauka, Moscow, pp 150–152

Zenkovich BA (1970) Whales and Plankton in Antarctic waters. Antarctic Ecol 1:183–190
Zhivotovsky LA (1979) The indicator of similarity of populations by polymorphic characters. J Gen Biol 39(5):453–460
Zhivotovsky LA (1982) The indicator of population variability by polymorphic characters. Phenetics of populations. Publishing house Science, Moscow, pp 38–44
Zimushko VV (1968) Some gray whale biology data – marine mammals. Publishing House Nauka, Moscow, pp 93–97
Zinchenko VL (1986) Changes in the population structure of the minke whale of the Southern Hemisphere during the feeding period in the Antarctic. In: Marine mammals: abstracts of the 9th All-Union conference on the study, protection and rational use of marine mammals. Arkhangelsk, pp 159–160
Zinchenko VL, Mikhalev YA (1986) The rate of fouling of minke whales with diatoms. In: Marine mammals. Abstracts of the 9th All-Union conference. Arkhangelsk, pp 164–165